student book volume 1

Cultural, Social and Technical

MATHEMATICS

Secondary Cycle Two, Year Two

Claude Boivin
Dominique Boivin
Antoine Ledoux
Nathalie Ricard

9001, boul. Louis-H.-La Fontaine, Anjou, Québec Canada H1J 2C5
Telephone: 514 351-6010 • Fax: 514 351-3534

ORIGINAL VERSION

Publishing Manger
Katie Moquin

Production Manager
Danielle Latendresse

Coordination Manager
Rodolphe Courcy

Project Manager
Diane Karneyeff

Linguistic Review (LES)
Denis Desjardins

Proofreader
Viviane Deraspe

Graphic Design
Dessine-moi un mouton

Technical Illustrations
Stéphan Vallières

General Illustrations
Yves Boudreau

Maps
Les Studios Artifisme

Iconographic Research
Jean-François Beaudette
Perrine Poiron

These programs are funded by Quebec's Ministère de l'Éducation, du Loisir et du Sport, through contributions from the Canada-Québec Agreement on Minoroty-Language Education and Second-Language Instruction.

Visions, Cultural, Social and Technical, Student Book, Volume 1, Secondary Cycle Two, Year Two
© 2009, Les Éditions CEC inc.
9001, boul. Louis-H.-La Fontaine
Anjou, Québec H1J 2C5

Translation of *Visions, Culture, Société et Technique, manuel de l'élève, volume 1,*
(ISBN 978-2-7617-2594-1) © 2009, Les Éditions CEC inc.

Legal Deposit : 2009
Bibliothèque et Archives nationales du Québec
Library and Archives Canada

ISBN 978-2-7617-2796-9

Printed in Canada
1 2 3 4 5 13 12 11 10 09

The authors and publisher wish to thank the following people for their collaboration in the evolution of this project.

Collaboration
Alain Bombardier, Collège Mont-Sacré-Cœur
Jocelyn Dagenais, Teacher, École Secondaire André-Laurendeau, CS Marie-Victorin
Isabelle Gendron, Teacher, Collège Mont-Royal

Scientific Consultation
Driss Boukhssmi, Professor, Université du Québec en Abitibi-Témiscamingue
Matthieu Dufour, Professor, Université du Québec à Montréal

Pedagogical Consultation
Stéphane Brosseau, Teacher, École Secondaire l'Horizon, CS des Affluents
Richard Cadieux, Teacher, École Jean-Baptiste Meilleur, CS des Affluents
Teodora Nadu, Teacher, École Jeanne-Mance, CS de Montréal
Dominic Paul, Teacher, École Pierre-Bédard, CS des Grandes-Seigneuries

ENGLISH VERSION

Translators and Linguistic Review
Communications McKelvey, inc.
Donna Aziz
Don Craig
Gordon Cruise
Cecilia Delgado
Shona French
Jennifer McCann
Gary Spiller

Pedagogical Consultant
Joanne Malowany

Pedagogical Review
Don Craig
Steve Element
Vilma Scattolin
Peggy Drolet

Project Management
Patrick Bérubé
Rite De Marco
Stephanie Vucko
Valerie Vucko

A special thank you to the following people for their collaboration in the evolution of this project.

Collaboration
Donna Boychuk
Michael J. Canuel
Robert Costain
Sylvie Desrochers
Margaret Dupuis
Christiane Dufour
Jean-Guy Dufour
Rosie Himo
Doris Kerec
Nancy Kerec
Louis-Gilles Lalonde
Linda Monette
Denis Montpetit
Shelley Orr
Mital Patel
Bev White

TABLE OF CONTENTS

volume 1

VISION 1

VISION 2

VISI3N

PRESENTATION OF STUDENT BOOK

This *Student Book* contains three chapters each called "Vision." Each "Vision" presents various "Learning and evaluation situations (LES)" sections and special features "Chronicle of the past," "In the workplace" and "Overview." At the end of the *Student Book*, there is a "Reference" section.

Revision

The "Revision" section helps to reactivate prior knowledge and strategies that will be useful in each "Vision" chapter. This feature contains one or two activities designed to review prior learning, a "Knowledge summary" which provides a summary of the theoretical elements being reviewed and a "Knowledge in action" section consisting of reinforcement exercises on the concepts involved.

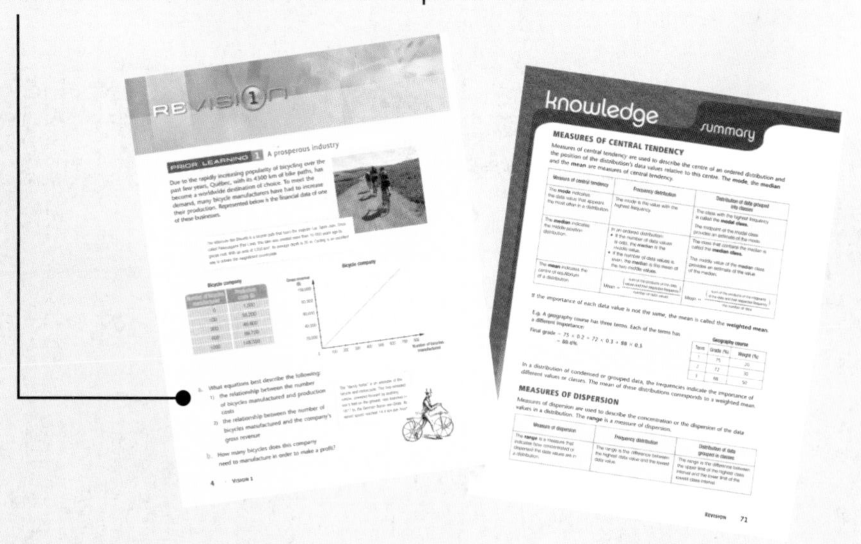

The sections

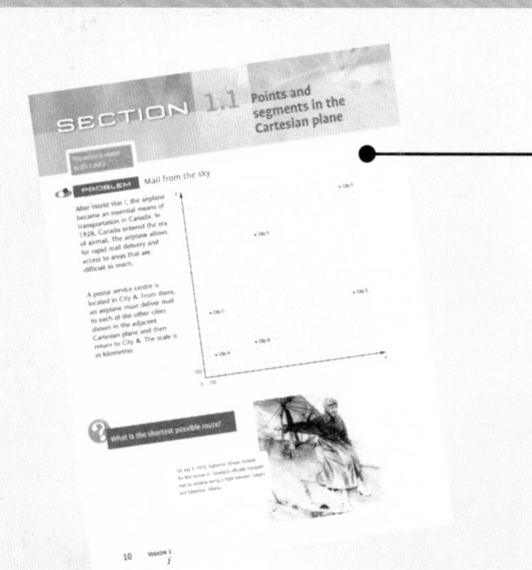

A "Vision" chapter is divided into sections, each starting with a problem and a few activities, followed by the "Technomath," "Knowledge" and "Practice" features. Each section is related to a LES that contributes to the development of subject-specific and cross-curricular competencies, as well as to the integration of mathematical concepts that underscore the development of these competencies.

Problem

The first page of a section presents a problem that serves as a launching point and is made up of a single question. Solving the problem engages several competencies and various strategies while calling upon the mobilization of prior knowledge.

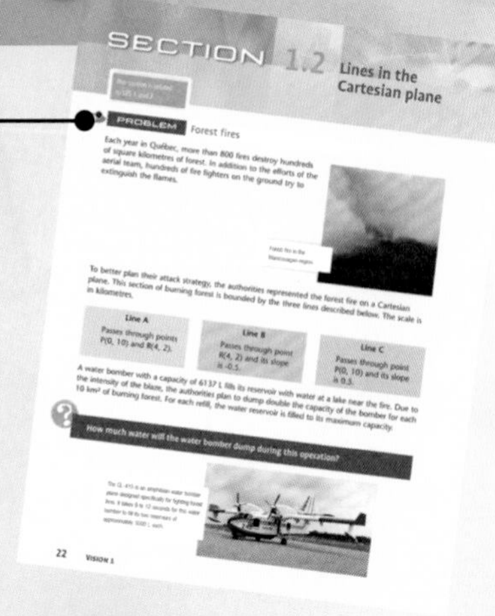

SECTION 1.2 Lines in the Cartesian plane

PROBLEM Forest fires

Each year in Québec, more than 800 fires destroy hundreds of square kilometres of forest. In addition to the efforts of the aerial team, hundreds of fire fighters on the ground try to extinguish the flames.

To better plan their attack strategy, the authorities represented the forest fire on a Cartesian plane. This section of burning forest is bounded by the three lines described below. The scale is in kilometres.

Line A	Line B	Line C
Passes through points P(0, 10) and R(4, 2).	Passes through point R(4, 2) and its slope is -0.5.	Passes through point P(0, 10) and its slope is 0.5.

A water bomber with a capacity of 6137 L fills its reservoir with water at a lake near the fire. Due to the intensity of the blaze, the authorities plan to dump double the capacity of the bomber for each 10 km² of burning forest. For each refill, the water reservoir is filled to its maximum capacity.

How much water will the water bomber dump during this operation?

22 VISION 1

Activity

The activities contribute to the development of subject-specific and cross-curricular competencies, require the use of various strategies, mobilize knowledge and further the understanding of mathematical notions. These activities can take on several forms: questionnaires, material manipulation, construction, games, stories, simulations, historical texts, etc.

ACTIVITY 1 Tricky marks!

a. If a represents the mark on the first test and b the mark on the second test, what system of equations is formed by the two statements made by the teacher?

b. What do you observe concerning the position of the variables in relation to the equals sign for the equation corresponding to:
1) the first statement?
2) the second statement?

c. What is the one-variable equation obtained when you replace variable a with the expression 2b − 92 into the equation from the first statement?

d. Solve the equation obtained in c). Explain what the solution represents in relation to the context.

e. What is this student's mark on the first test?

f. Graph the system of equations corresponding to this situation and find the coordinates of the intersection point. What did you notice?

SECTION 1.3 35

Technomath

The "Technomath" section allows students to use technological tools such as a graphing calculator, dynamic geometry software or a spreadsheet program. In addition, the section shows how to use these tools and offers several questions in direct relation to the mathematical concepts associated with the content of the chapter.

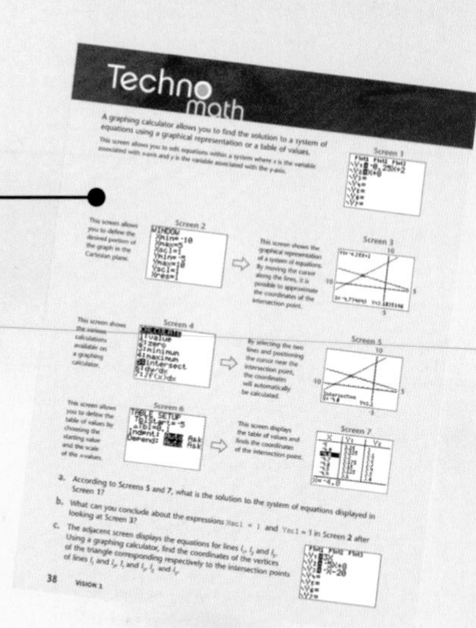

Technomath

A graphing calculator allows you to find the solution to a system of equations using a graphical representation or a table of values.

This screen allows you to edit equations within a system where x is the variable associated with x-axis and y is the variable associated with the y-axis.

This screen allows you to define the desired portion of the graph in the Cartesian plane.

This screen shows the graphical representation of a system of equations. By moving the cursor along the lines, it is possible to approximate the coordinates of the intersection point.

This screen shows the various calculations available on a graphing calculator.

By selecting the two lines and positioning the cursor near the intersection point, the coordinates will automatically be calculated.

This screen allows you to define the table of values by choosing the starting value and the scale of the x-values.

This screen displays the table of values and finds the coordinates of the intersection point.

a. According to Screens 5 and 7, what is the solution to the system of equations displayed in Screen 1?

b. What can you conclude about the expressions Xscl = 1 and Yscl = 1 in Screen 2 after looking at Screen 3?

c. The adjacent screen displays the equations for lines l_1, l_2 and l_3. Using a graphing calculator, find the coordinates of the vertices of the triangle corresponding respectively to the intersection points of lines l_1 and l_2, l_1 and l_3, l_2 and l_3.

38 VISION 1

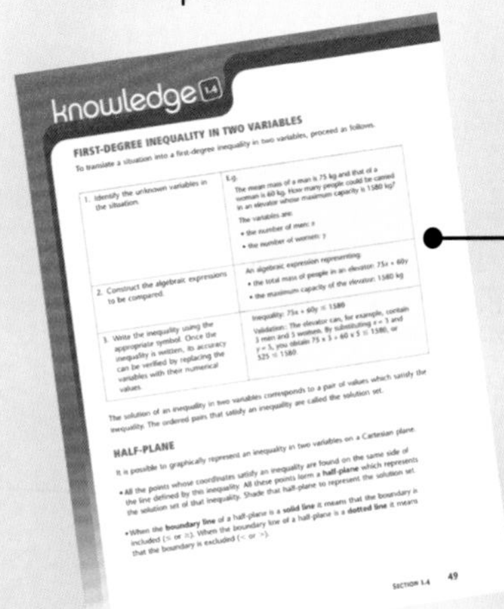

knowledge 1.4

FIRST-DEGREE INEQUALITY IN TWO VARIABLES

To translate a situation into a first-degree inequality in two variables, proceed as follows.

1. Identify the unknown variables in the situation.	E.g. The mean mass of a man is 75 kg and that of a woman is 60 kg. How many people could be carried in an elevator whose maximum capacity is 1580 kg? The variables are: • the number of men: x • the number of women: y
2. Construct the algebraic expressions to be compared.	An algebraic expression representing: • the total mass of people in an elevator: 75x + 60y • the maximum capacity of the elevator: 1580 kg
3. Write the inequality using the appropriate symbol. Once the inequality is written, its accuracy can be verified by replacing the variables with their numerical values.	Inequality: 75x + 60y ≤ 1580 Validation: The elevator can, for example, contain 3 men and 5 women. By substituting x = 3 and y = 5, you obtain 75 × 3 + 60 × 5 ≤ 1580, or 525 ≤ 1580.

The solution of an inequality in two variables corresponds to a pair of values which satisfy the inequality. The ordered pairs that satisfy an inequality are called the solution set.

HALF-PLANE

It is possible to graphically represent an inequality in two variables on a Cartesian plane.

• All the points whose coordinates satisfy an inequality are found on the same side of the line defined by this inequality. All these points form a **half-plane** which represents the solution set of that inequality. Shade that half-plane to represent the solution set.

• When the **boundary line** of a half-plane is a **solid line** it means that the boundary is included (≤ or ≥). When the boundary line of a half-plane is a **dotted line** it means that the boundary is excluded (< or >).

SECTION 1.4 49

Knowledge

The "Knowledge" section presents a summary of the theoretical elements encountered in the section. Theoretical statements are supported with examples in order to foster students' understanding of the various concepts.

Practice

The "Practice" section presents a series of contextualized exercises and problems that foster the development of the competencies and the consolidation of what has been learned throughout the section.

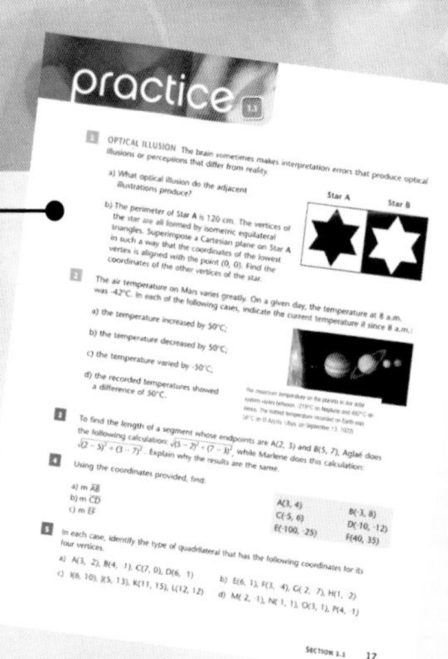

Special Features

Chronicle of the past

The "Chronicle of the past" feature recalls the history of mathematics and the lives of certain mathematicians who have contributed to the development of mathematical concepts that are directly related to the content of the "Vision" chapter being studied. This feature includes a series of questions that deepen students' understanding of the subject.

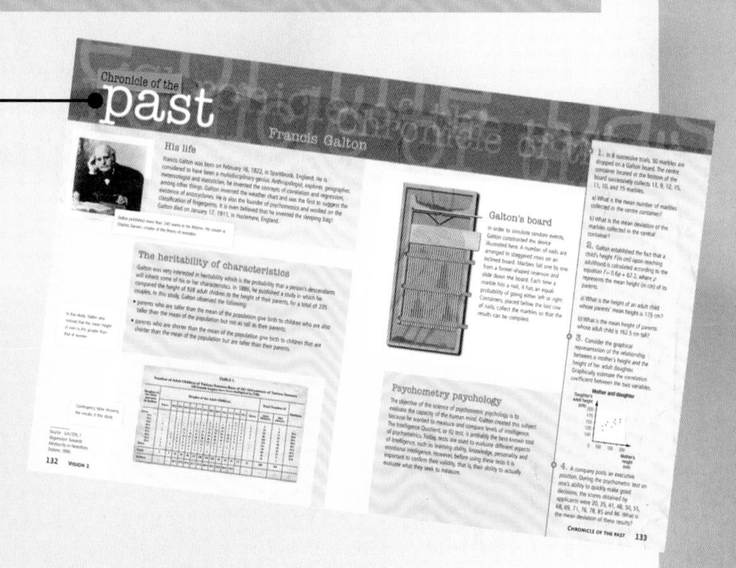

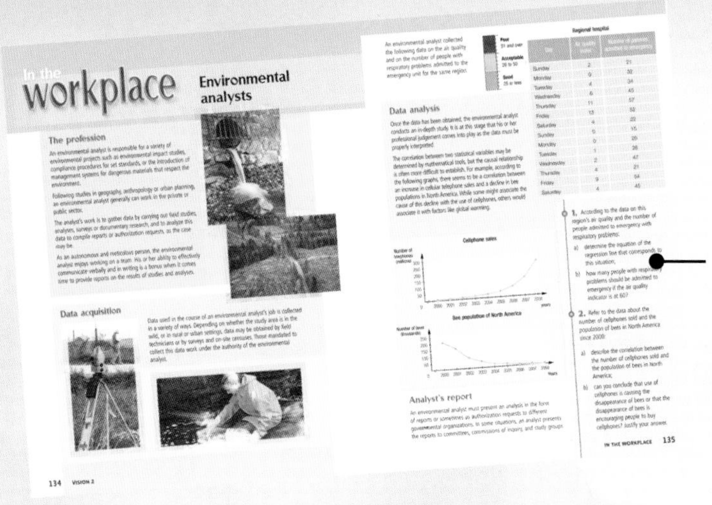

In the workplace

The "In the workplace" feature presents a profession or a trade that makes use of the mathematical notions studied in the related "Vision" chapter. This feature includes a series of questions designed to deepen students' understanding of the subject.

Overview

The "Overview" feature concludes each "Vision" chapter and presents a series of contextualized exercises and problems that integrate and consolidate the competencies that have been developed and the mathematical notions studied. This feature ends with a bank of problems, each of which focuses on solving, reasoning or communicating.

The "Practice" and "Overview" features, include the following:

- A number in a blue square refers to a Priority **1** and a number in an orange square a Priority **2**.
- When a problem refers to actual facts, a keyword written in red uppercase indicates the subject with which it is associated.

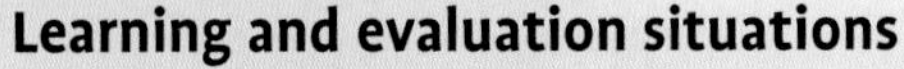

Learning and evaluation situations

The "Learning and evaluation situations" (LES) are grouped according to a common thematic thread; each focuses on a general field of instruction, a subject-specific competency and two cross-curricular competencies. The knowledge acquired through the sections helps to complete the tasks required in the LES.

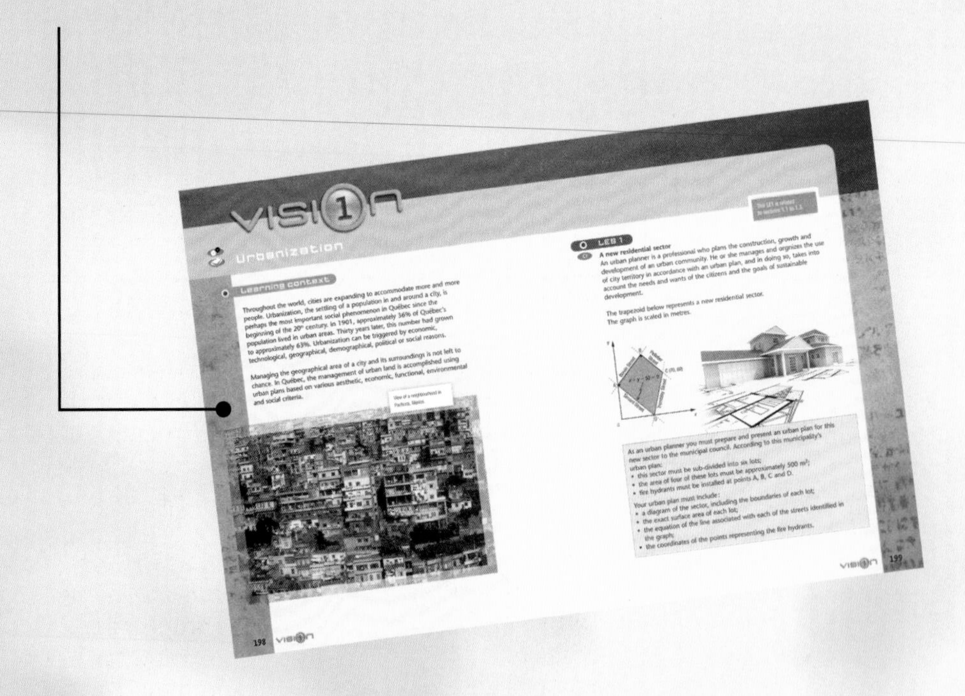

Reference

Located at the end of the *Student Book*, the "Reference" section contains several tools that support the student-learning process. It consists of two distinct parts.

The "Technology" part provides explanations pertaining to the functions of a graphing calculator, the use of a spreadsheet program as well as the use of dynamic geometry software.

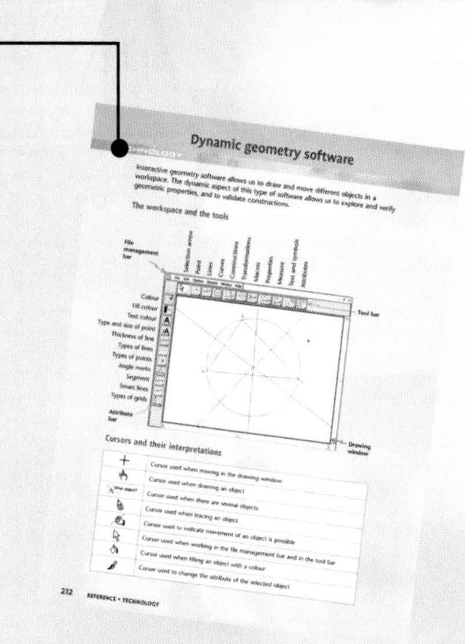

Dynamic geometry software

Interactive geometry software allows us to draw and move different objects in a workspace. The dynamic aspect of this type of software allows us to explore and verify geometric properties, and to validate constructions.

The workspace and the tools

Cursors and their interpretations

+	Cursor used when moving in the drawing window
	Cursor used when drawing an object
	Cursor used when there are several objects
	Cursor used when tracing an object
	Cursor used to indicate movement of an object is possible
	Cursor used when working in the file management bar and in the tool bar
	Cursor used when filling an object with a colour
	Cursor used to change the attribute of the selected object

212 REFERENCE • TECHNOLOGY

The "Knowledge" part presents notations and symbols used in the *Student Book*. Geometric principles are also listed. This part concludes with a glossary and an index.

Icons

Indicates that a worksheet is available in the *Teaching Guide*.

Indicates that some key features of subject-specific competency 1 are mobilized.

Indicates that subject-specific competency 1 is being targeted in the LES.

Indicates that the activity can be performed in teams. Details on this topic are provided in the *Teaching Guide*.

Indicates that some key features of subject-specific competency 2 are mobilized.

Indicates that subject-specific competency 2 is being targeted in the LES.

Indicates that some key features of subject-specific competency 3 are mobilized.

Indicates that subject-specific competency 2 is being targeted in the LES.

VISION 1

From lines to systems of equations

What equation best describes the trajectory of an airplane? What is the distance between two cities? What are the coordinates of a point associated with the position of a ship? Which company offers the best rates?In “Vision 1,” you will express situations using first-degree equations and inequalities in two variables. You will also learn certain concepts related to analytical geometry to calculate the distance between two points and to find the coordinates of points on a line segment in a given ratio.

Arithmetic and algebra	Geometry	Statistics	Probabilities
• Solving first-degree equations in two variables using substitution and elimination methods • First-degree inequalities in two variables	• Distance between two points • Point of division • Slope of a segment and slope of a line • Equation of a line • Parallel and perpendicular lines • Half-planes		

REVISION 1

PRIOR LEARNING 1 A prosperous industry

Due to the rapidly increasing popularity of bicycling over the past few years, Québec, with its 4300 km of bike paths, has become a worldwide destination of choice. To meet the demand, many bicycle manufacturers have had to increase their production. Represented below is the financial data of one of these businesses.

The Véloroute des Bleuets is a bicycle path that tours the majestic Lac Saint-Jean. Once called Piékouagami (Flat Lake), this lake was created more than 10 000 years ago by glacial melt. With an area of 1350 km², its average depth is 20 m. Cycling is an excellent way to admire the magnificent countryside.

Bicycle company

Number of bicycles manufactured	Production costs ($)
0	1,500
100	16,200
300	45,600
600	89,700
1000	148,500

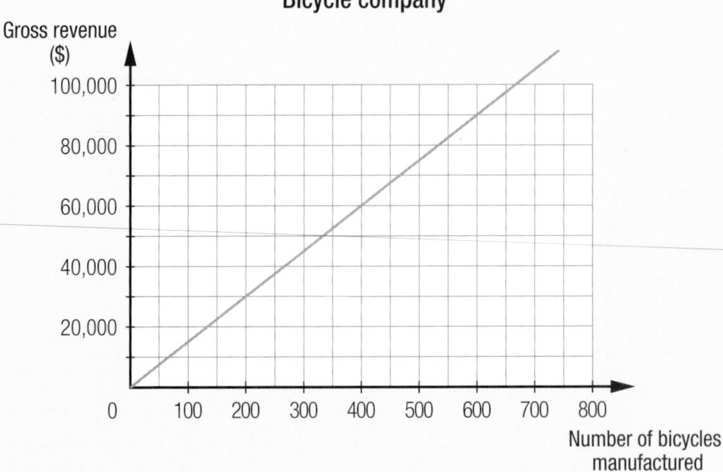

a. What equations best describe the following:
   1) the relationship between the number of bicycles manufactured and production costs
   2) the relationship between the number of bicycles manufactured and the company's gross revenue

b. How many bicycles does this company need to manufacture in order to make a profit?

The "dandy horse" is an ancestor of the bicycle and motorcycle. This two-wheeled vehicle, powered forward by pushing one's feet on the ground, was invented in 1817 by the German Baron von Drais. Its record speed reached 14.4 km per hour!

PRIOR LEARNING 2 The great trek

The Traversée de la Gaspésie is an annual event for amateur cross-country skiers. Over the course of a week, approximately 150 skiers cover at least 305 km of trails surrounded by magnificent landscapes from Rimouski to Gaspé-Forrilon.

Finish line: Gaspé-Forillon

The information below is in relation to this cross country ski trek:

- The distance travelled on the second day is equal to 55 km less than twice the distance covered on the first day.
- The distance travelled on the third day is equal to 10 km more than the distance travelled on the first day.
- The distance travelled on the fourth day is the same as the distance travelled on the second day.
- The distance travelled on the fifth day is the same as the distance travelled on the first day.
- The distance travelled on the sixth day is equal to 5 km more than the distance travelled on the first day.

a. Let d be the distance travelled on the first day. For each of the following days, find an algebraic expression that represents the distance travelled.

b. What is the minimum distance travelled each day by the skiers?

The Chic-Choc Mountains, which cross the Gaspé Peninsula, have 25 summits over 1000 m high. Mount Jacques-Cartier reaches a height of 1268 m. In winter, the Chic-Choc Mountains receive the most snow in all of Québec.

SOLVING SYSTEMS OF EQUATIONS

Different strategies can be used to solve systems of first-degree equations in two variables, in other words, to find the values of the variables that simultaneously satisfy both equations.

Graphical representation

In a graphical representation, the coordinates of the intersection point of two lines represent the solution to the system of equations. Graphical representation often provides only an approximation of the solution.

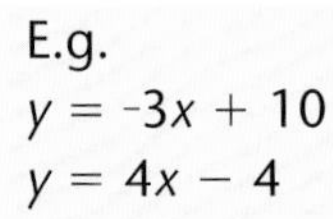

E.g.
$y = -3x + 10$
$y = 4x - 4$

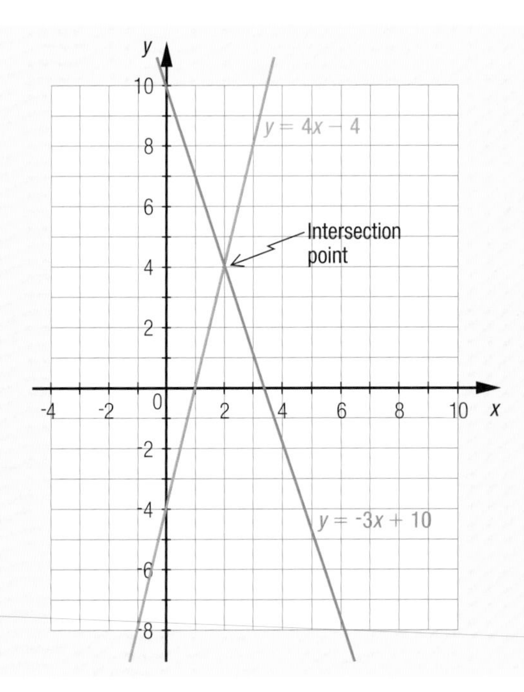

The solution is (2, 4).

Table of values

It is possible to solve a system of equations by constructing a table of values. Find the value of the independent variable for which the dependent variables are equal.

E.g.
$y = 2x + 1$
$y = -4x + 7$

x	-2	-1	0	**1**	2	3	4
y	-3	-1	1	**3**	5	7	9
y	15	11	7	**3**	-1	-5	-9

The solution is (1, 3).

Comparison method

The comparison method allows you to solve a system of equations by comparing algebraic expressions when they are of the form $\begin{array}{l} y = a_1x + b_1 \\ y = a_2x + b_2 \end{array}$.

E.g. To solve the system $y = -110x + 1900$, $y = -150x + 2400$ using the comparison method, do the following:	
1. Compare the two algebraic expressions containing the variable that is not isolated.	$-110x + 1900 = -150x + 2400$
2. Solve the equation resulting from Step **1**.	$-110x + 1900 = -150x + 2400$ $40x = 500$ $x = 12.5$
3. In one of the initial equations, replace the variable solved with the value obtained in Step **2** in order to find the value of the other variable.	$y = -150 \times \mathbf{12.5} + 2400$ $y = 525$ The solution is (12.5, 525).
4. Validate the solution by substituting 12.5 for x and 525 for y in each of the initial equations. $\mathbf{525} = -110 \times \mathbf{12.5} + 1900$ $\mathbf{525} = -150 \times \mathbf{12.5} + 2400$	

FIRST-DEGREE INEQUALITIES IN ONE VARIABLE

Finding the set of values that satisfy an inequality means that you have **solved** this inequality. Sometimes it is necessary to use inequalities to find the solution to a problem. Follow the procedure below.

1. Identify the unknowns.	E.g. The perimeter of a rectangular plot of land is at least 178 m. The length is 5 m more than triple its width. Calcuate the possible dimensions of the land. The unknowns are: • the width of the plot • the length of the plot
2. Represent each unknown quantity with a variable or an algebraic expression involving variables.	The width of the plot (in m) = x The length of the plot (in m) = $3x + 5$
3. Construct an inequality that represents the situation.	$2(x + 3x + 5) \geq 178$
4. Solve the inequality following the rules of inequality transformations.	$2(x + 3x + 5) \geq 178$ $2(4x + 5) \geq 178$ $8x + 10 \geq 178$ $8x \geq 168$ $x \geq 21$
5. State the solution taking the context into account.	You can deduce that the width of the plot must be at least 21 m. For example, the plot's dimensions could be 21 m by 68 m.

knowledge in action

1 Express each of the following statements as an inequality.

a) x is no more than 2.

b) x is at least 14.

c) x is less than 9.

d) x is no less than 6.

e) x is smaller than y.

f) x is less than or equal to y.

2 In each case, find the solution to the system of equations.

a) System of equations 1

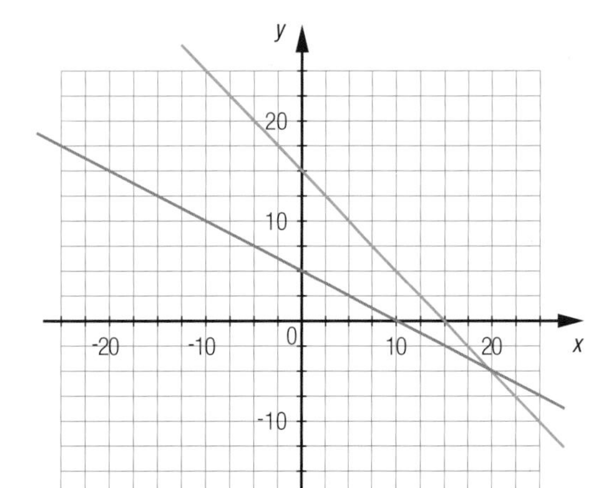

b) System of equations 2

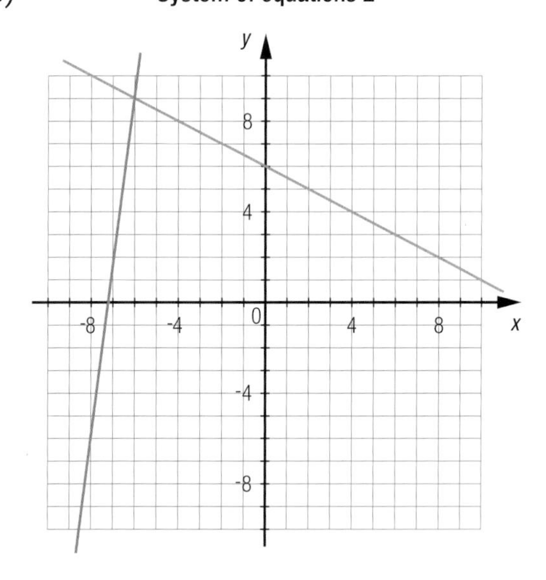

c) System of equations 3

x	-30	-15	0	30	60
y_1	160	130	100	40	-20
y_2	85	130	175	265	355

d) System of equations 4

x	-38	-36	-34	-32	-30
y_1	78	76	74	72	70
y_2	84	80	76	72	68

3 Match each of the expressions in the left-hand column with one of the graphical representations in the right-hand column.

A $t < -3$

B $t \geq 3$

C $-3 \leq t < 3$

D $t \leq 3$

1 

2 -3

3 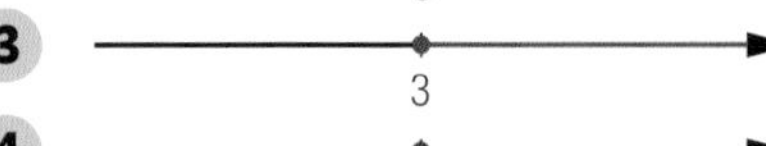

4 3

4 Solve the following inequalities.

a) $24 + 3x > 18$

b) $3a - 12 \leq -2a + 28$

c) $-(t - 1) + 7 \geq -11$

d) $-14 + 0.2b < 8$

e) $-3.4m - 7.2 \leq 7.08$

f) $18 \geq 36 - 3(c + 11)$

g) $-\frac{5n + 1}{2} < 9$

h) $\frac{2x - 4}{3} > -\frac{x - 2}{2}$

5 Solve the following systems of equations.

a) $y = 30 - x$
$y = x + 2$

b) $y = -7x + 10$
$y = x - 10$

c) $y = -x$
$y = x + 1$

d) $y = 17 - 3x$
$y = -x - 3$

e) $y = \frac{x}{2}$
$y = -3x + 4$

f) $y = \frac{2x - 17}{25}$
$y = \frac{x - 6}{15}$

6 The base of the adjacent parallelogram measures four times its height. For what values of x:

a) does the parallelogram exist?

b) is the perimeter of the parallelogram less than or equal to 60 cm?

c) is the area of the parallelogram greater than 169 cm^2?

7 Eve and Myriam go cross-country skiing together. Eve has 1800 mL of water in a container, and she consumes it at a rate of 500 mL/h, whereas Myriam drinks 100 mL/h from her own one-litre supply.

a) Identify the unknowns in this situation and represent them using different variables.

b) Express the situation above as a system of equations.

c) When is the amount of water in both containers equal?

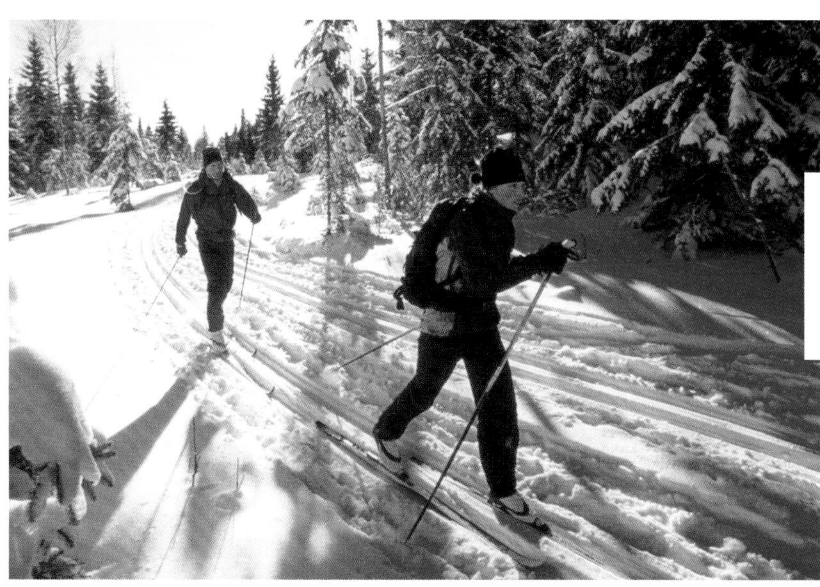

In cross-country skiing, the classic style consists of moving forward in two parallel tracks. A new technique called skating or free technique resembles ice-skating.

SECTION 1.1 Points and segments in the Cartesian plane

This section is related to LES 1 and 2.

PROBLEM Mail from the sky

After World War I, the airplane became an essential means of transportation in Canada. In 1928, Canada entered the era of airmail. The airplane allows for rapid mail delivery and access to areas that are difficult to reach.

A postal service centre is located in City **A**. From there, an airplane must deliver mail to each of the other cities shown in the adjacent Cartesian plane and then return to City **A**. The scale is in kilometres.

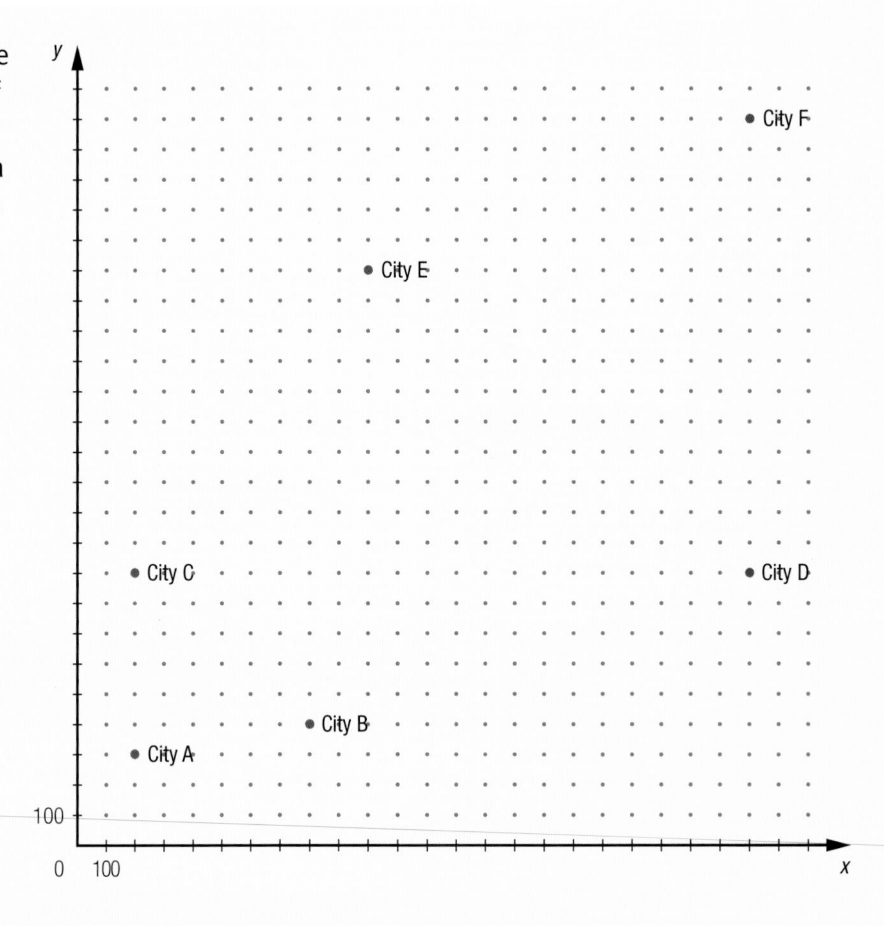

What is the shortest possible route?

On July 9, 1918, Katherine Stinson became the first woman in Canada to officially transport mail by airplane during a flight between Calgary and Edmonton, Alberta.

ACTIVITY 1 Caution! Steep hill

Road signs, similar to this one, warn drivers of an upcoming steep hill. The percentage posted corresponds to the steepness of this portion of the road in relation to a horizontal plane. This particular sign indicates a drop of 12 m for each 100 m of flat road.

In 1923, J. Omer Martineau, an assistant chief engineer with the Ministère de la Voirie du Québec, implemented the use of symbols (pictograms or images) on signs rather than words because at that time, few people knew how to read.

A car travels from point A to point B on the inclined portion of the road shown in the adjacent graph. The scale of the graph is in metres.

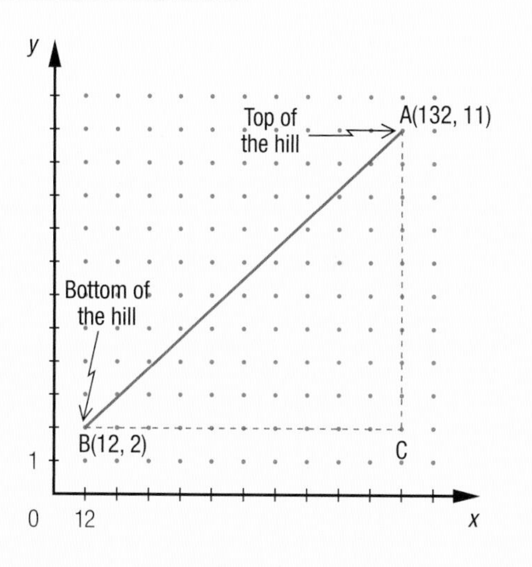

a. What is the change in the y-coordinate, or the number associated with the change in the vertical position of the vehicle?

b. What is the change in the x-coordinate, or the number associated with the change in the horizontal position of the vehicle?

c. A road sign indicating a steep hill is installed near this section of the road. What percent of inclination should be posted on the sign?

d. What are the coordinates of point C?

e. What type of triangle is ΔABC?

f. Find:

1) the distance between point A and point C
2) the distance between point B and point C
3) the distance travelled by the car on the inclined portion of the road

Road signs, as old as the road system itself, were mainly developed after the introduction of cars. Certain signs, for example, those indicating a steep hill, are fairly similar throughout the world. Others, however, are more unique. For example:

Australia

China

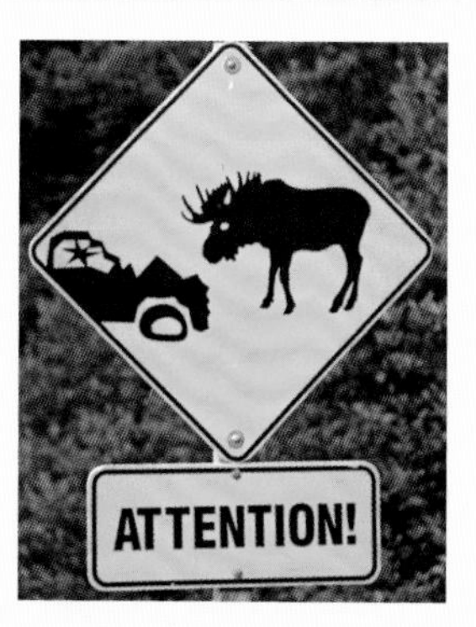

Newfoundland & Labrador

ACTIVITY 2 Land surveying

In 1626, Samuel de Champlain, considered Canada's first land surveyor, surveyed the first three seigneuries. Today, land surveying is required to delimit or describe a region or any piece of land used for specific purposes such as wildlife reserves or power transmission corridors.

Samuel de Champlain (1570-1635) Cartographer, explorer and Governor of New France

On the land shown in the adjacent graph, a surveyor must place survey marker M at the midpoint of $\overline{AB}$.

The beginning of the surveyor's work is illustrated in red. The scale is in metres.

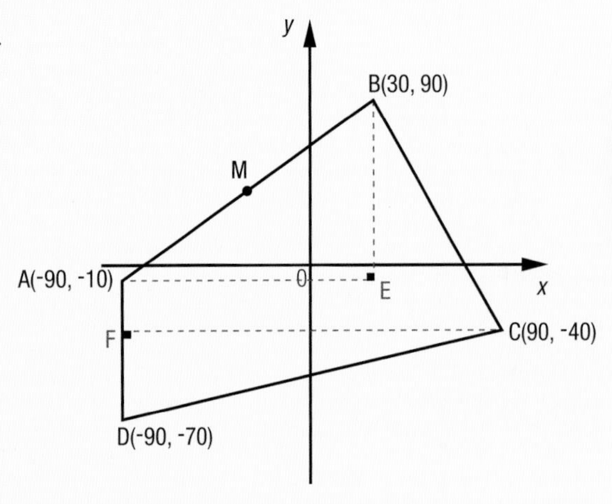

a. What are the coordinates of point E?

b. Find the coordinates of:

1) the midpoint of $\overline{AE}$
2) the midpoint of $\overline{BE}$
3) the survey marker M

On the same piece of land, survey marker P must be positioned $\frac{2}{3}$ along segment DC, i.e. at $\frac{2}{3}$ of the segment starting at point D.

c. Will the survey marker P be situated closer to point C or point D? Explain your answer.

d. Point P divides $\overline{DC}$ into what ratio?

e. Find the coordinates of:

1) the point situated $\frac{2}{3}$ along $\overline{FC}$
2) the point situated $\frac{2}{3}$ along $\overline{DF}$
3) survey marker P

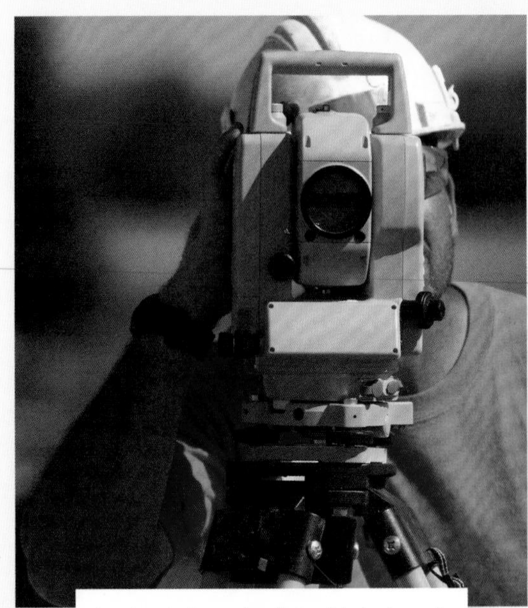

Land surveying makes it possible to draw all kinds of maps, precisely locate buildings, mark the property boundaries and determine the position of satellites or oil-drilling rigs.

ACTIVITY 3 Algebra or geometry?

Analytic geometry makes it possible to solve geometric problems using algebra.

French mathematicians René Descartes and Pierre de Fermat made many discoveries related to analytic geometry. However, they independently developed certain concepts related to the field of mathematics.

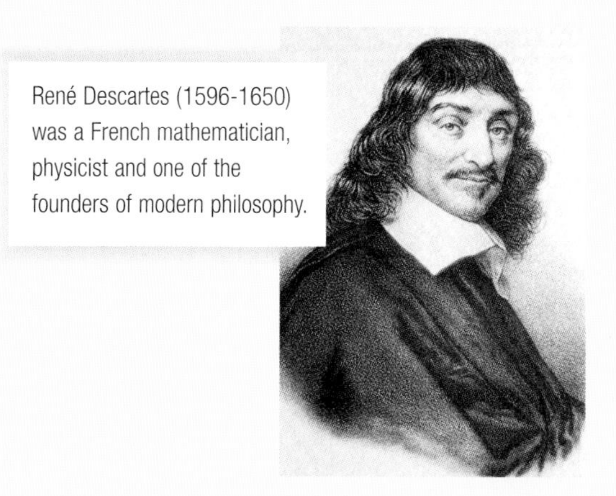

René Descartes (1596-1650) was a French mathematician, physicist and one of the founders of modern philosophy.

Pierre de Fermat (1601-1665) was a French jurist and mathematician. He was known as the "prince of amateurs."

Descartes used algebra to support geometry in proving that the midpoint of the hypotenuse of a right triangle is equidistant from each of the three vertices of the triangle.

To prove this, right triangle ABC is drawn on the Cartesian plane below.

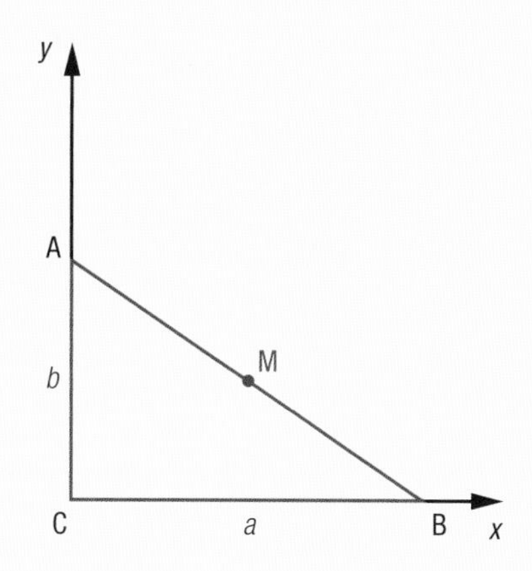

a. What needs to be proven?

b. Find the coordinates of each of the vertices of ABC.

c. Find the coordinates of the midpoint M.

d. Find an algebraic expression representing the length of:

1) $\overline{AM}$ 2) $\overline{BM}$ 3) $\overline{CM}$

e. What can you conclude after comparing the three lengths found in **d.**?

f. Why is it preferable to align two sides of triangle ABC with the axes of the Cartesian plane?

Techno math

Dynamic geometry software allows you to draw and manipulate figures in a Cartesian plane. By using the tools: SHOW AXES, LINE SEGMENT, MIDPOINT, COORDINATES, PERPENDICULAR LINE, TRIANGLE, DISTANCE and SLOPE, you can draw a segment in the Cartesian plane and display its slope.

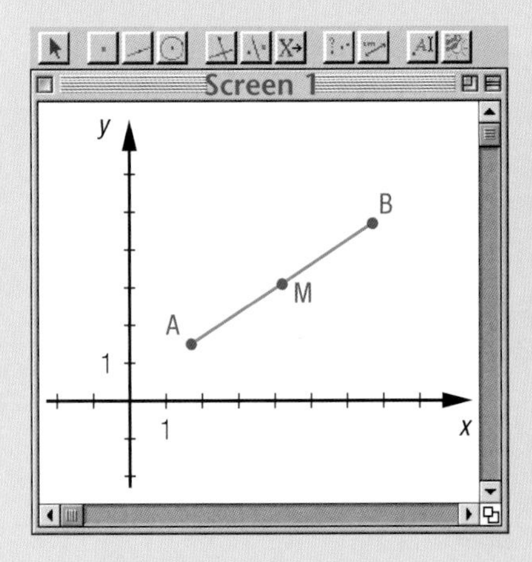

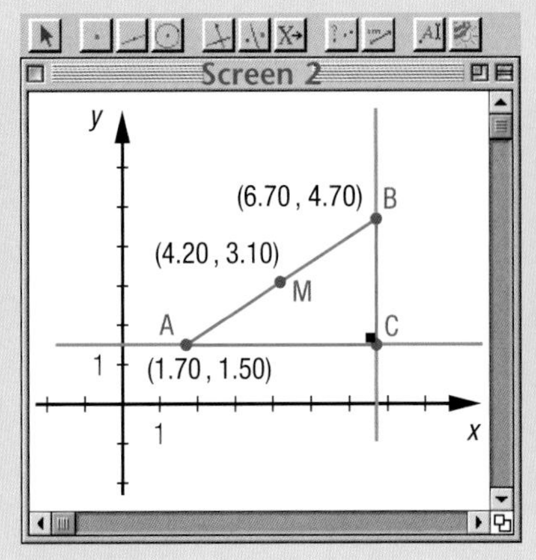

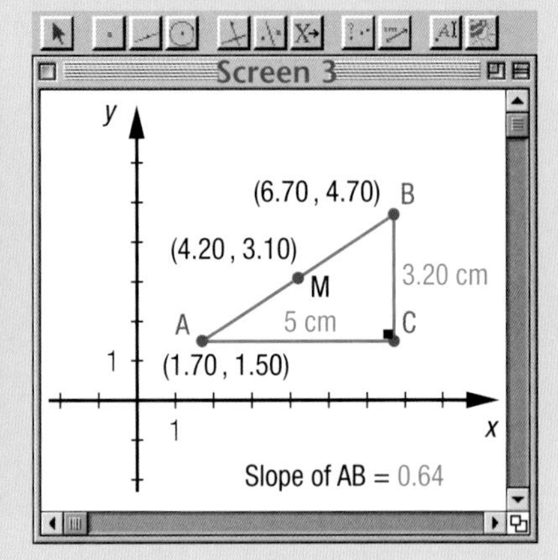

By modifying the position or the inclination of the segment AB, you can observe changes to the coordinates of the points and of the slope.

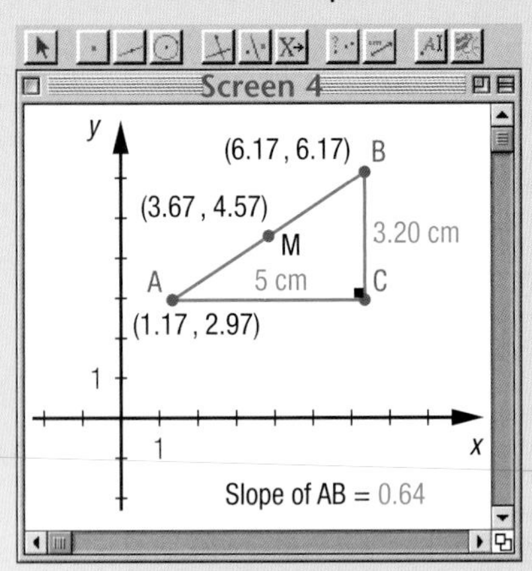

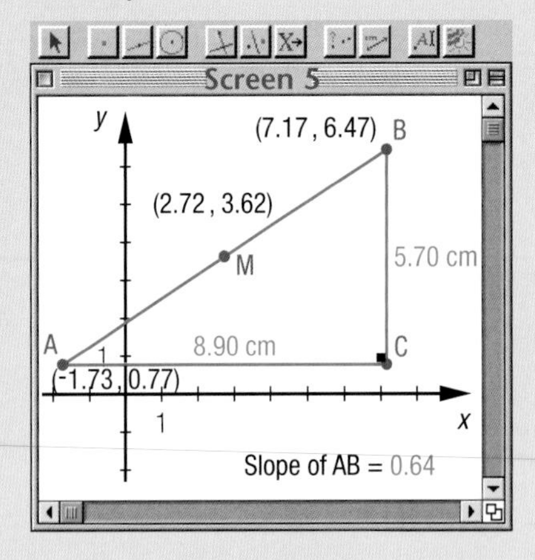

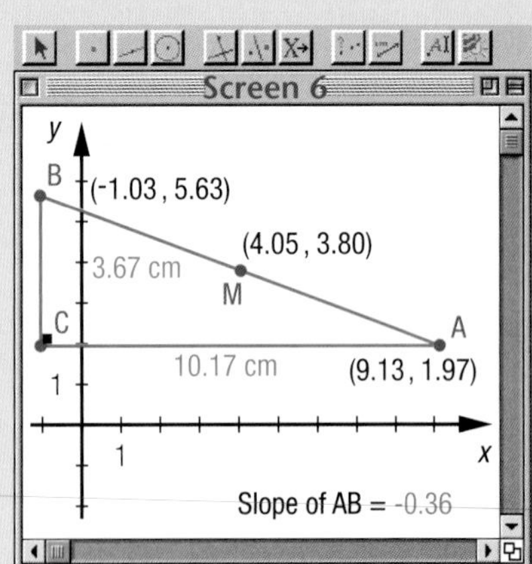

a. Compared to Screen **3**, what changes have been made to:

1) Screen **4**? 2) Screen **5**? 3) Screen **6**?

b. For Screens **3** through **6**, verify the following:

1) The *x*-value of point M equals the mean of the *x* values of points A and B.
2) The *y*-value of M equals the mean of the *y* values of points A and B.

c. For Screens **4** through **6** do the following:

1) Compare the ratio: $\frac{\text{change in } y\text{-values from B to A}}{\text{change in } x\text{-values from B to A}}$ to the slope of segment AB.
2) Find the length of segment AB.

d. Using dynamic geometry software, determine:

1) if there is a relationship between the slope and the length of a segment
2) the slope of a segment that is parallel to the *x*-axis
3) the slope of a segment that is parallel to the *y*-axis

knowledge 1.1

CHANGE ON THE AXES

For a point $A(x_1, y_1)$ and a point $B(x_2, y_2)$, note the following:

- The **change in x-values** from A to B is: $\Delta x = x_2 - x_1$.
- The **change in y-values** from A to B is: $\Delta y = y_2 - y_1$.

SLOPE OF A SEGMENT

The slope of a segment whose endpoints are $A(x_1, y_1)$ and $B(x_2, y_2)$ is a value that defines its slope, rate of change, gradient, inclination or steepness. The slope equals the ratio of the change in y-value to the change in x-value. The slope of a segment joining these two points is calculated by using the following formula.

$$\text{Slope of } \overline{AB} = \frac{\Delta y}{\Delta x} = \frac{y_2 - y_1}{x_2 - x_1}$$

E.g. The slope of segment AB whose endpoints are A(1, 6) and B(-7, 12) is calculated as follows.

$$\text{Slope of } \overline{AB} = \frac{y_2 - y_1}{x_2 - x_1} = \frac{12 - 6}{-7 - 1} = \frac{3}{-4}$$

DISTANCE BETWEEN TWO POINTS

The distance between point A and point B is equal to the length of the segment joining these two points. The length is expressed as a positive number.

The distance d between point $A(x_1, y_1)$ and point $B(x_2, y_2)$ is calculated using the following formula.

The absolute value of a real number allows you to ignore the sign of the number. Absolute value is indicated by putting the actual number between two vertical bars like so: $|3| = 3$ and $|-3| = 3$.

$$d(A, B) = \sqrt{(x_2 - x_1)^2 + (y_2 - y_1)^2}$$

E.g. The distance between point A(3, 4) and point B(-2, 6) is calculated as follows.

$$d(A, B) = \sqrt{(x_2 - x_1)^2 + (y_2 - y_1)^2} = \sqrt{(-2 - 3)^2 + (6 - 4)^2} = \sqrt{29} \text{ or } \approx 5.39 \text{ u}$$

POINT OF DIVISION

The position of any division point on a segment can be found using a fraction or a ratio.

E.g.

In the adjacent graphical representation, note the following:

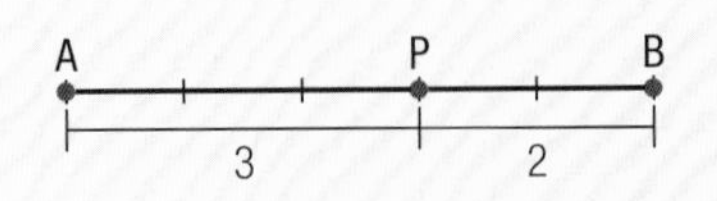

- Point P is situated $\frac{3}{5}$ along segment AB.
- Point P divides segment AB into a ratio of 3:2.
- Point P is situated $\frac{2}{5}$ along segment BA.
- Point P divides segment BA into a ratio of 2:3.

Point of division P is on segment AB, whose endpoints are $A(x_1, y_1)$ and $B(x_2, y_2)$.

If point P is located at a fraction $\frac{a}{b}$ of the distance between points A and B, then its coordinates are:

$$\left(x_1 + \frac{a}{b} \times \Delta x, y_1 + \frac{a}{b} \times \Delta y\right)$$

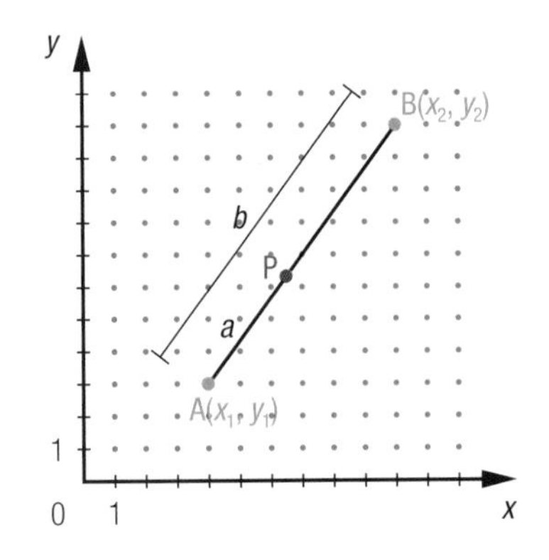

E.g.

1. The coordinates of the midpoint M of segment AB whose endpoints are A (-2, -4) and B(4, 6) can be found by doing the following calculations:

$$\left(x_1 + \frac{a}{b} \times \Delta x, y_1 + \frac{a}{b} \times \Delta y\right)$$

$$\left(-2 + \frac{1}{2}(4 + 2), -4 + \frac{1}{2}(6 + 4)\right)$$

(1, 1)

The coordinates of the midpoint M are (1, 1).

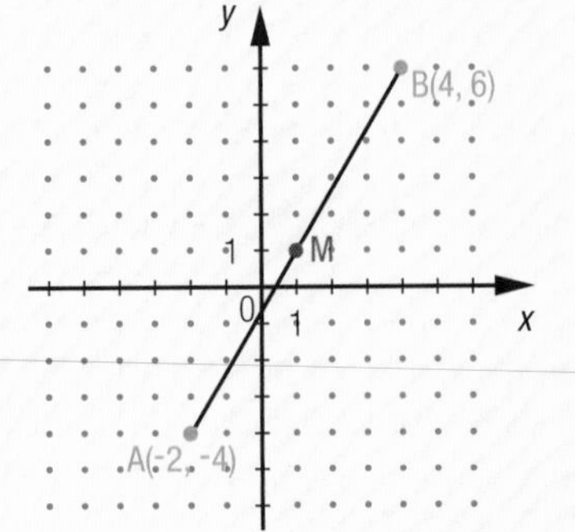

2. Point P divides segment AB in a ratio of 3:1 whose endpoints are A(3, 7) and B(-4, -10). The coordinates of point P can be found by doing the following calculations.

The ratio 3:1 corresponds to the fraction $\frac{3}{4}$.

$$\left(x_1 + \frac{a}{b} \times \Delta x, y_1 + \frac{a}{b} \times \Delta y\right)$$

$$\left(3 + \frac{3}{4}(-4 - 3), 7 + \frac{3}{4}(-10 - 7)\right)$$

(-2.25, -5.75)

The coordinates of point P are (-2.25, -5.75).

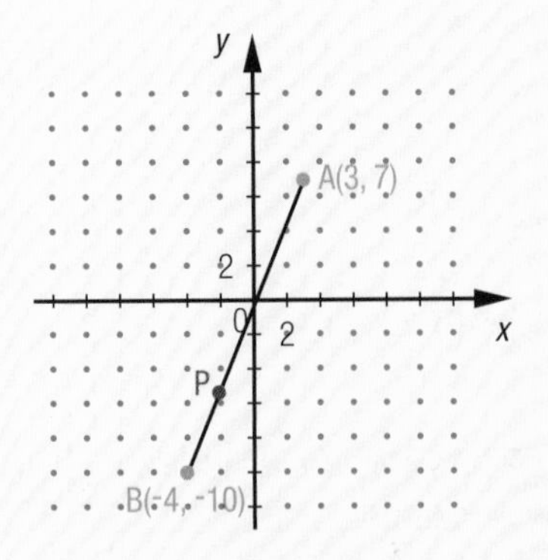

practice 1.1

1 **OPTICAL ILLUSION** The brain sometimes makes interpretation errors that produce optical illusions or perceptions that differ from reality.

Star A Star B

a) What optical illusion do the adjacent illustrations produce?

b) The perimeter of Star **A** is 120 cm. The vertices of the star are all formed by isometric equilateral triangles. Superimpose a Cartesian plane on Star **A** in such a way that the coordinates of the lowest vertex is aligned with the point (0, 0). Find the coordinates of the other vertices of the star.

2 The air temperature on Mars varies greatly. On a given day, the temperature at 8 a.m. was -42°C. In each of the following cases, indicate the current temperature if since 8 a.m. the following occured:

a) The temperature increased by 50°C.

b) The temperature decreased by 50°C.

c) The temperature varied by -50°C.

d) The recorded temperatures showed a difference of 50°C.

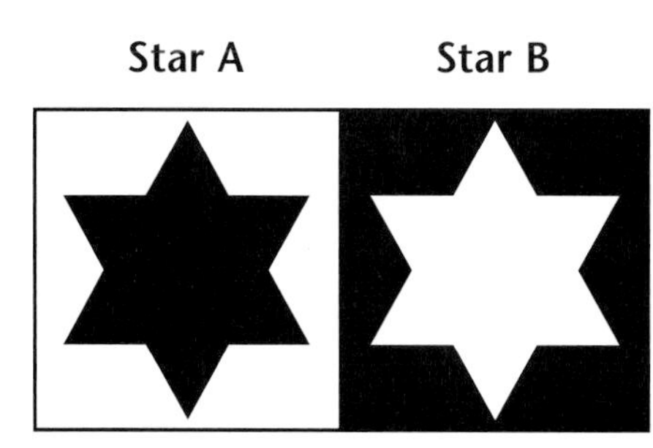

The maximum temperature on the planets in our solar system varies between -219°C on Neptune and 482°C on Venus. The hottest temperature recorded on Earth was 58°C (in El Azyzia, Libya, on September 13, 1922).

3 To find the length of a segment whose endpoints are A(2, 3) and B(5, 7), Aglaé does the following calculation: $\sqrt{(5-2)^2+(7-3)^2}$, while Marlene does this calculation: $\sqrt{(2-5)^2+(3-7)^2}$. Explain why the results are the same.

4 Using the coordinates provided, find:

a) m $\overline{AB}$
b) m $\overline{CD}$
c) m $\overline{EF}$

A(3, 4)	B(-3, 8)
C(-5, 6)	D(-10, -12)
E(-100, -25)	F(40, 35)

5 In each case, identify the type of quadrilateral that has the following coordinates for its four vertices.

a) A(3, -2), B(4, -1), C(7, 0), D(6, -1)
b) E(6, 1), F(3, -4), G(-2, -7), H(1, -2)
c) I(6, 10), J(5, 13), K(11, 15), L(12, 12)
d) M(-2, -1), N(-1, 1), O(3, 1), P(4, -1)

6 A seamstress sews buttons on a shirt. Below is the distance between each button and the collar.

1st button	2nd button	3rd button	4th button	5th button
2 cm	12 cm	22 cm	32 cm	42 cm

If the first button corresponds to point A and the fifth button to point B:

a) into what ratio does the second button divide segment AB?

b) into what ratio does the fourth button divide segment BA?

c) state the position of the 3rd button by using a fraction of $\overline{AB}$.

In the Middle Ages, the button had not yet been invented. A buckle or metal brooch called a "fermail" was used to hold or fasten together two sides of garments.

7 Identify the type of triangle defined by the following vertices:

a) A(10, 20), B(10, 60), C(40, 20)

b) D(0, 0), E(1, $\sqrt{3}$), F(2, 0)

c) G(0, -3), H(-1, 1), I(3, 2)

d) J(2, 1), K(1, -1), L(-2, -1)

8 Determine if the slope of a segment whose endpoints are A(x_1, y_1) and B(x_2, y_2) is greater, less than or equal to 0, given that from A to B, the following occur:

a) The change in *x*-values is greater than 0 and the change in *y*-values is less than 0.

b) $\Delta x < 0$ and $\Delta y > 0$.

c) $\Delta x < 0$ and $\Delta y < 0$.

d) The change in *x*-values is less than 0, and $\Delta y = 0$.

9 Calculate the slope of a segment whose endpoints are:

a) A(3, 5), B(5, 13)

b) B(-2, 4), C(3, 19)

c) C(-2, -3), D(24, -3)

d) D(2, 3), E(-2,-8)

10 Find the coordinates of a point:

a) located in the middle of segment AB if the endpoints are A(2, 3) and B(4, -11)

b) situated $\frac{3}{5}$ along segment CD if the endpoints are C(-3, -1) and D(2, 14)

c) situated $\frac{2}{3}$ along segment EF if the endpoints are E(3, 4) and F(1, -5)

d) that divides segment BC if the endpoints are B(3, -2) and C(-2, 3), in a ratio of 2:3

11 Find the circumference of a pool whose diameter has endpoints A(-1, -9) and B(-5, 5).

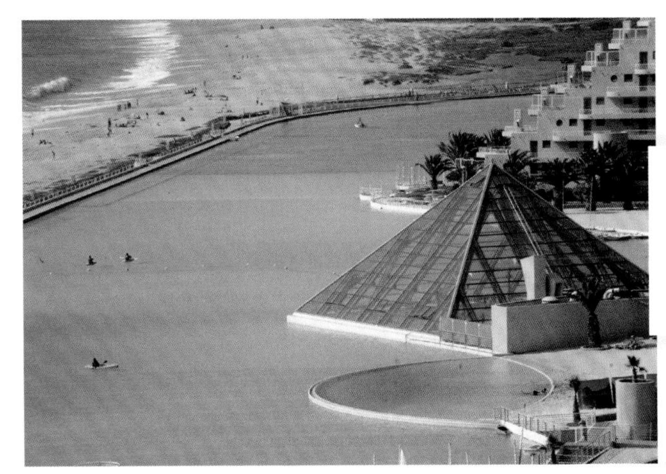

The Guinness Book of World Records recognizes San Alfonso del Mar, in Algarrobo, Chile, as the largest swimming pool in the world. The lagoon has a length of more than 1 km and has a depth of 3.5 m. It holds approximately 250 million litres of salt water.

12 In the adjacent Cartesian plane, right trapezoid ABCD has been drawn and the coordinates of its vertices are provided.

a) Find the midpoints of $\overline{AD}$ and $\overline{BC}$ and label these points M_1 and M_2 respectively.

b) Calculate the slope of segment:

1) AB
2) M_1M_2
3) DC

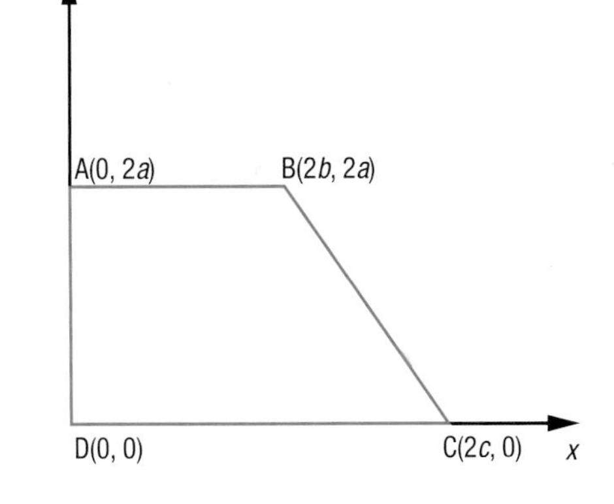

c) Using an algebraic expression, define the length of:

1) AB
2) M_1M_2
3) DC

d) Is it correct to state that the segment connecting the midpoints of the non-parallel sides of a right trapezoid is parallel to the bases and that the length of this segment is equal to half the sum of the bases? Justify your answer.

13 Point P(5, 8) is situated $\frac{3}{4}$ along segment EH. What are the coordinates of H if those of point E are (-1, -7)?

14 Below is some information on two neighbouring cities:

- City **A** is located at (15, 6).
- City **B** is located on the *y*-axis.
- The distance between the two cities is 17 km.

Find the coordinates of City **B**.

15 The adjacent figure represents the work of a landscaper. The perimeter is edged with a decorative border that costs \$5.95/m. The scale is in metres. Find:

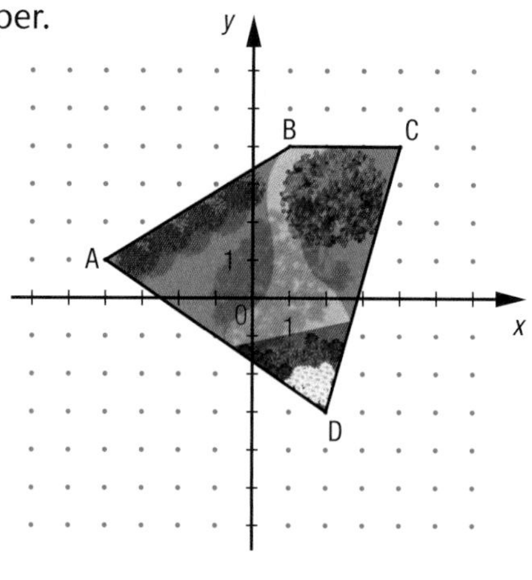

a) the total cost of the decorative border

b) the coordinates of an automatic sprinkler if it were situated at the midpoint of $\overline{AB}$

c) the coordinates of an automatic sprinkler if it were situated $\frac{4}{7}$ along segment DC

16 The slope of the ladder leaning against this building is -4. The scale is in metres. Find:

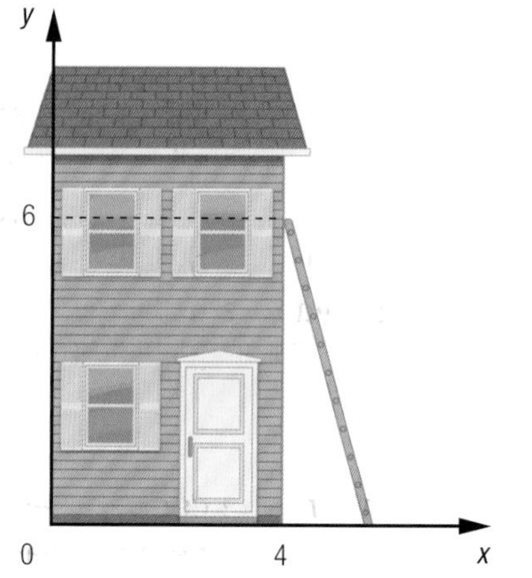

a) the coordinates corresponding to the foot of the ladder

b) the distance between the foot of the ladder and the building

c) the length of the ladder

17 **COLLEGE STUDIES** Enrollment in post-secondary schools varies according to various factors including those that are economic. Given that the situation from 2005 to 2009 can be graphically represented by a straight line, find the number of college students enrolled in 2007.

College enrollment

Year	Number of students
2000	159 617
2004	154 026
2005	153 290
2009	176 473

18 The slope of a line is a value that defines its inclination.

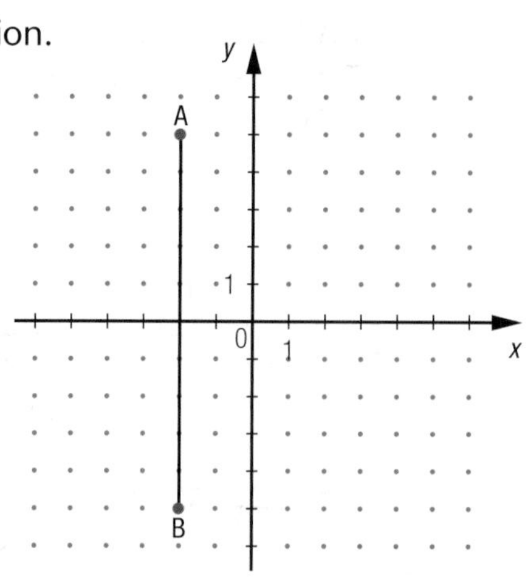

a) What is the angle formed by line AB and the x-axis?

b) What is the change in x-values from A to B?

c) What is the change in y-values from A to B?

d) What do you notice when creating a ratio between the change of y-values to x-values?

e) What is your conclusion about the slope of a vertical line?

19 Show that the midpoints of the sides of the adjacent quadrilateral ABCD are the vertices of a parallelogram.

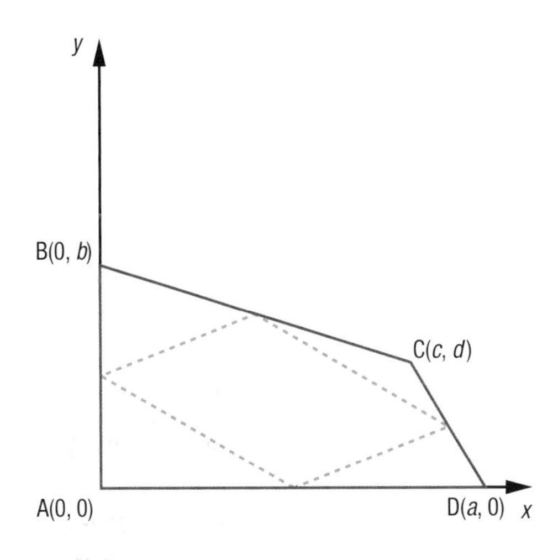

20 In the adjacent Cartesian plane, Cities **A**, **B** and **C** are joined to show the roads connecting these cities. To free up the roads during rush hour, a secondary road is constructed equivalent to the median stemming from City **C**. The scale is in kilometres.

a) What are the coordinates of the intersection point where the new road meets road AB?

b) What is the length of the new road?

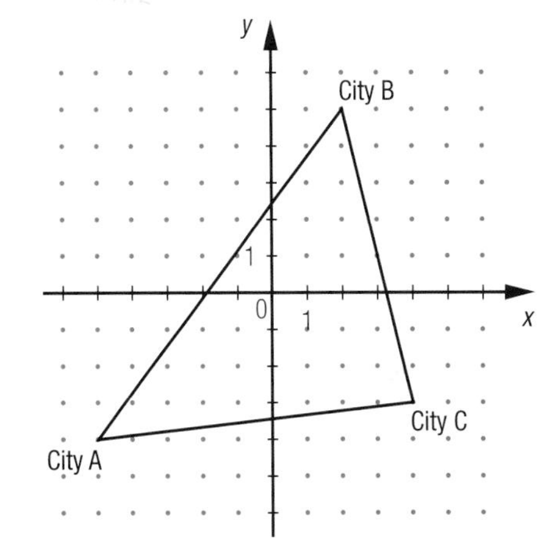

21 This graph shows two towers, a cable that joins them, and a cable car. The graph scale is in metres.

a) What is the slope of the cable?

b) What is the length of the cable?

c) What is the distance between the cable car and Tower **B**, if the tram's position on the cable divides the cable in a ratio of 5:2?

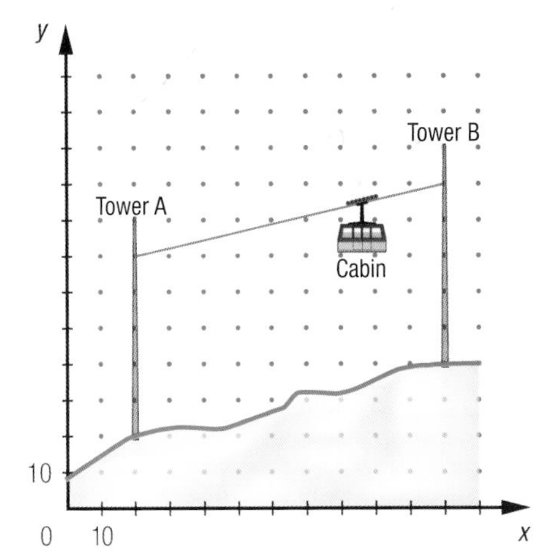

An aerial tramway or cable car is a type of aerial lift in which one or two cabins are suspended from a cable. It is often used at ski resorts.

SECTION 1.2 Lines in the Cartesian plane

This section is related to LES 1 and 2.

PROBLEM Forest fires

Each year in Québec, more than 800 fires destroy hundreds of square kilometres of forest. In addition to the efforts of the aerial team, hundreds of fire fighters on the ground try to extinguish the flames.

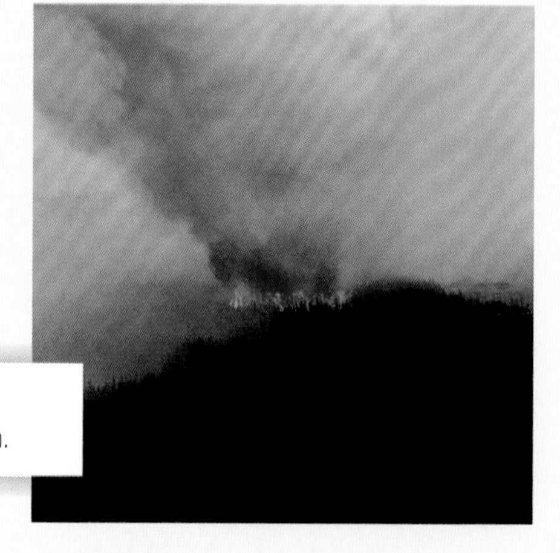

Forest fire in the Manicouagan region.

To better plan their attack strategy, the authorities represented the forest fire on a Cartesian plane. This section of burning forest is bounded by the three lines described below. The scale is in kilometres.

Line A	Line B	Line C
Passes through points P(0, 10) and R(4, 2).	Passes through point R(4, 2) and its slope is -0.5.	Passes through point P(0, 10) and its slope is 0.5.

A water bomber with a capacity of 6137 L fills its reservoir with water at a lake near the fire. Due to the intensity of the blaze, the authorities plan to dump double the capacity of the bomber for each 10 km² of burning forest. For each refill, the water reservoir is filled to its maximum capacity.

How much water will the water bomber dump during this operation?

The CL-415 is an amphibian water bomber plane designed specifically for fighting forest fires. It takes 9 to 12 seconds for this water bomber to fill its two reservoirs of approximately 3000 L each.

ACTIVITY 1 Cabinet making

During the 17th century, cabinet making was a branch of the carpentry trade; cabinet makers worked mainly on veneers and parquetry. Today, the term "carpentry" is mainly used to refer to building construction and "cabinet making" to refer to the furniture industry.

Token from the 1748 Carpenters and Cabinet Makers Community.

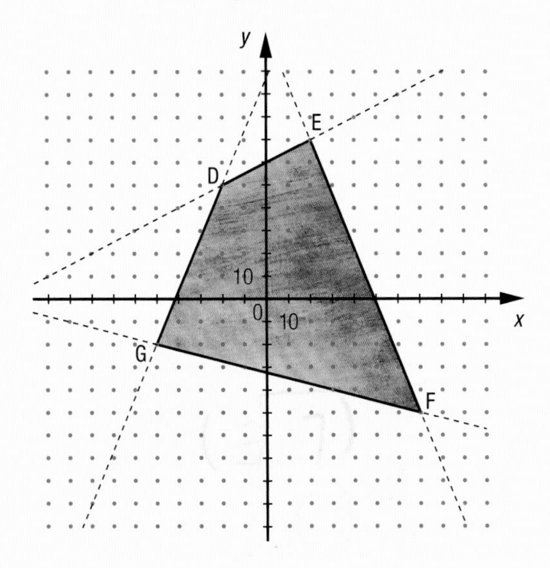

When designing furniture, a cabinet maker sometimes represents pieces on a Cartesian plane. The adjacent graph represents a tabletop. The scale is in centimetres.

a. Define the equation of EF, in function form: $y = ax + b$.

b. What are the slope and the y-intercept of this line?

c. What algebraic manipulations must be made to the equation found in **a.** in order to:

1) find the x-intercept?
2) express the equation in its general form: $Ax + By + C = 0$?

According to the cabinet maker's calculations, the general form of the equation of line DE is: $x - 2y + 120 = 0$.

d. In this equation, what is:

1) the value of **A**?
2) the value of **B**?
3) the value of **C**?

e. What algebraic manipulation must be made to the equation $x - 2y + 120 = 0$ in order to express it in function form?

f. Establish a relationship between the values found in **d.** and:

1) the slope of DE
2) the y-intercept of line DE
3) the x-intercept of line DE

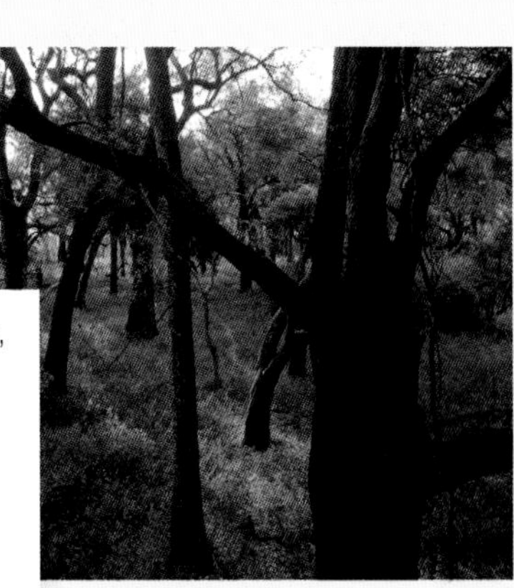

The term for cabinet making in French, "*ébénisterie*," comes from the word "*ébène*" which means ebony, the rare and valuable black wood used mainly for decorative furniture as opposed to more utilitarian pieces.

ACTIVITY 2 Negative reciprocal

Pierre de Fermat stated that two lines in the same plane are parallel if they never meet. Based on this idea, a specific relationship between the slopes of two parallel lines and of two perpendicular lines was established.

Pierre de Fermat (1601-1665)

Pierre de Fermat, the Toulousian lawyer, was a well-known lawyer who was interested in science. Along with Descartes, he was the inventor of analytic geometry. He, like Pascal, is considered a father of probability theory. Using arithmetic, he discovered the properties of many numbers and his work is central to modern number theory. In optical physics, he developed the Fermat principle.

The right triangle ABC is defined by lines l_1, l_2 and l_3.

a. Use the Pythagorean Theorem to prove that this is a right triangle.

b. Find the slope of the line:

1) l_1
2) l_3

c. Verify that the product of the slopes found in **b.** is equal to -1.

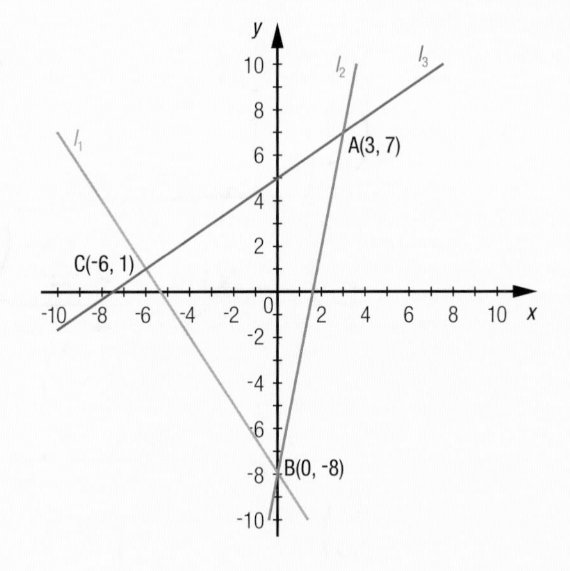

Quadrilateral DEFG is formed by lines l_4, l_5, l_6 and l_7.

d. Find the slope of the line:

1) l_4
2) l_5

e. Compare the slopes obtained in **d.** What do you observe?

f. What can be said of the positions of lines l_4 and l_5 in relation to each other?

g. What type of quadrilateral is DEFG?

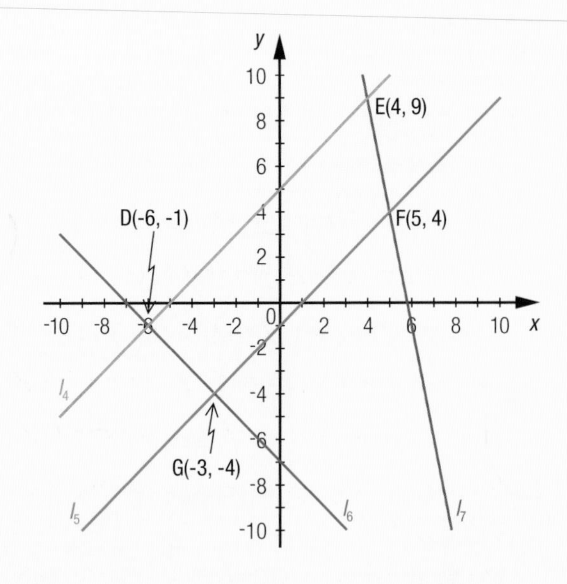

Techno math

Dynamic geometry software allows you to draw and manipulate figures in a Cartesian plane. By using the tools: SHOW AXES, LINES, PARALLEL LINES, PERPENDICULAR LINES, you can draw parallel and perpendicular lines and present their equations in various forms.

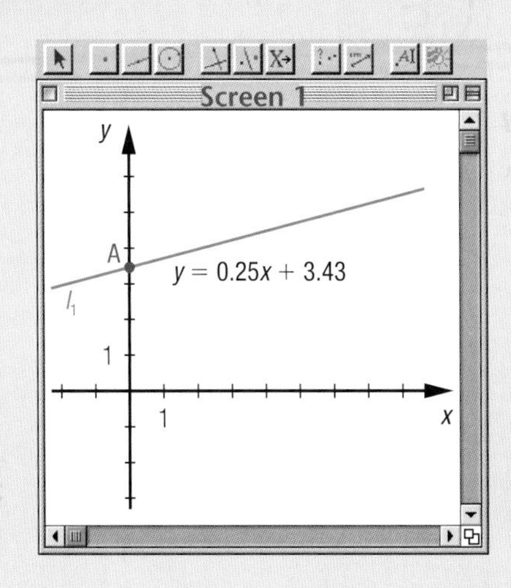

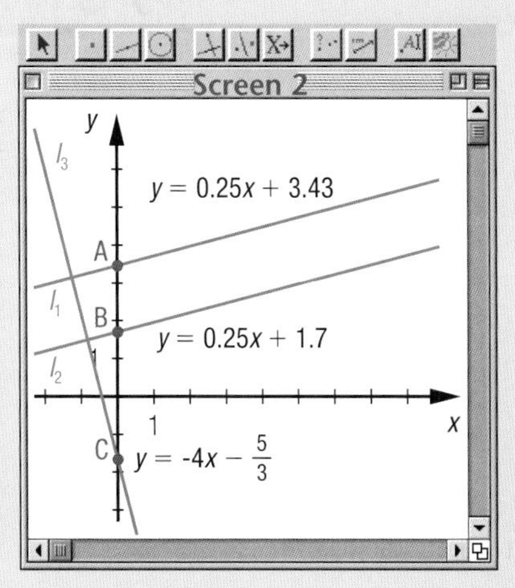

By modifying the inclination of line l_1 or the position of points A, B and C, you can observe certain changes that occur to the equations of straight lines.

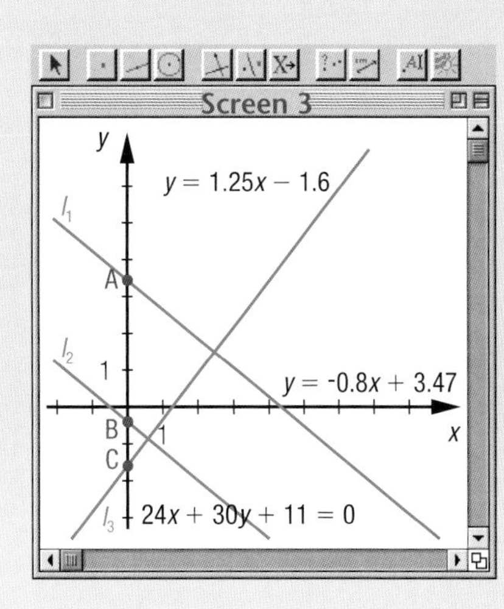

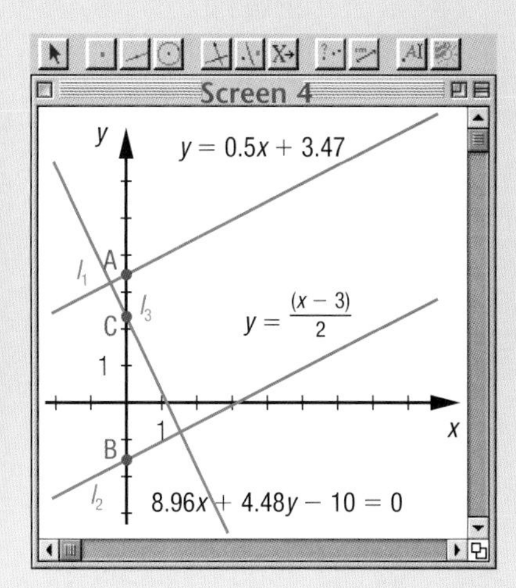

a. Find the coordinates of four points on the line displayed in Screen **1**.

b. On Screen **2**, how do the equations of the lines l_1 and l_2 prove that $l_1 // l_2$?

c. On Screen **2**, verify that the product of the slopes of lines l_1 and l_3 is equal to -1.

d. Write the equation of line l_2 from Screen **3** in function form.

e. Find the *y*-intercept of each of the lines displayed in Screen **4**.

f. In Screen **4**, given that $l_1 \perp l_3$ what can be said about the position of line l_2 in relation to the other lines?

g. Using dynamic geometry software, what happens to the equation of a line that is:

1) parallel to the *x*-axis

2) parallel to the *y*-axis

EQUATION OF A LINE

There are various ways to write the equation of a line. Below are two:

Equation type	Equation	Relationship between parameters	Characteristics
Function Form	$y = ax + b$	Slope: a y-intercept: b x-intercept: $-\frac{b}{a}$	Can be used to describe any non-vertical line.
General Form	$Ax + By + C = 0$	Slope: $-\frac{A}{B}$ y-intercept: $-\frac{C}{B}$ x-intercept: $-\frac{C}{A}$	Can be used to describe any line.

Using algebraic manipulations, it is possible to convert the equation of a line from general form to function form and vice versa.

E.g.

1) The equation $y = -2x + 4$ can be expressed in general form as follows.

$$y - y = -2x + 4 - y$$
$$0 = -2x - y + 4 \text{ therefore } -2x - y + 4 = 0$$

2) The equation $3x + 4y - 4 = 0$ can be expressed in function form as follows.

$$3x + 4y - 4 - 3x = 0 - 3x$$
$$4y - 4 = -3x$$
$$4y - 4 + 4 = -3x + 4$$
$$4y = -3x + 4$$
$$\frac{4y}{4} = -\frac{3x}{4} + \frac{4}{4}$$
$$y = -\frac{3x}{4} + 1$$

It is also possible to express the equation $3x + 4y - 4 = 0$ in function form using the parameters $A = 3$, $B = 4$ and $C = -4$.

Slope: $-\frac{A}{B} = -\frac{3}{4}$ y-intercept: $-\frac{C}{B} = -\frac{-4}{4} = 1$

The equation is therefore: $y = -\frac{3}{4}x + 1$.

PARALLEL AND PERPENDICULAR LINES

In a Cartesian plane, two lines with the same slope are parallel.

E.g. The linear equation $y = -5x - 7$ and linear equation $10x + 2y - 12 = 0$ are parallel because their slope is the same, -5.

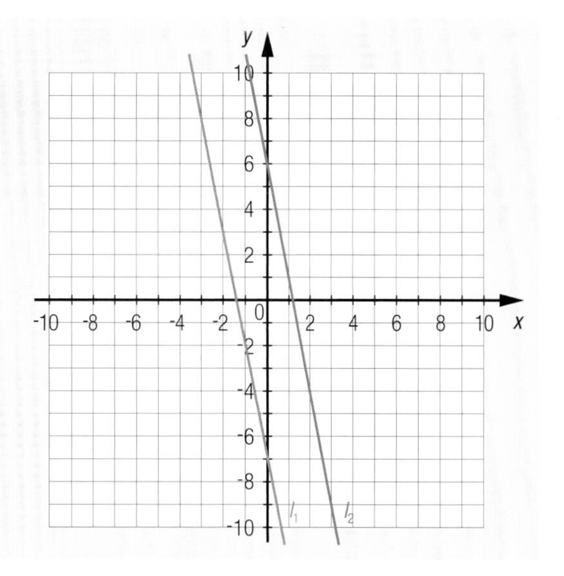

In a Cartesian plane, two lines with slopes that are the negative reciprocal of each other are perpendicular. When this is the case, the product of the slopes of these two lines is equal to -1.

E.g.

1) Linear equation $y = -\frac{1}{3}x + 9$ and linear equation $y = 3x - 1$ are perpendicular because the slope of one line is the negative reciprocal of the slope of the other line: $-\frac{1}{3} \times 3 = -1$.

2) To verify that the two lines l_1 and l_2 are perpendicular, find the slope of each line.

Slope of $l_1 = \frac{-3 - 0}{0 - 6} = \frac{1}{2}$

Slope of $l_2 = \frac{0 - 8}{4 - 0} = -2$

Lines l_1 and l_2 are perpendicular because the slope of l_1 is the negative reciprocal of the slope of l_2 and their product is $\frac{1}{2} \times -2 = -1$.

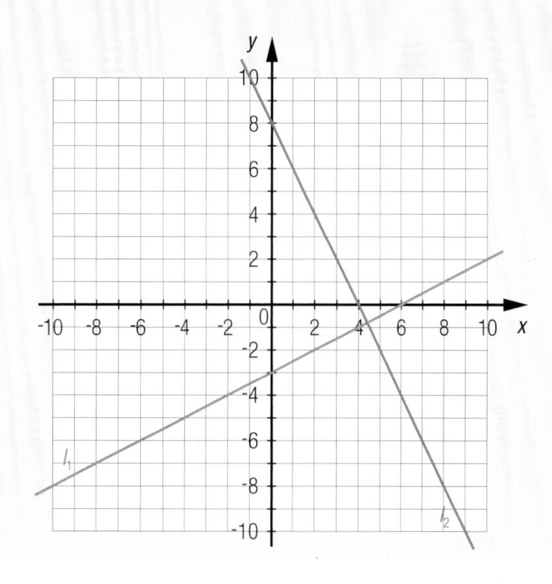

practice 1.2

1 Find the equation for each of the following lines.

a)

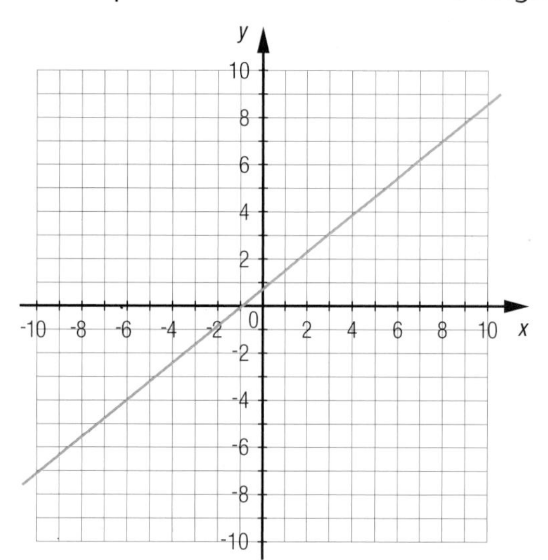

b)

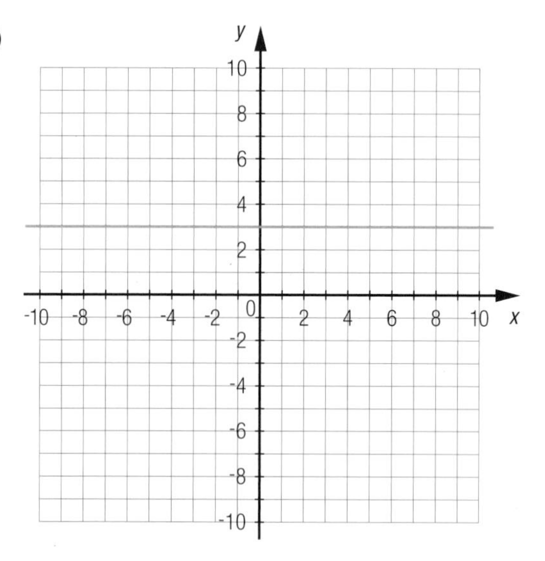

c)

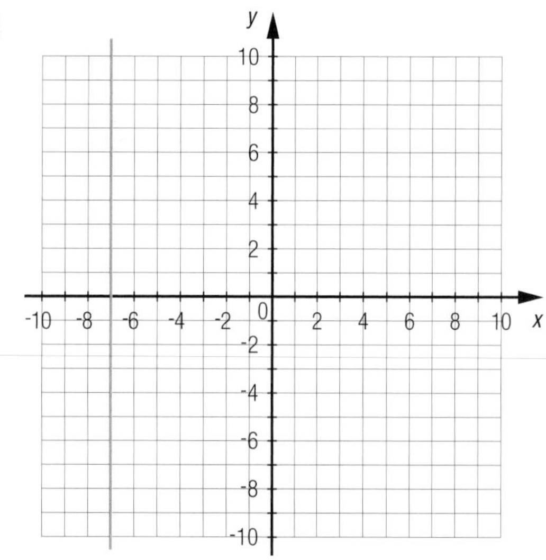

d)

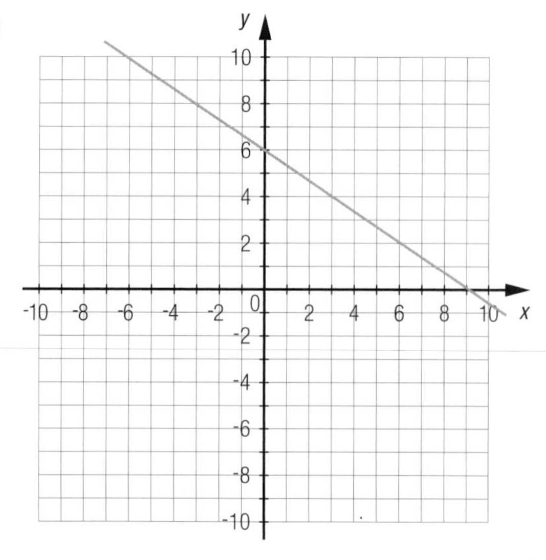

2 For each of the following cases, find the equation of each line:

1) in function form 2) in general form

a) The line whose slope is 4 and y-intercept is -13.

b) The line whose slope is 5 and passes through point P(2, 5).

c) The line whose slope is $\frac{4}{5}$ and passes through point R(3, 5).

d) The line that passes through points S(-1, 25) and T(3, -7).

3 Find the slope, the y-intercept and the x-intercept for the lines with the following equations:

a) $y = ^{-}x + 23$ b) $y = ^{-}12x + 5$ c) $x + y = 15$

d) $6x - 2y = 19$ e) $x + y - 7 = 0$ f) $3x - 4y + 5 = 0$

4 Find the equation of the line that:

a) passes through point E(0, 3) and is parallel to $y = 2x + 5$

b) passes through point F(3, 4) and is parallel to $y = 5$

c) passes through point H(-2, 8) and is parallel to $x = 2$

d) passes through point J(2, 3) and is perpendicular to $y = 2x + 5$

e) passes through point K(4, 12) and is perpendicular to $4x - 5y + 3 = 0$

5 In each case, draw the line with the following characteristics.

a) x-intercept: 7
y-intercept: 5

b) parallel to the y-axis and passes through P(6, 4)

c) slope is 0.5 and x-intercept is 3

d) slope is $\frac{2}{3}$ and y-intercept is -2

6 Express each equation:

a) in general form

1) $y = ^{-}2x - 1$ 2) $y = \frac{x}{7} + 2$ 3) $y = ^{-}\frac{4}{5}x - 33$

b) in function form

1) $0 = x - y + 15$ 2) $x + y = 112$ 3) $10x - 2y + 5 = 0$

7 Find the equation for line l_2, given that line l_2 is perpendicular to line l_1 and that both lines pass through the origin.

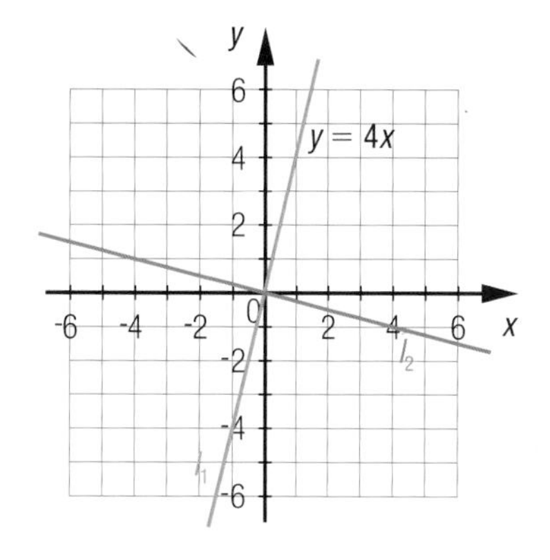

8 For the tables of value below, state which ones graphically represent a line that is parallel to the line with equation $3x - 4y - 68 = 0$.

1

X	Y1
2	-15.5
3	-14.75
4	-14
5	-13.25
6	-12.5
7	-11.75
8	-11

X=2

2

X	Y2
-18	10.5
-17	11.25
-16	12
-15	12.75
-14	13.5
-13	14.25
-12	15

X= -18

3

X	Y3
21	30
22	31.333
23	32.667
24	34
25	35.333
26	36.667
27	38

X=21

9 Among these equations, find those that represent:

a) parallel lines

b) perpendicular lines

1. $y = -7x - 2$
2. $21x + 3y + 6 = 0$
3. $y = -\frac{2}{5}x + 8$
4. $-\frac{7}{2}x - \frac{y}{2} - 1 = 0$
5. $2x + 5y - 40 = 0$

10 Three lines intersect in this Cartesian plane forming triangle ABC. Determine:

a) the equation of line BC in general form

b) the equation of the perpendicular bisector of BC in function form

c) if this is a right triangle

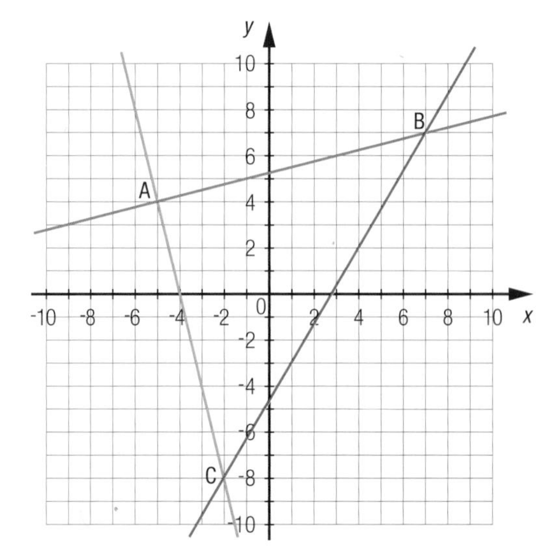

11 What is the value of k if line l_1 passes through points A(-1, 20) and B(k, 5) and it is parallel to the line whose slope is -3?

12 The adjacent illustration shows the front view of a piece of furniture in the shape of an isosceles trapezoid. Find the area, given that the slopes of each of its non-parallel sides are -7.5 and 7.5 respectively.

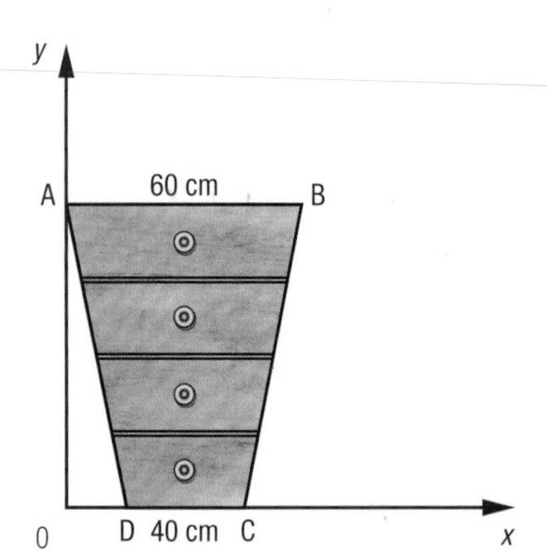

13 The x-intercept of a line is -12 and its slope 2.5.

a) What is the y-intercept of this line?

b) Find the equation of this line in general form.

c) Find the equation of a line that is perpendicular to this line.

14 The following information is about the adjacent road map:

- The line representing Point Road has x-intercept of -4 and y-intercept of 5.
- Point Road is perpendicular to Post Office Road.

Using general form, write the equation representing Post Office Road.

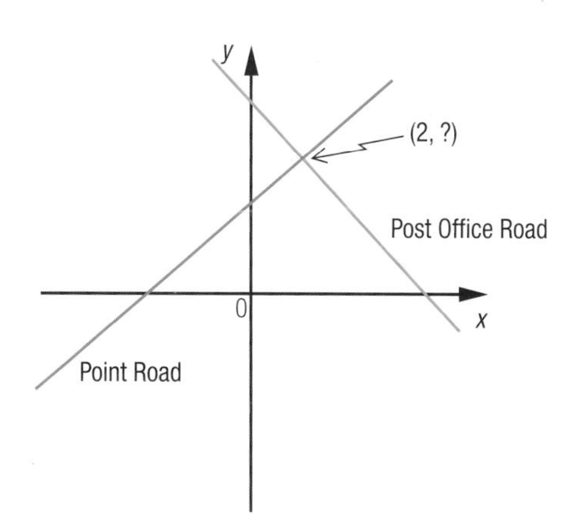

15 This adjacent graph provides information about an airplane's approach toward a runway.

a) What slope did the pilot choose for the airplane's descent?

b) Using general form, write the equation representing the trajectory of the airplane's descent.

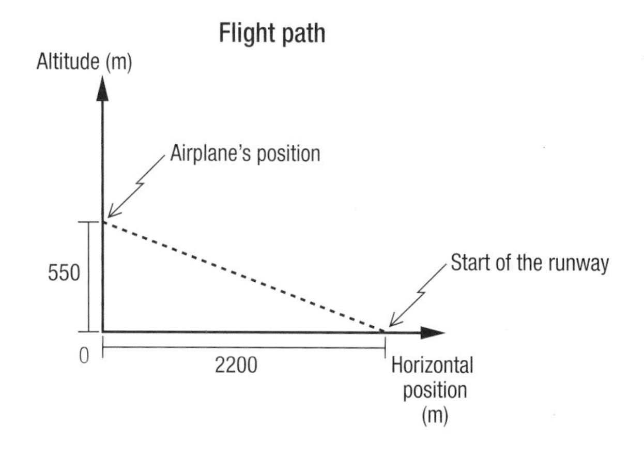

Most runways are used for both take-offs and landings. Runways are generally aligned with the dominant winds. This allows air currents to help lift the planes during take-offs and to slow them down as they land into the wind, using the air's resistance as an additional brake.

16 CANADIAN DOLLAR Below is some information regarding the mean annual value of the American dollar in Canadian currency:

Canadian dollar

Year	2003	2004	2005	2006	2007
Value ($)	1.40	1.30	1.21	1.13	1.07

a) Represent the situation with a scatter plot.

b) Draw the line of best fit.

c) Find the equation for this line.

d) Using this equation, what was the mean value of the American dollar in Candian currency for 2002?

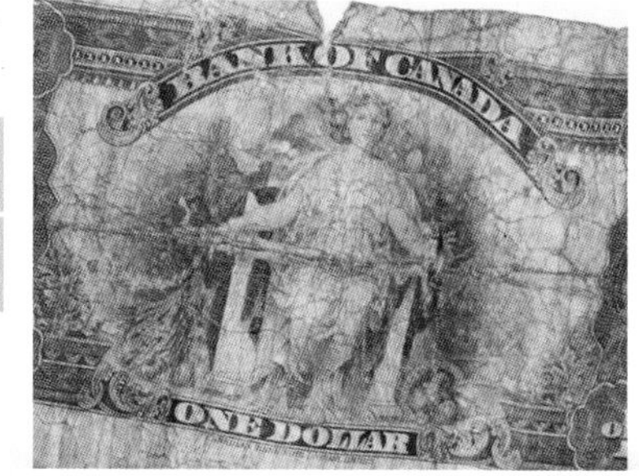

The word "dollar" comes from the German "*thaler*," a silver coin used in 1519. The word "cent" has its roots in the Latin "*centum*" meaning one hundred. The Canadian dollar has been the official currency of Canada since 1858.

17 **SEIGNEURY** Around 1626, the land of New France was distributed according to the seigneurial system of landownership. Land was divided in long plots aligned with the St. Lawrence River or its tributaries. The adjacent illustration represents a seigneury. The scale is in kilometres.

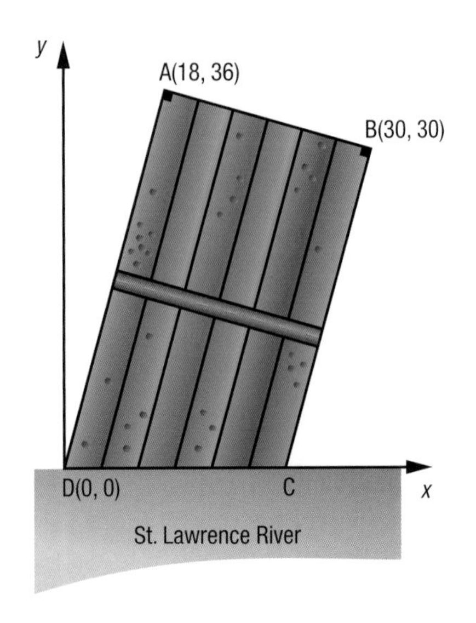

a) Find the equations of the lines that pass through points:

1) A and B
2) B and C
3) C and D
4) A and D

b) Find the coordinates of point C.

c) Calculate:

1) the perimeter of the seigneury
2) the area of the seigneury

18 **SINGLE SLOPE ROOF** The single slope roof made its appearance in Montréal around 1860. The adjacent drawing illustrates a building with this type of roof.

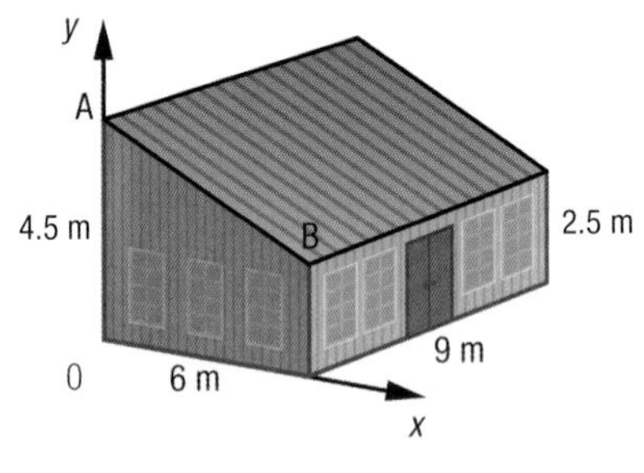

a) What is the equation of the line that passes through points A and B?

b) What are the dimensions of the roof?

c) If a person 1.77 m tall stands on the floor in the centre of this building, what is the distance between the top of his or her head and the roof?

Ecological Row Housing (Manchester, Great Britain)

19 The adjacent diagram shows the plan of a barn roof. On this plan, point B is situated $\frac{1}{3}$ along $\overline{AC}$ and D is situated $\frac{2}{3}$ along $\overline{CE}$.

a) Find the equation for:
   1) line AC
   2) line CE
   3) line BH
   4) line DF

b) Find the coordinates for:
   1) point H
   2) point F
   3) point B
   4) point D

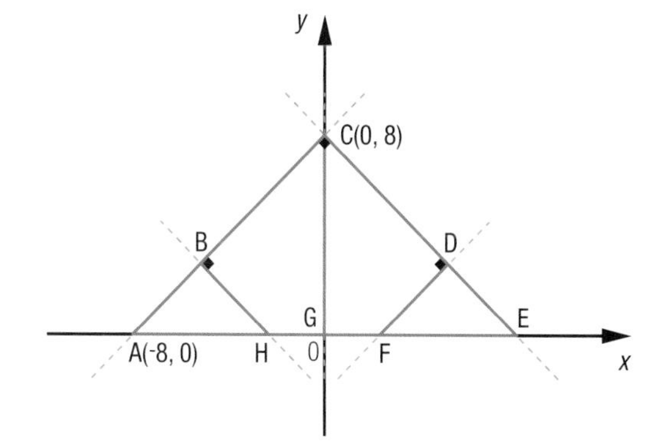

20 During a fire alarm, the fire fighters must first find the exact location of the fire on a map in order to determine the shortest way to get there. The graph below shows the three main roads of a city. The scale is in kilometres.

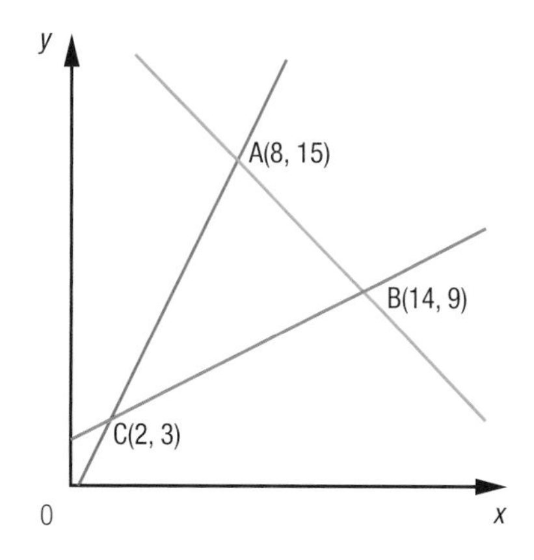

A fire breaks out at the intersection of the two secondary roads in this area. This fire is located at the intersection between the median of vertex B and the altitude of vertex A.

a) Find the coordinates of the location of the fire.

b) Find the amount of time required for the firefighting team to arrive on the scene if the fire station is located at point B and the fire truck's average speed is 65 km/h.

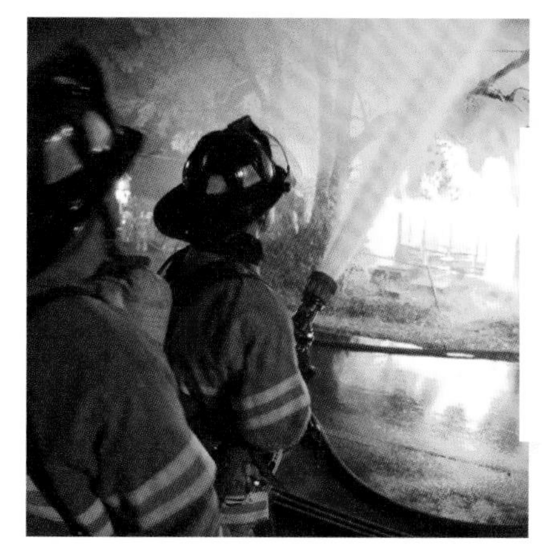

Established in 2000, the École nationale des pompiers du Québec (ÉNPQ) strives to improve and standardize the skills of all firefighting personnel. A 275-hour training session "Firefighter 1" is required for first-level firefighters in order for them to work in communities of 25 000 people or fewer. The school also offers specialized training in automobile extrication and operating automatic water trucks and lifts.

SECTION 1.3 Systems of equations

This section is related to LES 1 and 2.

PROBLEM Horticulture

Horticulture is a branch of agriculture that includes the cultivation of vegetables, flowers, shrubs and decorative trees. Specialists in this field usually work outdoors or in greenhouses.

The Métis Gardens in the Matane region of Gaspésie are a world-renowned horticultural work of art. Within an area of 17 hectares, more than 3000 indigeneous and exotic plants are cultivated. The Métis gardens also host the International Festival of Gardens, a unique event in North America that is held every summer from June to September.

Below is data that was collected while observing the growth of two plants of the same species using different fertilizers.

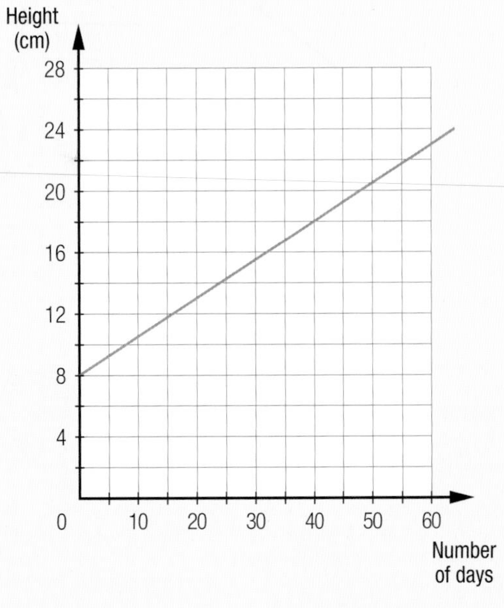

Fertilizer B

Number of days	Height (cm)
0	0
10	4.5
24	10.8
32	14.4
46	20.7
50	22.5

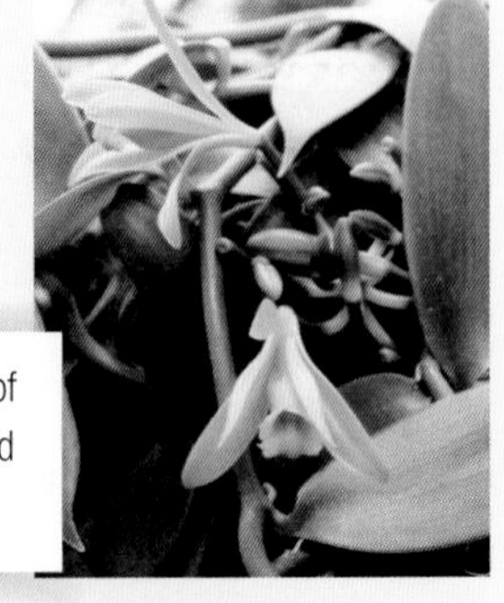

Vanilla planifolia is the scientific name of a tropical Mexican orchid. Its fruit, called beans, produce a spice called "vanilla."

When will the two plants be the same height?

ACTIVITY 1 Tricky marks!

a. If a represents the mark on the first test and b the mark on the second test, what system of equations is formed by the two statements made by the teacher?

b. What do you observe concerning the position of the variables in relation to the equals sign for the equation corresponding to:

1) the first statement?

2) the second statement?

c. What is the one-variable equation obtained when you replace variable a with the expression $2b - 92$ into the equation from the first statement?

d. Solve the equation obtained in **c**. Explain what the solution represents in relation to the context.

e. What is this student's mark on the first test?

f. Graph the system of equations corresponding to this situation and find the coordinates of the intersection point. What did you notice?

ACTIVITY 2 Burning calories

Regular physical activity helps to lower body fat and increase muscle mass by burning calories. The number of calories burned depends on a person's gender, mass and age, and also on the length, intensity, and the type of physical activity or sport.

The calorie is an old measuring unit of heat. There is a distinction between calorie (cal) and the kilocalorie (kcal) equivalent to 1000 calories. In nutrition, the calorie is used as a measurement of energy contained in certain food and corresponds to the old *grand calorie*. Today, the word "calorie" in common vernacular is misused because it actually refers to kilocalories (kcal).

Below is some information about Guillaume's last two workouts:

- At the beginning of the week, 30 minutes of cross-country skiing and 60 minutes of swimming burned a combined total of 1275 calories.
- At the end of the week, 90 minutes of cross-country skiing and 30 minutes of swimming burned a combined total of 1950 calories.

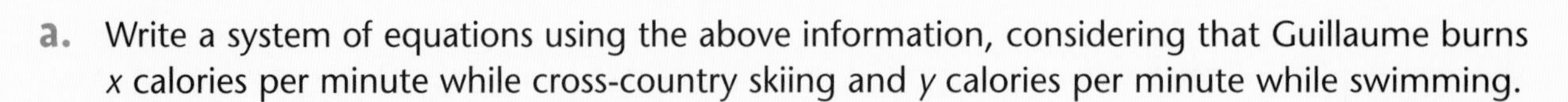

a. Write a system of equations using the above information, considering that Guillaume burns x calories per minute while cross-country skiing and y calories per minute while swimming.

b. What do you observe concerning the position of the variables in relation to the equals sign in the equation representing Guillaume's workout:

1) at the beginning of the week? 2) at the end of the week?

c. 1) What would be the system of equations if Guillaume had doubled his workout time at the end of the week?

2) Would this change the number of calories burned per minute for each sport? Explain your answer.

Next week, Guillaume will triple his workout times for both activities at the beginning of the week and keep his end-of-week workouts the same as now. This situation can be represented by the following system of equations.

❶ $90x + 180y = 3825$

❷ $90x + 30y = 1950$

d. What do you notice about the coefficient of the x variable in each equation?

e. What equation is obtained by subtracting each like term in equation ❷ from equation ❶?

f. How many calories does Guillaume burn per minute when he:

1) swims?

2) cross-country skis?

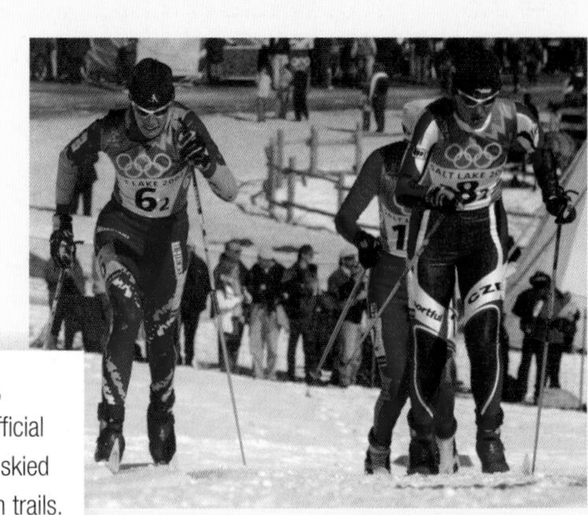

At the 1924 Winter Olympics in Chamonix, France, cross-country skiing became an official international event, open to men only who skied the classic style on both 18 km and 50 km trails.

ACTIVITY 3 Floods

In Québec, floods can happen at any time of the year. A flood can be caused by heavy rains or rapid snow thaw causing, among other things, sudden high river water levels.

From July 18 to 21, 1996, more than 200 mm of rain fell on the Saguenay–Lac-Saint-Jean region causing the evacuation of 16 000 people and destroying more than 1350 buildings.

Below is information regarding the water levels of two rivers within a specific precipation zone:

Bark River

- Water level before precipitation: 150 mm
- Mean variation of the water level since the start of precipitation: 2.5 mm/h

Salmon River

- Water level before precipitation: 210 mm
- Mean variation of the water level since the start of precipitation: 2.5 mm/h

a. Write a system of equations expressing *n* as the water level of each river and *t* as the time elapsed since the start of the precipitation.

b. Graph the lines corresponding to the system of equations.

c. Describe the position of the two lines in relation to each other.

d. At what point is the water level in the two rivers the same? Explain your answer.

The authorities noted that the water level in the Red River varies according to the rule $n = 2.5t + 210$, where n represents the water level (in mm) and t represents the time elapsed (in h) since the start of the precipitation.

e. Write a sytem of equations expressing *n* as the water level of the Red and Salmon rivers and *t* as the time elapsed since the start of the precipitation.

f. Graph the lines corresponding to the system of equations.

g. Describe the position of the two lines in relation to each other.

h. At what time is the water level of the two rivers the same? Explain your answer.

Techno math

A graphing calculator allows you to find the solution to a system of equations using a graphical representation or a table of values.

This screen allows you to edit equations within a system where x is the variable associated with horizontal axis and y is the variable associated with the vertical axis.

Screen 1

```
Plot1 Plot2 Plot3
\Y1 = -0.25X+2
\Y2 = X+8
\Y3 =
\Y4 =
\Y5 =
\Y6 =
\Y7 =
```

This screen allows you to define the desired portion of the graph in the Cartesian plane.

Screen 2

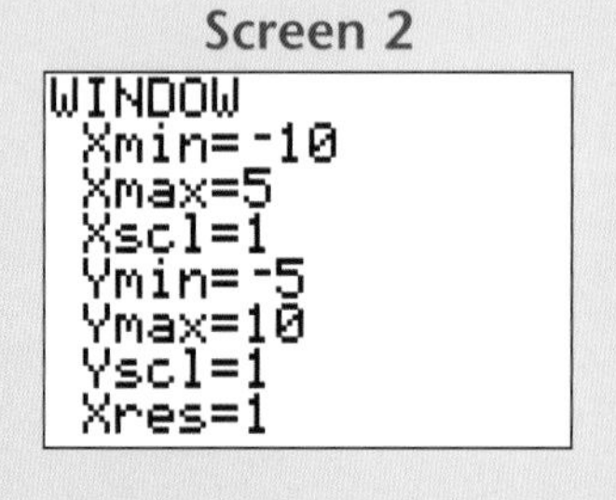

This screen shows the graphical representation of a system of equations. By moving the cursor along the lines, it is possible to approximate the coordinates of the intersection point.

Screen 3

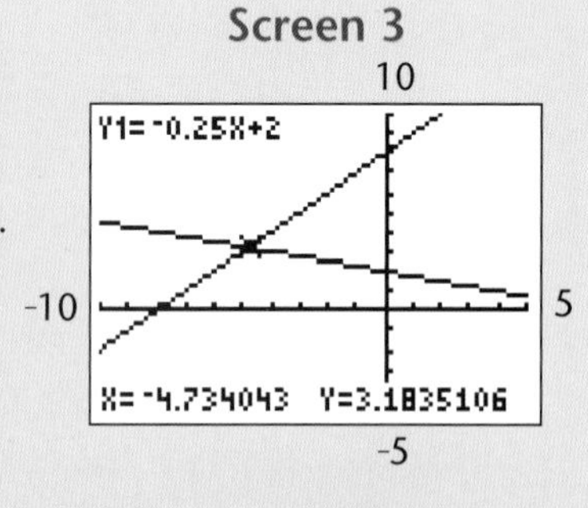

This screen shows the various calculations available on a graphing calculator.

Screen 4

By selecting the two lines and positioning the cursor near the intersection point, the coordinates will automatically be calculated.

Screen 5

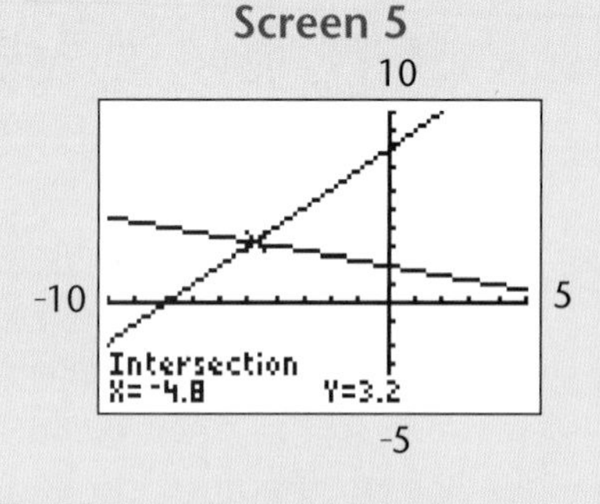

This screen allows you to define the table of values by choosing the starting value and the scale of the x-values.

Screen 6

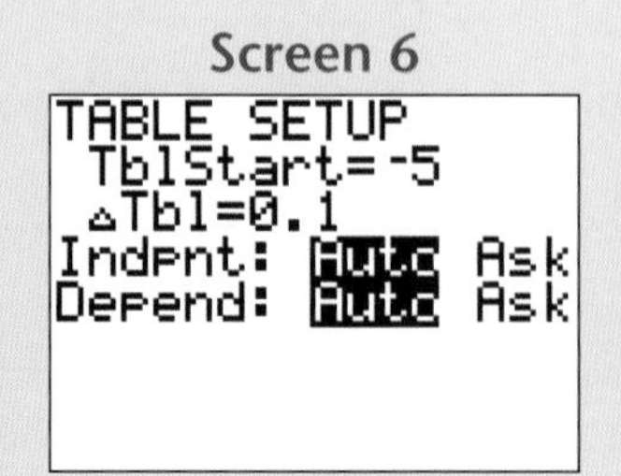

This screen displays the table of values and finds the coordinates of the intersection point.

Screen 7

X	Y1	Y2
-5	3.25	3
-4.9	3.225	3.1
-4.8	3.2	3.2
-4.7	3.175	3.3
-4.6	3.15	3.4
-4.5	3.125	3.5
-4.4	3.1	3.6

X=-4.8

a. According to Screens **5** and **7**, what is the solution to the system of equations displayed in Screen **1**?

b. What can you conclude about the expressions `Xscl = 1` and `Yscl = 1` in Screen **2** after looking at Screen **3**?

c. The adjacent screen displays the equations for lines l_1, l_2 and l_3. Using a graphing calculator, find the coordinates of the vertices of the triangle corresponding respectively to the intersection points of lines l_1 and l_2, l_1 and l_3, l_2 and l_3.

```
Plot1 Plot2 Plot3
\Y1 = 3X
\Y2 = -5X+8
\Y3 = -X-20
\Y4 =
\Y5 =
\Y6 =
\Y7 =
```

SOLVING SYSTEMS OF EQUATIONS

There are different strategies used to solve a system of two first-degree equations in two variables, in other words, to find a set of values (x, y) that satisfies both equations.

Substitution method

The substitution method uses algebraic manipulations to solve a system of equations of the form $\begin{matrix} a_1x + b_1y = c_1 \\ y = a_2x + b_2 \end{matrix}$; one of the variables is isolated in one of the equations.

E.g. To solve the system $\begin{matrix} 3x + 2y = 5 \\ y = -x - 4 \end{matrix}$ using the substitution method:	
1. If necessary, isolate a variable in one of the equations.	$3x + 2y = 5$ $y = -x - 4$
2. Substitute this variable in the other equation with the equivalent expression,creating one equation with one variable.	$3x + 2(-x - 4) = 5$
3. Solve the resulting equation.	$3x - 2x - 8 = 5$ $x = 13$
4. Substitute the value obtained in the initial equation in order to solve for the second variable.	$y = -13 - 4$ $y = -17$ The solution is therefore (13,-17).
5. Verify the solution by substituting $x = 13$ and $y = -17$ for each of the equations: $3 \times \mathbf{13} + 2 \times \mathbf{-17} = 5$; $\mathbf{-17} = \mathbf{-13} - 4$	

Elimination Method

The elimination method uses algebraic manipulations to solve a system of equations of the form $\begin{matrix} a_1x + b_1y = c_1 \\ a_2x + b_2y = c_2 \end{matrix}$; both variables are located on the same side of the equals sign.

E.g. To solve the system $\begin{array}{l} -3x + 7y = 8 \\ -4x + y = -6 \end{array}$ using the elimination method:	
1. If necessary, create a system of equations where the coefficients of one variable are equal or opposite.	$\begin{array}{l} -3x + 7y = 8 \\ -4x + y = -6 \end{array} \Rightarrow \begin{array}{l} -3x + 7y = 8 \\ -28x + 7y = -42 \end{array}$ (× 7)
2. Create an equation with one variable by adding or subtracting the equations.	$\begin{array}{r} -3x + 7y = 8 \\ - \quad -28x + 7y = -42 \\ \hline 25x = 50 \end{array}$
3. Solve the resulting equation.	$25x = 50$ $x = 2$
4. Substitute the value obtained into either of the original equations in order to solve for the second variable.	$-3 \times \mathbf{2} + 7y = 8$ $y = 2$ The solution is therefore (2, 2).
5. Verify the solution by substituting $x = 2$ and $y = 2$ in each of the equations: $-3 \times \mathbf{2} + 7 \times \mathbf{2} = 8$; $-4 \times \mathbf{2} + \mathbf{2} = -6$	

SPECIAL SYSTEMS OF EQUATIONS

Non-coinciding parallel lines

Two lines are parallel and non-coinciding when their equations have the same slope but have different *y*-intercepts. Solving this type of system algebraically leads to an impossibility and therefore has no solution.

E.g.

The solution to the system $\begin{array}{l} 21x - 3y = -3 \\ y = 7x + 5 \end{array}$ leads to an impossibility and has no solution.

$$\begin{aligned} 21x - 3(7x + 5) &= -3 \\ 21x - 21x - 15 &= -3 \\ \mathbf{-15} &= \mathbf{-3} \end{aligned}$$

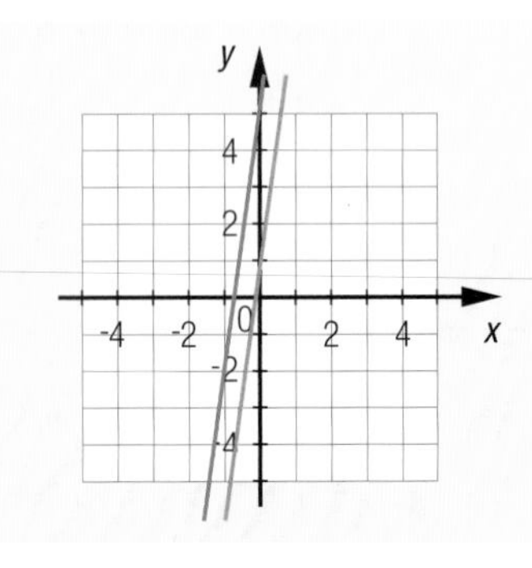

Coinciding parallel lines

Two parallel lines are called coinciding when their equations have the same slope and the same *y*-intercept. Solving this type of system algebraically leads to a true equality and has an infinite number of solutions.

E.g.

The solution to the system $\begin{array}{l} -3x - 3y = 12 \\ y = -x - 4 \end{array}$ leads to a true equality and has an infinite number of solutions.

$$\begin{aligned} -3x - 3(-x - 4) &= 12 \\ -3x + 3x + 12 &= 12 \\ \mathbf{12} &= \mathbf{12} \end{aligned}$$

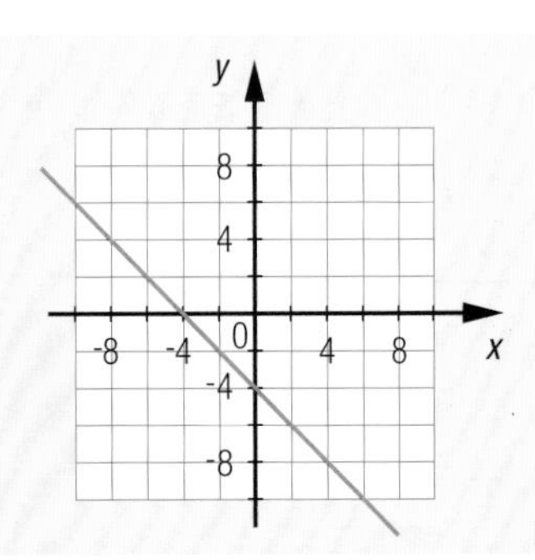

practice 1.3

1 In each case, solve the system of equations.

a) System of equations 1

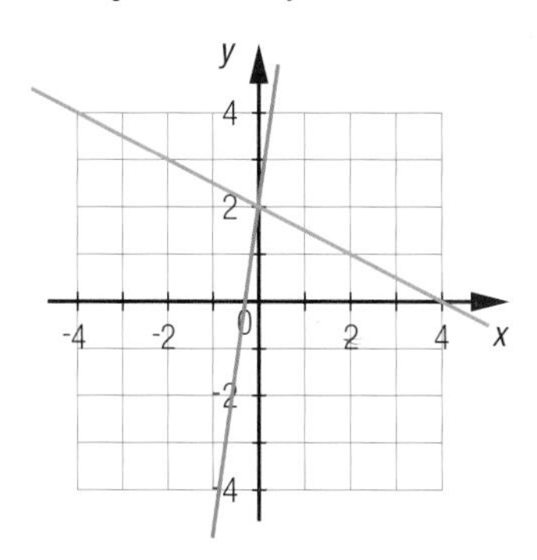

b) System of equations 2

x	25	30	35	40	45
y_1	14.25	15.5	16.75	18	19.25
y_2	11.25	13.5	15.75	18	20.25

c) System of equations 3

x	0	1	2	3	4
y_1	-15	-14.2	-13.4	-12.6	-11.8
y_2	-15	-14.2	-13.4	-12.6	-11.8

d) System of equations 4

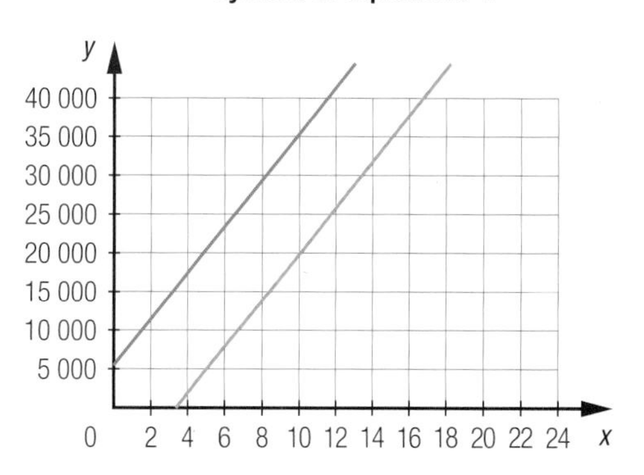

2 Solve the system of equations below using the most appropriate method.

a) $y = -x + 10$
$y = x - 4$

b) $-2x - 2y + 30 = 0$
$y = x - 1$

c) $y = \frac{3x}{7}$
$y = \frac{5x - 1}{12}$

d) $-x + y = 0$
$x + y = 1$

e) $2x - y = 8$
$x = y + 7$

f) $-2x + 25y = -17$
$-x + 15y = -6$

3 Find the equation that graphically represents a non-coinciding parallel line to the linear equation: $-5x + 2y - 7 = 0$.

4 Find the equation that graphically represents a coinciding parallel line to the linear equation: $y = \frac{4x + 5}{3}$.

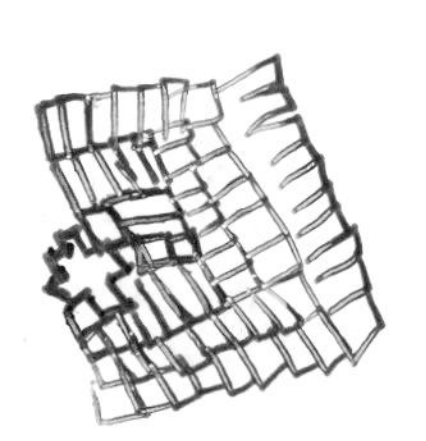

5 In each case, do the following:

1) Identify the unknowns and represent them using different variables.
2) Express the situation as a system of equations.
3) Find the solution.

a) Thomas has to choose between two photocopying stores to have his documents printed. Store **A** charges \$0.06 for each document and an additional \$25 for assembly. Store **B** charges \$0.31 for each document, including assembly. How many documents must Thomas print in order for the total cost to be the same at both stores?

b) MaryLou buys three sweaters and two pairs of pants for a total of \$125 in one store. In the same store, Sarah purchased four sweaters and three pairs of pants similar to MaryLou's for a total of \$180. What is the cost of each sweater and each pair of pants, given that the sweaters are all the same price and that the pants are all the same price?

c) The perimeter of a rectangular field is 248 m. The length of the field is equal to three times its width. What are the field's dimensions?

d) The tray on Scale **A** is holding two bottles and five glasses while the tray on Scale **B** is holding three bottles and three glasses. What is the mass of a glass if Scale **A** indicates a total mass of 440 g and Scale **B** indicates 534 g? All the bottles are identical and all the glasses are identical.

6 The slope of line l_1 is $\frac{3}{2}$ and the y-intercept is -5. Line l_2 is perpendicular to line l_1 and its y-intercept is 47. What is the intersection point of these two lines?

7 Find angle C given that the measure of angle A is 6° less than twice the measure of angle C.

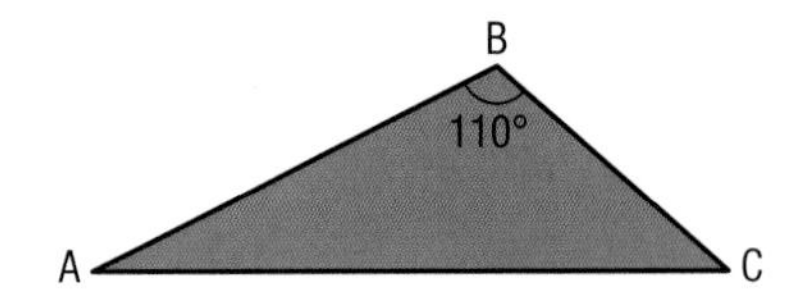

8 **MERCURY** The diameter of the planet Mercury is equivalent to $\frac{1220}{3189}$ of that of Earth. Three times the diameter of Earth plus twice the diameter of Mercury is equal to 48 028 km. What is Mercury's diameter?

In Roman mytholgy, Mercury was the god of business and travel as well as the messenger of other gods. Wednesday ("*mercredi*"), the fourth day of the week, was dedicated to him. The planet Mercury was probably named after him because of its speedy movement through the sky.

9 In the system of equations $\begin{array}{l} y = -x + 4 \\ x + y = k \end{array}$:

a) for what value of k is there an infinite number of solutions to this system?

b) for what value of k is there no solutions to this system?

10 The solution to a system of equations is $\left(-5, \frac{11}{2}\right)$. The slope of the first line is $\frac{1}{2}$, while the other is $\frac{7}{2}$. Find the equations of this system.

11 Given that these scales are balanced, do the following:

1) Express the situation as a system of equations.
2) Find the mass of each object.

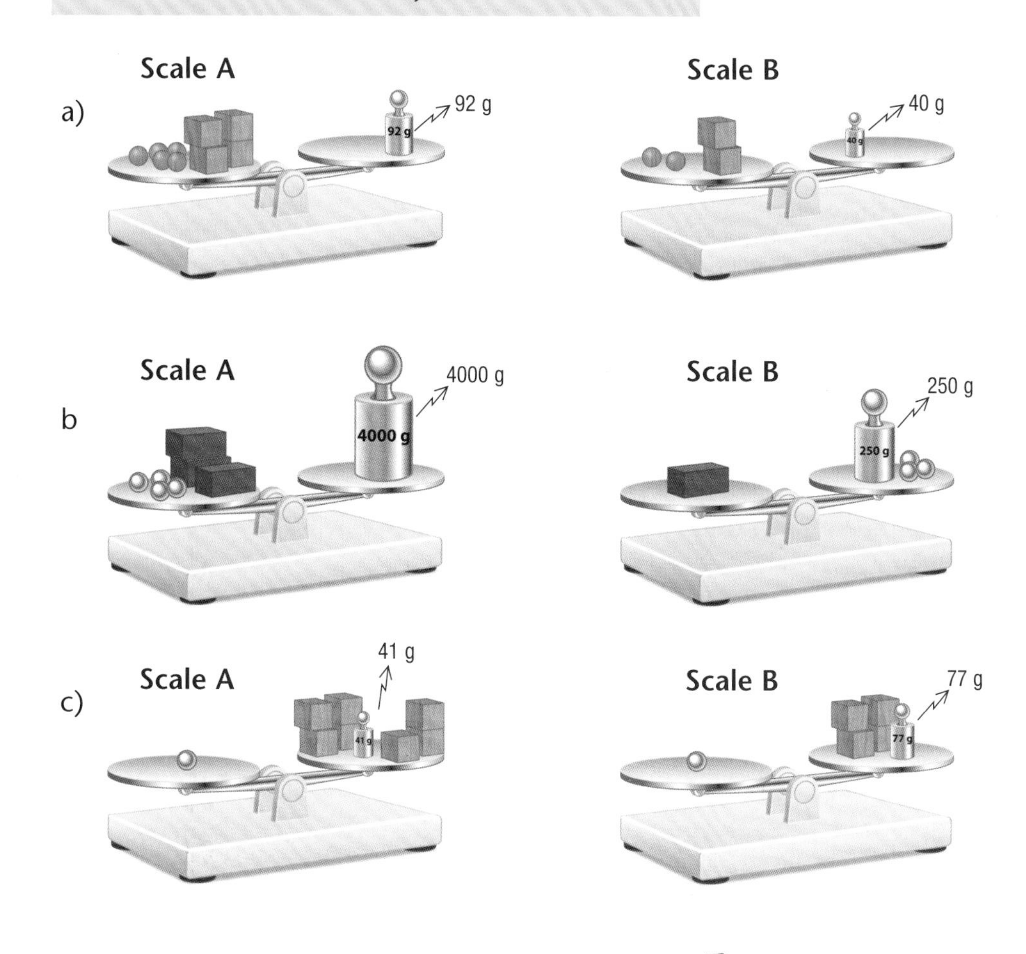

12 At Good Morning restaurant, two croissants and three cups of coffee cost \$6.15; five croissants and two cups of coffee cost \$9.05. What is the price of each item?

13 A piggybank contains loonies (\$1) and toonies (\$2). Find the number of coins of each denomination if there are 59 coins and the total value of the bank is \$90.

14 The adjacent illustration shows the side view of a truncated pyramid with a square base. Considering that the area of the side shown is 792 m² and the length of side AD is equal to five times the length of side BC, find:

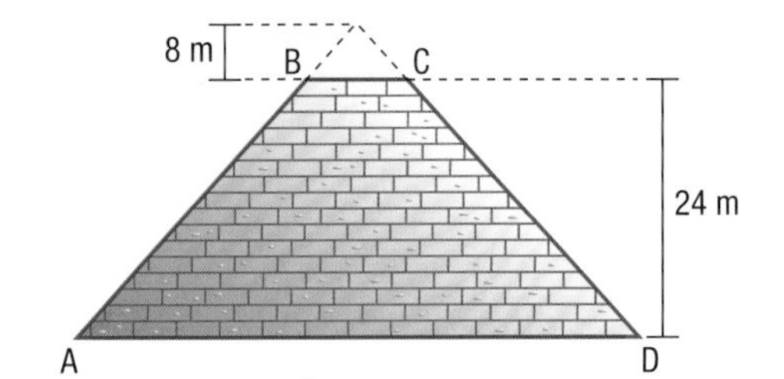

a) the length of AD and BC

b) the area of the visible side of the pyramid

c) the volume of the truncated pyramid

15 The rules below indicate the assets *A* (in $) of three companies working in the same field, where *m* equals the number of months gone by since the start of the year.

Company ❶	Company ❷	Company ❸
$A = 1000m + 12\ 000$	$A = -1000m + 28\ 000$	$A = 2000m + 4000$

Which of the three companies is in better financial health? Explain your answer.

16 The adjacent plan illustrates the triangular garden that Christopher planted behind his house. The scale is in metres.

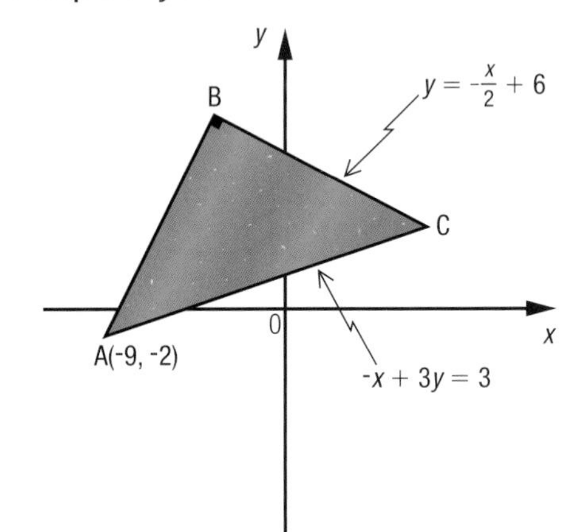

a) What are the coordinates of the automatic sprinklers located at vertices B and C?

b) What is the volume of earth that Christopher needs to purchase in order to spread a layer 25 cm thick over his garden?

17 **METEOROLOGY** To convert temperature in degrees Fahrenheit (F) to the temperature in degrees Celsius (C), use the following equation: $C = \frac{5}{9}(F - 32)$. The temperatures below were recorded during the past two days at a meteorological station:

Yesterday

The difference between the temperature measured in Fahrenheit and the temperature measured in Celsius was 52.

The day before yesterday

The sum of the temperature measured in Fahrenheit and the temperature measured in Celsius was equal to 74.

During which of these two days was the temperature hotter?

SECTION 1.4 Half-planes in the Cartesian plane

This section is related to LES 2.

PROBLEM Electricity consumption

The electricity bill represents a big part of an average family's budget. Changing certain patterns of electricity consumption can result in substansial savings. Electricity is measured in kilowatt hours (kWh). One kilowatt hour corresponds to 1000 watts consumed per hour.

Below is some information about the Tremblay family's electricity consumption:

Electricity consumption for July

Appliance	Power (W)	Time Consumption (t)
Water heater	3500	120
Lighting	15	280
Refrigerator	300	186
Stove	4500	30
Washer/dryer	2650	15
TV	300	95
Other	350	100

The rule $C = 0.0529k + 17$ represents the cost of electricity C (in \$) as a function of the number of kilowatt hours k consumed during the month of July.

Did the Tremblay family respect their budget if they expected to pay less than \$55/month for electricity?

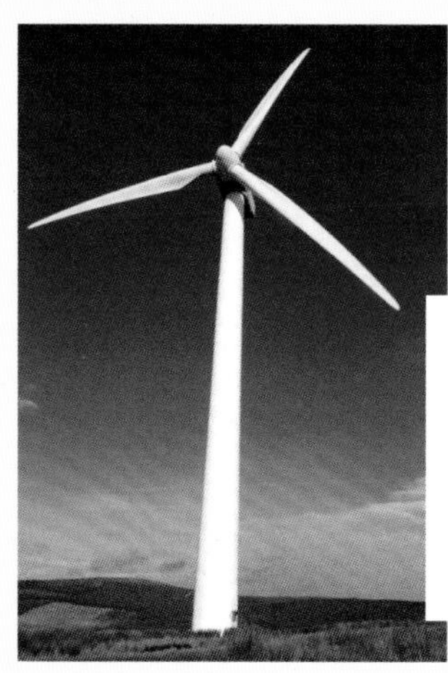

The amount of wind energy produced depends mainly on the force of the wind, the surface of the windmill blades and air density. Electricity can be produced at wind speeds of at least 12 to 14 km/h with full production at 50-60 km/h. Wind speeds over 90 km/h damage the equipment, so in those conditions electricity production has to be interrupted.

ACTIVITY 1 One situation, many solutions

Everyday situations can often be described with the help of inequalities. Following are four situations:

Situation 

In a duathlon the distance b travelled by bicycle is greater than three times the distance r travelled by running.

$b > 3r$

Situation 2

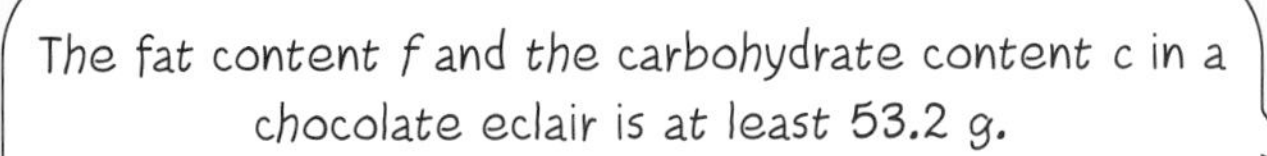

The fat content f and the carbohydrate content c in a chocolate eclair is at least 53.2 g.

$f + c \geq 53.2$

Situation 

In 2007, four times the population q of Québec was less than the population c of Canada.

$4q < c$

Situation 

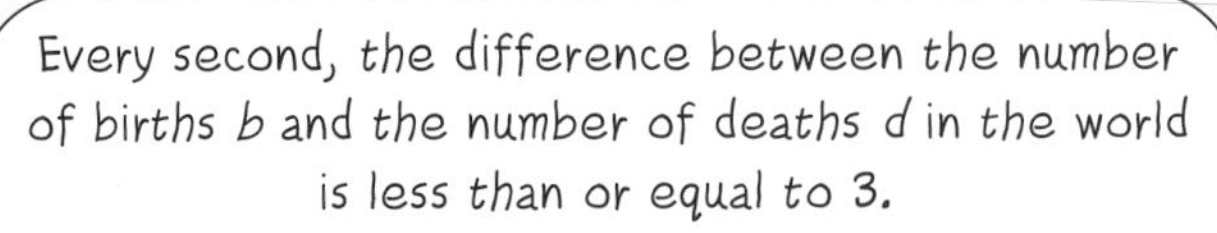

Every second, the difference between the number of births b and the number of deaths d in the world is less than or equal to 3.

$b - d \leq 3$

a. In Situation ❶, can a duathlon participant cycle 15 km and run 5 km? Explain your answer.

b. In Situation ❷, how much fat can there be in a chocolate eclair that contains 40 g of carbohydrates?

c. In Situation ❸, what is the minimum population of Canada if Québec has a population of 7.7 million people?

d. In Situation ❹, is it possible to record, at a given moment in the world, eight births and two deaths? Explain your answer.

e. How many possible solutions are there for each of these situations?

ACTIVITY 2 A safe distance

When fighting a fire, firefighters must respect safety rules to avoid being injured by a collapsing outside wall of a building engulfed in flames. Thus, the distance between the firefighters and the building must be at least $\frac{4}{3}$ the height of the building.

a. If x is the height of the building and y the safe distance to be respected by the firefighters, represent this statement as an inequality.

b. If the height of a burning building is 6 m, are firefighters safe if they are located:

1) 6 m from the building?
2) 8 m from the building?
3) 10 m from the building?
4) 12 m from the building?

c. Firefighters are located 12 m away from the burning building. What is the possible height of the building if they are in the safety zone?

The adjacent graph represents this situation:

d. What connection can you establish between:

1) the shaded half-plane and the inequality sign representing this situation?
2) the dotted line and the inequality sign representing this situation?

e. Can you make the following statements:

1) The points located below the dotted line are solutions to this situation. Explain your answer.
2) The points located above the dotted line are solutions to this situation. Explain your answer.
3) The points located on the dotted line are solutions to this situation. Explain your answer.

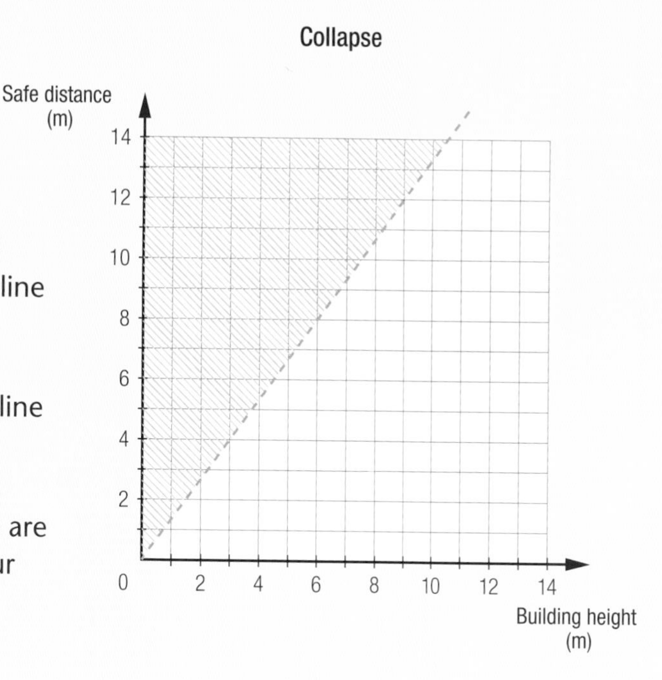

After various studies, specialists reached the conclusion that the distance between the firefighters and the burning building must be **greater than** or **equal to** $\frac{4}{3}$ the height of the building.

f. How does this change affect:

1) the inequality that represents this situation?
2) the graphical representation?
3) the solution set?

Techno math

A graphing calculator allows you to view the solution set of an inequality on a Cartesian plane.

This screen allows you to enter and edit the equations of one or more curves. The type of line used to draw the curves can also be modified.

: normal line

: dotted line

: thick line

Screen 1 Screen 2

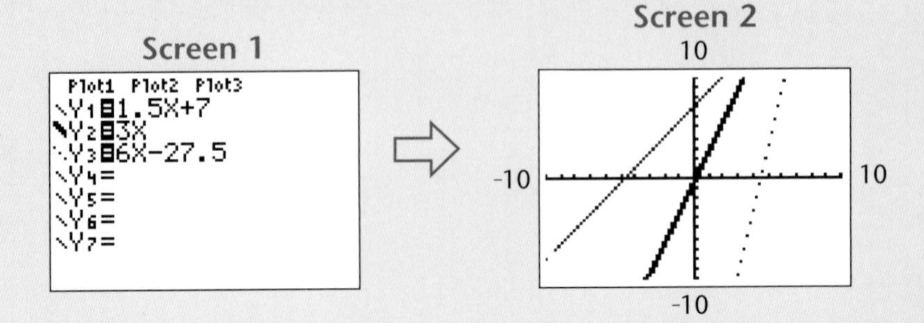

It is possible to represent the solution to an inequality by shading the appropriate region on a Cartesian plane.

Screen 3

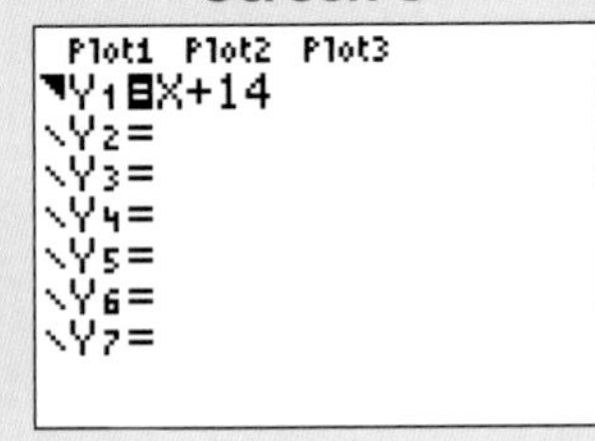

Screen 4

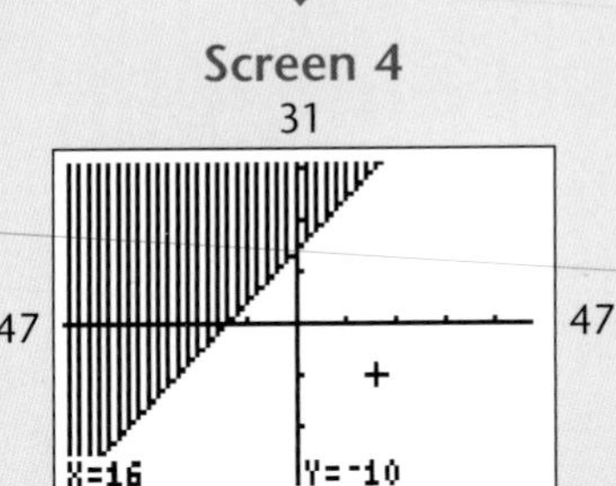

By moving the cursor, on the graphic screen, it is possible to view the coordinates that may or may not be part of the solution set.

Screen 5

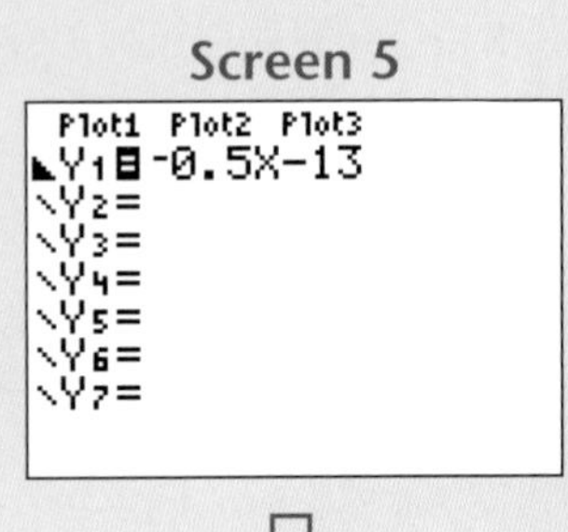

Screen 6

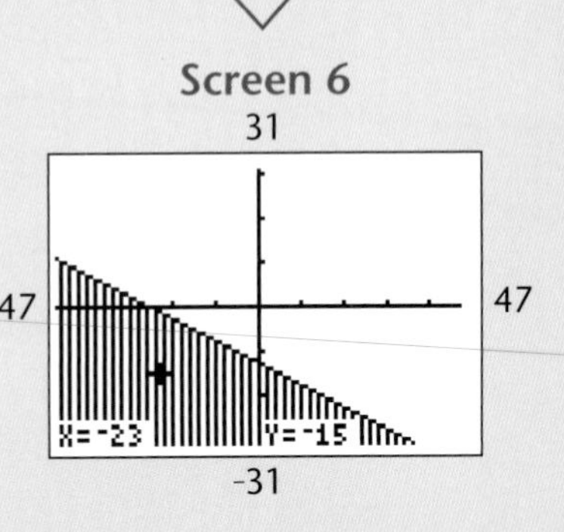

a. What is the inequality represented in:

1) Screens **3** and **4**?

2) Screens **5** and **6**?

b. Use algebra to show that the coordinates of the point shown:

1) on Screen **4** does not belong to the solution set of the inequality

2) on Screen **6** does belong to the solution set of the inequality

c. Find the coordinates of a point located:

1) in the 3rd quadrant of Screen **4**, which is not in the shaded region

2) in the 2nd quadrant of Screen **6**, which is in the shaded region

d. Using a graphing calculator, graph the solution set for each of the following inequalities:

1) $y \geq 2x + 3$

2) $y \leq -0.25x + 1$

3) $y \geq -3x + 8$

FIRST-DEGREE INEQUALITY IN TWO VARIABLES

To translate a situation into a first-degree inequality in two variables, proceed as follows.

1. Identify the unknown variables in the situation.	E.g. The mean mass of a man is 75 kg and that of a woman is 60 kg. How many people could be carried in an elevator whose maximum capacity is 1580 kg? The variables are: • the number of men: x • the number of women: y
2. Construct the algebraic expressions to be compared.	An algebraic expression representing: • the total mass of people in an elevator: $75x + 60y$ • the maximum capacity of the elevator: 1580 kg
3. Write the inequality using the appropriate symbol. Once the inequality is written, its accuracy can be verified by replacing the variables with their numerical values.	Inequality: $75x + 60y \leq 1580$ Validation: The elevator can, for example, contain 3 men and 5 women. By substituting $x = 3$ and $y = 5$, you obtain $75 \times 3 + 60 \times 5 \leq 1580$, or $525 \leq 1580$.

The solution of an inequality in two variables corresponds to a pair of values which satisfy the inequality. The ordered pairs that satisfy an inequality are called the solution set.

HALF-PLANE

It is possible to graphically represent an inequality in two variables on a Cartesian plane.

- All the points whose coordinates satisfy an inequality are found on the same side of the line defined by this inequality. All these points form a **half-plane** which represents the solution set of that inequality. Shade that half-plane to represent the solution set.

- When the **boundary line** of a half-plane is a **solid line** it means that the boundary is included ($\leq$ or $\geq$). When the boundary line of a half-plane is a **dotted line** it means that the boundary is excluded ($<$ or $>$).

E.g. 1) Solution set for the inequality $y \leq x + 3$.

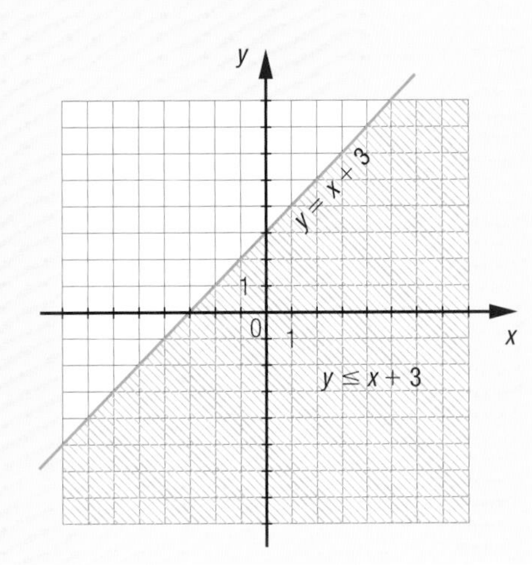

2) Solution set for the inequality $y > -0.5x + 6$.

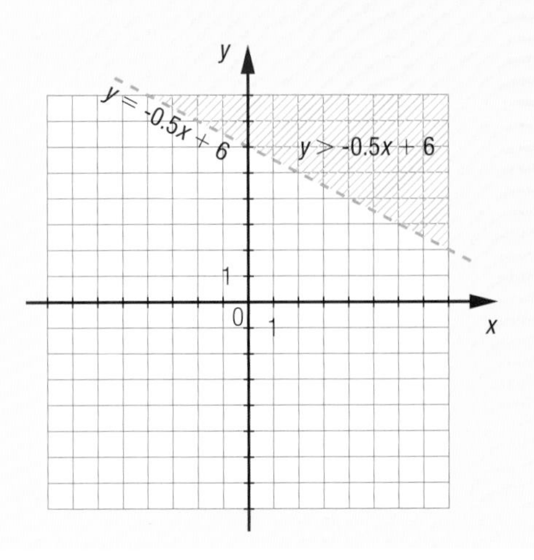

To graphically represent the solution set of a first-degree inequality in two variables, proceed as follows.

<table>
<tr>
<td>1. Write the inequality in the form $y < ax + b$, $y > ax + b$, $y \leq ax + b$ or $y \geq ax + b$.</td>
<td>E.g.
You want to graphically represent the solution set of the inequality $-x + 4y < -4$.
$-x + 4y < -4$
$4y < x - 4$
$y < 0.25x - 1$</td>
</tr>
<tr>
<td>2. Draw the boundary line of the equation $y = ax + b$ using a solid or dotted line according to whether or not the equation is part of the inequality.</td>
<td>The equation of the boundary line is $y = 0.25x - 1$.
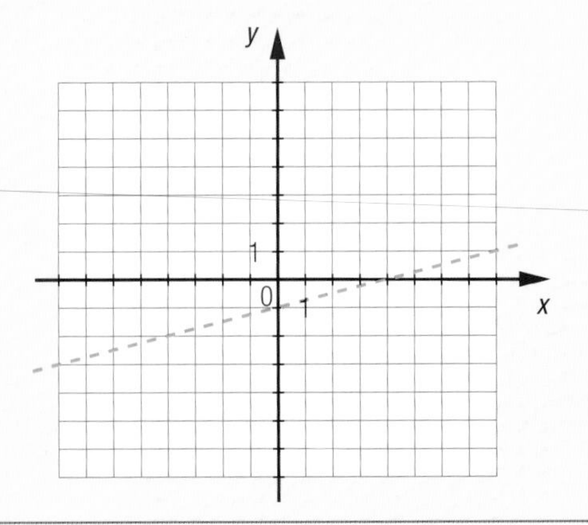
</td>
</tr>
<tr>
<td>3. Colour or shade the half-plane below the boundary line if the symbol is $<$ or $\leq$, or above the boundary line if the symbol is $>$ or $\geq$.</td>
<td>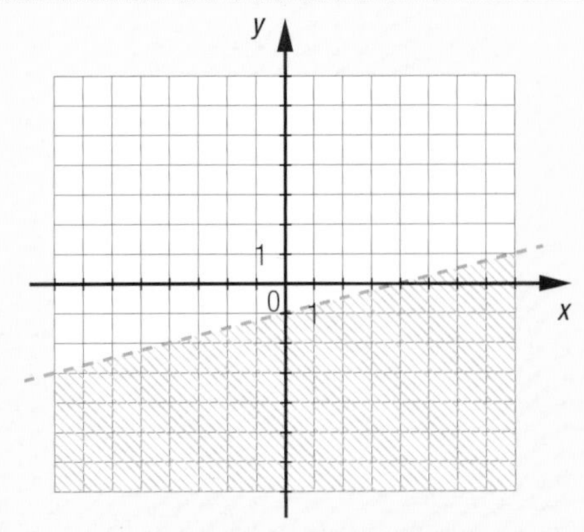
</td>
</tr>
</table>

practice 1.4

1 Represent each of the following statements as a first-degree inequality in two variables.

a) The difference between length l and width w of a rectangular field is greater than or equal to 250 m.

b) In a school, the number of girls g is more than twice the number of boys b.

c) A truck transports x containers of 100 kg and y containers of 50 kg. It cannot transport more than 6000 kg.

d) A room with less than 500 seats contains x children and y adults.

e) On a farm, the ratio between the number of cows c and the number of chickens d is greater than 12.

2 Write the following inequalities in the form $y \leq ax + b$ or $y \geq ax + b$.

a) $4x + 3y - 6 \geq 0$ b) $5x - 10y + 2 \geq 0$ c) $-\frac{y}{7} \leq -x + 11$

d) $\frac{x}{2} + \frac{y}{3} \leq 1$ e) $2(x - 5y) \geq 3x + 1$ f) $\frac{1 - x}{2} + \frac{3 - y}{5} \leq \frac{1}{2}$

3 For each situation below, do the following:

1) Identify the unknowns and represent them using different variables.
2) Express the situation as an inequality.

a) Mr. Pincher predicts a harvest this year of at least twice as much wheat as barley.

b) Simon and his father compare their ages. Simon says that three times his age plus his father's age equals a number less than 86.

c) The surface area of Canada is 14 times greater than that of France.

d) Of the five Great Lakes, Lake Superior has a surface area of at least 5000 km^2 more than four times the surface area of Lake Ontario which is the smallest of the Great Lakes.

In total, the Great Lakes contain about 23 000 km^3 of water and have a total surface area of 244 000 km^2. They form the largest freshwater surface reservoir in the world and represent approximately 18% of the world's fresh water reserves. Lake Superior is the largest, deepest and coldest of the lakes. Water stays in the lake for approximately 191 years.

4 Match each of the inequalities with one of the graphical representations.

A $3x - 5y + 20 \geq 0$ **B** $y > \frac{x + 1}{2}$ **C** $y \geq \frac{3}{5}x + 4$ **D** $x - 2y + 1 > 0$

Graph **1**

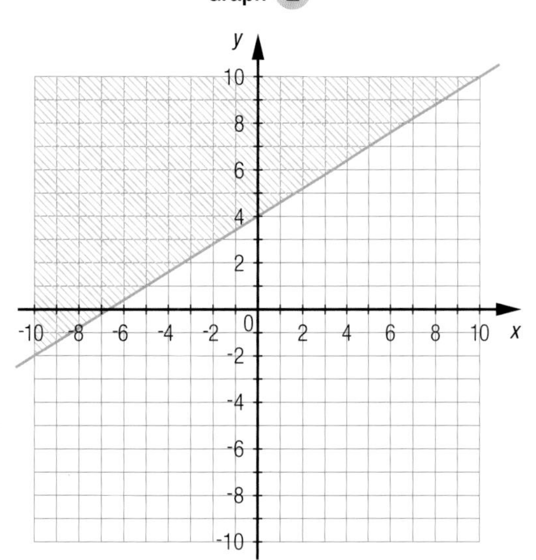

Graph **2**

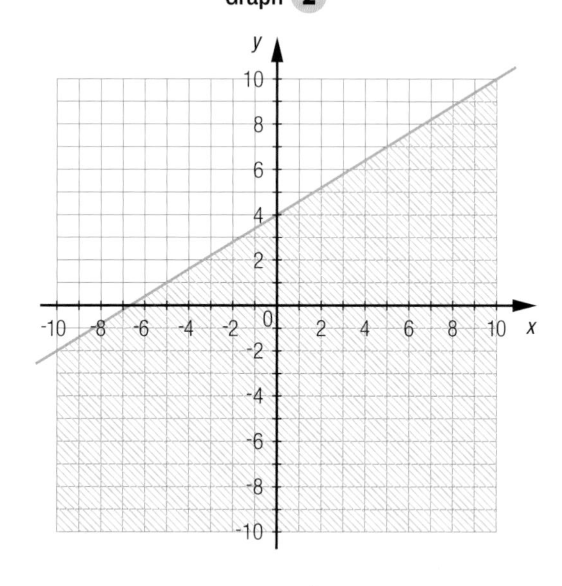

Graph **3**

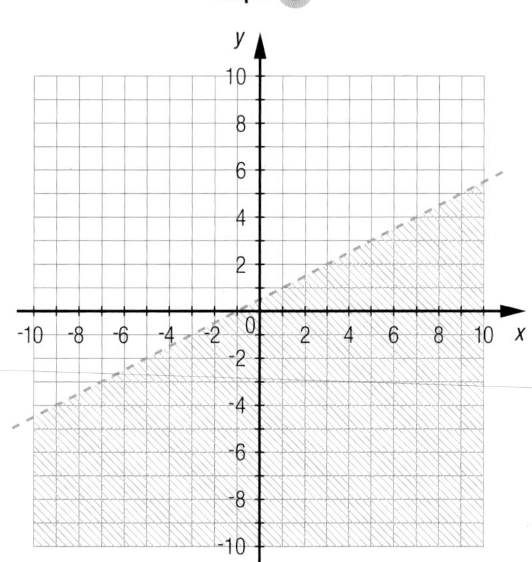

Graph **4**

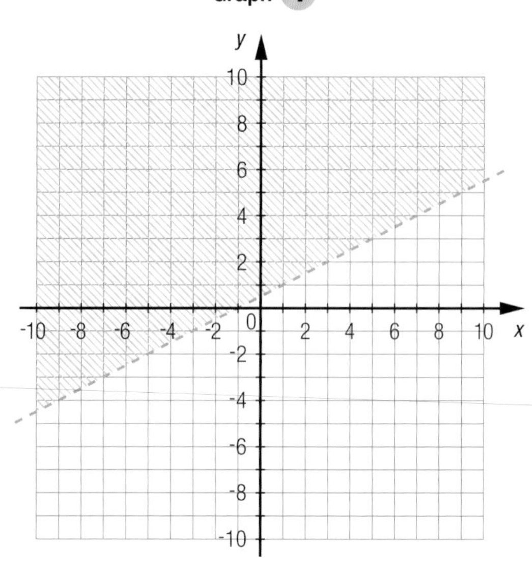

5 Graphically represent the solution set of the inequalities below.

a) $x \geq 3$ b) $y < 7$ c) $3x + y \leq 8$

d) $15x - 30y - 60 > 0$ e) $12x - 3y < 15$ f) $\frac{x - y}{3} \geq -1$

6 Determine the inequality represented in the adjacent graphs considering that the boundary line is not part of the solution set.

Screen 1

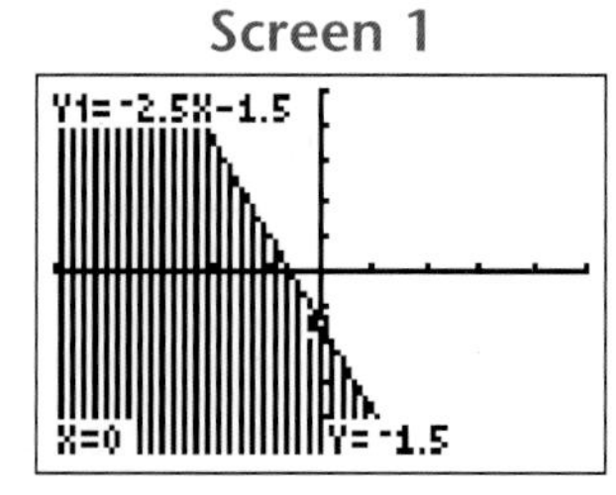

Screen 2

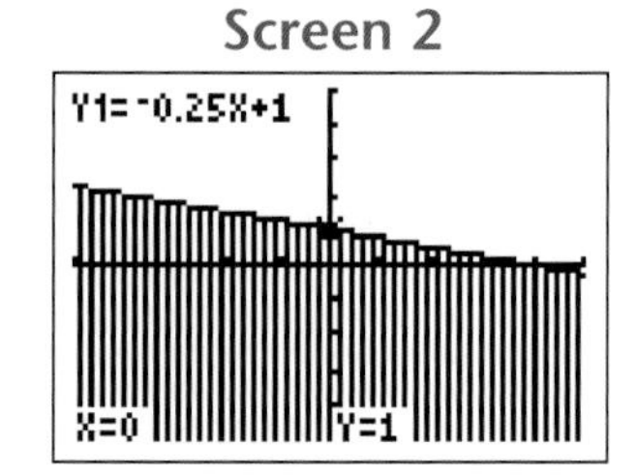

7 Represent each situation as an inequality.

a)

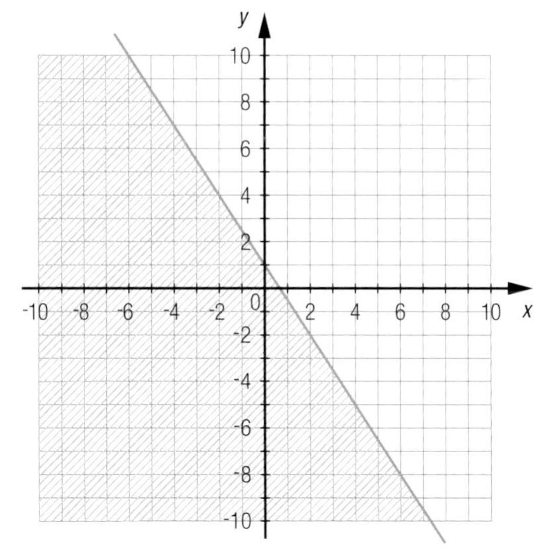

b)

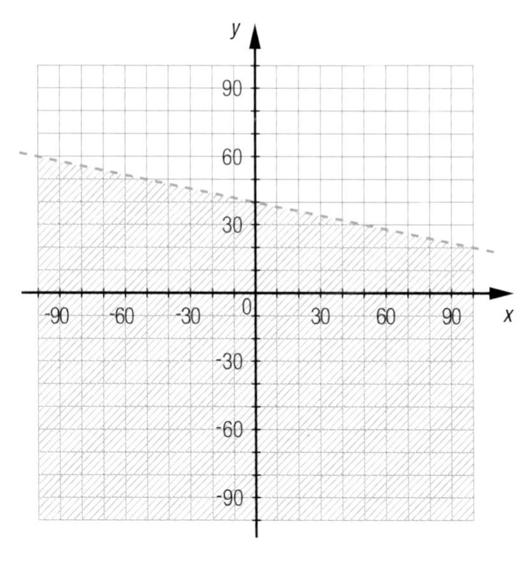

c)

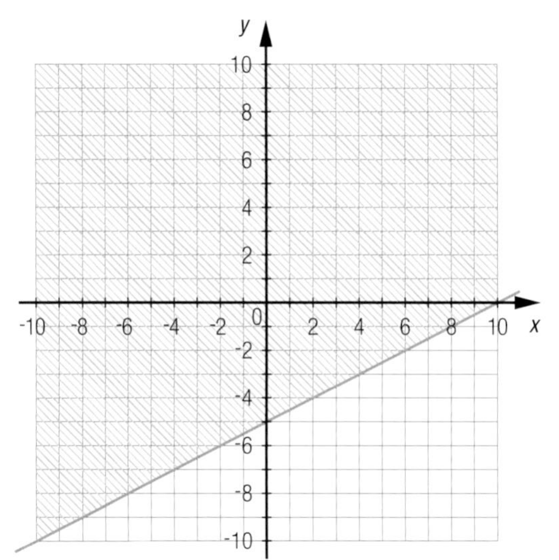

d)

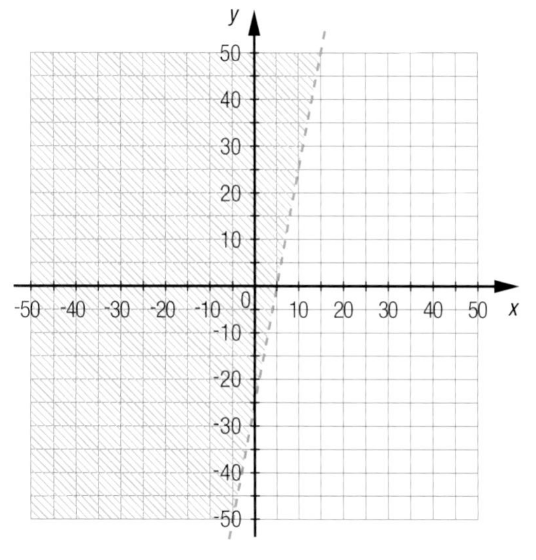

8 An amount of money a is invested at an annual rate of 5% while an amount of money b is invested at an annual rate of 6%. The combined annual interest is at most $207. This situation is represented in the adjacent graph:

a) Find the inequality that describes this situation.

b) What is the equation of the boundary line associated with the inequality found in **a)**?

c) Are the points situated on the boundary part of the solution set? Explain your answer.

d) Which Region (**A** or **B**) represents the solution set of the inequality associated with this situation?

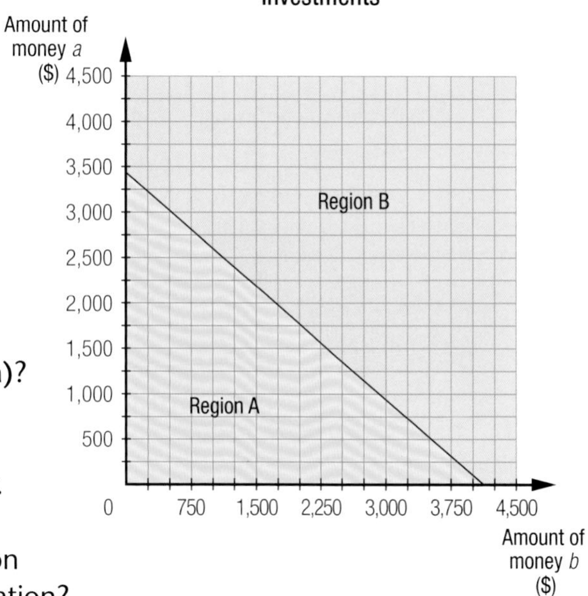

9 The area of the right trapezoid shown in the illustration below is less than 18 cm^2.

a) Describe this situation with an inequality.

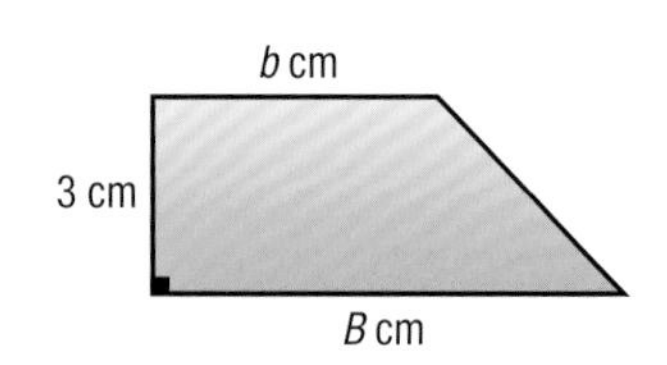

b) Find three possible lengths for the large base *B* and the small base *b* of the trapezoid.

c) Can the large base *B* and the small base *b* measure 7 cm and 5 cm respectively? Explain your answer.

10 At a campground, the area allocated for a tent or a tent trailer is 50 m^2 and 150 m^2 for a camper van. The total area of the campground is 15 750 m^2. If 151 sites are reserved for tents and tent trailers, what is the maximum number of camper vans that can be accommodated?

Most of Québec's 27 provincial parks provide campsites that are either completely serviced, semi-serviced, or primitive. Some parks also offer winter camping. All provincial parks offer hiking trails and nature interpretation programs.

11 For each situation below, do the following:

1) Translate the situation into a first-degree inequality in two variables.
2) Graphically represent the solution set for this inequality.

a) A lift contains 12 kg boxes and 20 kg boxes. The maximum capacity of the lift is 440 kg.

b) May's age substracted from twice Kevin's age is less than 25.

c) Hoang's mark on his last exam is greater than twice Matthew's mark.

d) The snowfall in February and March is at least 150 cm.

12 Find the number of solutions for the inequality $3x + 5y - 45 \leq 0$ where the coordinates are natural non-zero numbers.

13 If $(-4, y)$ is part of the solution set for the inequality $2x + 3y \leq 36$, what is the largest possible value of y?

14 Melanie estimates that her pool contains between 18 000 L and 24 000 L of water. She empties it with Pump **A** at a rate of 50 L/min and with Pump **B** at a rate of 30 L/min. How long will the operation take if both pumps are used at the same time?

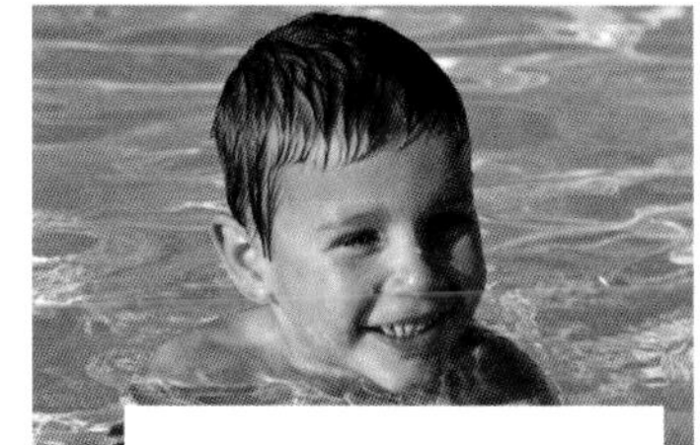

In 2007 there were almost 300 000 residential pools in Québec with 75% of these above ground. From 1986 to 2005, 222 of the 225 deaths from drowning occurred in these residential pools and children aged from 1 to 4 years old accounted for 50% of these deaths.

15 A field is defined on a Cartesian plane by the following inequalities.

$y \leq -\frac{2x}{5} + 8$ $\quad y \leq \frac{5x}{2} - 21$ $\quad y \geq 0$ $\quad x \geq 0$

Find the area of the field, given that the Cartesian plane is scaled in decametres.

16 Identify which of the following inequalities includes the most points among those shown in the adjacent graph.

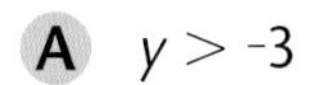

A $y > -3$

B $x + y - 1 < 0$

C $y \geq \frac{7x - 9}{3}$

D $-3x - 4y - 5 \leq 0$

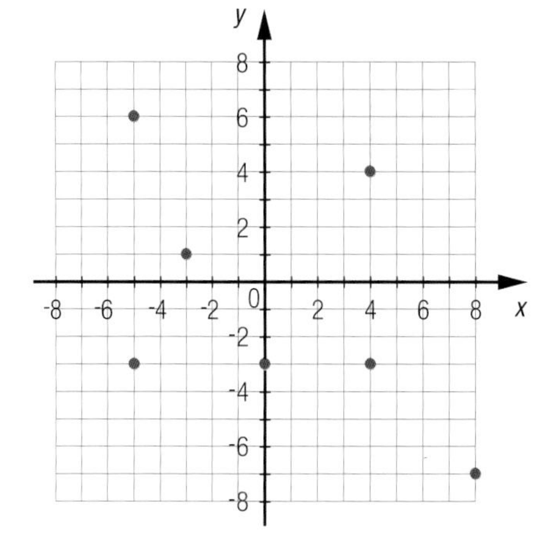

17 In each case, create an inequality where the given coordinates belong to the solution set.

a) (0, 0), (5, 2), (7, -1), (12, 9)

b) (-2, 0), (0, -6), (8, -8), (15, 1)

c) (15, 7), (20, 5), (25, 3), (30, 1)

d) (-10, -12), (17, -4), (50, 13), (90, 100)

18 Below is some information concerning a moving object:

- The speed between the departure point and point A is 30 m/s.
- The speed between point A and the destination point is 45 m/s.
- The distance travelled is at most 5400 m.

a) Identify the unknowns in this situation.

b) If point A represents the departure point, what is the maximum time the object travelled?

c) If point A represents the destination point, what is the maximum time the object travelled?

d) Find the maximum time required by the object to travel the distance between point A and the destination point given that it took 30 seconds to travel the distance between the departure point and point A.

Chronicle of the

past

René Descartes

His life

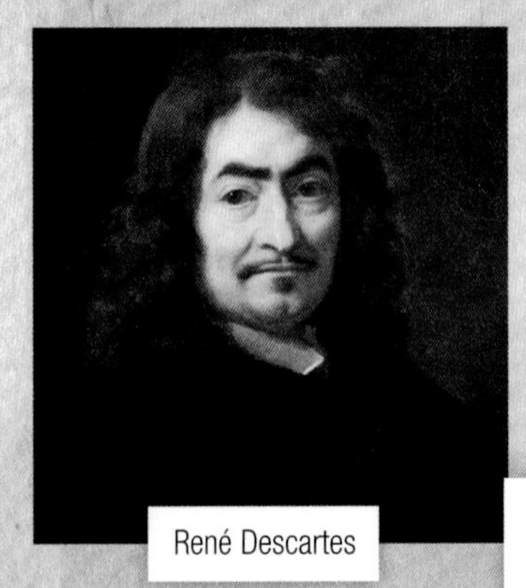

René Descartes

French philosopher, physician and mathematician, René Descartes made his mark in each of these disciplines. In mathematics he made several discoveries in geometry and algebra. An important law of optical physics bears his name, and in philosophy, he is the author of a famous sentence.

I think, therefore I am.

Born March 31, 1596 in Descartes, France; died at age 53 on February 11, 1650 in Stockholm, Sweden.

DISCOURS
DE LA METHODE
Pour bien conduire sa raison, & chercher la verité dans les sciences.
PLUS
LA DIOPTRIQVE.
LES METEORES.
ET
LA GEOMETRIE.
Qui sont des essais de cete METHODE.

A LEYDE
De l'Imprimerie de IAN MAIRE.
cIↄ Iↄ c XXXVII.
Auec Priuilege.

His *Discourse of Method* (1668) is his most famous work. He qualified it himself as his "discourse on how to think correctly and to seek the truth through science."

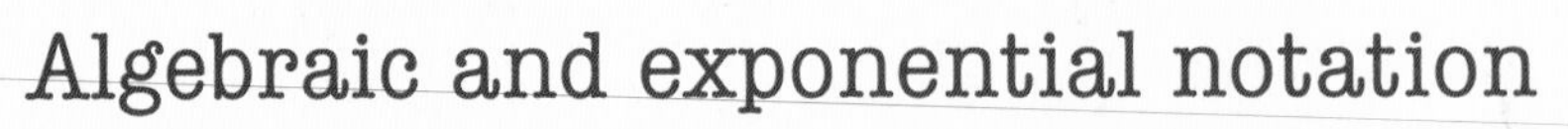

Algebraic and exponential notation

In Descartes' era, equations such as $4x^2 + 3x = 16$ were written as 4Aq + 3A equals 16 where A was the unknown, and q for quadratus meant "squared." The notation "4cc + 3c.16" was used where c was the unknown, and the period represented the equals sign.

René Descartes was the first to propose a simplified notation in which unknowns were expressed as the last few letters of the alphabet and known quantities as the first few. He developed a more practical notation for powers: exponents which considerably simplified written equations.

Using geometry for arithmetic operations

René Descartes demonstrated his creativity by using geometric constructions and to find the square roots of numbers and to carry out mathematical operations such as multiplication and division.

Multiplication and division

Descartes used a scale diagram similar to that shown below to determine the product or the quotient of two numbers. In the diagram below, $\overline{BD} // \overline{CE}$ and $m\,\overline{AD} \times m\,\overline{AC} = m\,\overline{AB} \times m\,\overline{AE}$. Assuming the length of $\overline{AB}$ to be 1, you can conclude that $m\,\overline{AC} \times m\,\overline{AD} = m\,\overline{AE}$, that $\frac{m\,\overline{AE}}{m\,\overline{AD}} = m\,\overline{AC}$ and that $\frac{m\,\overline{AE}}{m\,\overline{AC}} = m\,\overline{AD}$.

For example, Descartes could find the product of 3 and 2 with the help of the adjacent diagram.

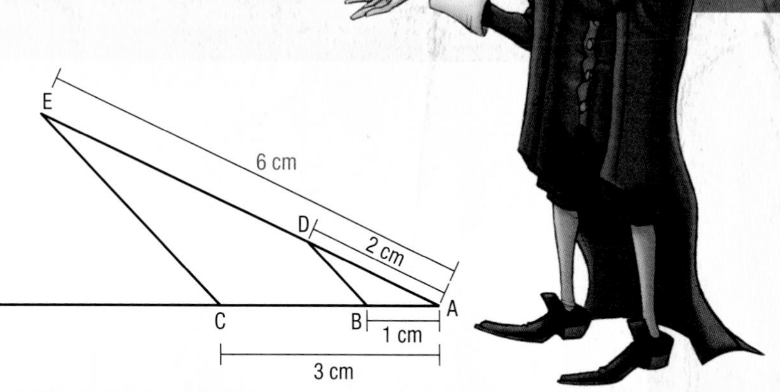

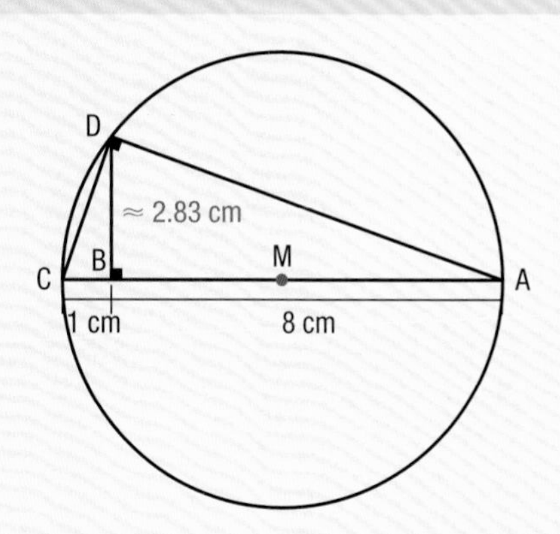

Finding square roots

Descartes used a method based on the adjacent diagram to find square roots. In this scale diagram, $\overline{AC}$ is the diameter of a circle with centre M and $\frac{m\,\overline{BC}}{m\,\overline{BD}} = \frac{m\,\overline{BD}}{m\,\overline{AB}}$. By letting the length of $\overline{BC}$ be equal to one unit, you can deduce that $(m\,\overline{BD})^2 = m\,\overline{AB}$; therefore $\overline{BD} = \sqrt{m\,\overline{AB}}$.

Thus Descartes could, for example, determine the square root of 8 with the help of this diagram.

The Cartesian Plane

In some situations, Descartes used algebra to develop geometric concepts. He used a grid to describe equations of lines. Legend has it that the idea came to him from watching a fly walk across the square panes of a window. He realized that he could define the position of the fly with the help of the squares which resulted in the creation of the coordinate system that we use today.

1. Rewrite and simplify this equation using modern notation. Let the unknown value be *x*.

6Aq – 15A – 3 – 9Aq + 17A + 5Aq + 7 equals 0.

2. Rewrite this equation using the notation from the 16th century.

$$9x^2 + 5x = 5$$

3. Using Descartes' geometric method:

a) show that $2 \times 4 = 8$

b) show that $10 \div 4 = 2.5$

c) find the square root of 10

In the workplace

Air traffic controllers

The profession

Air traffic controllers direct and supervise aircraft movement both in the air and on the runways within a predetermined area for the safety of all individuals and vehicles concerned.

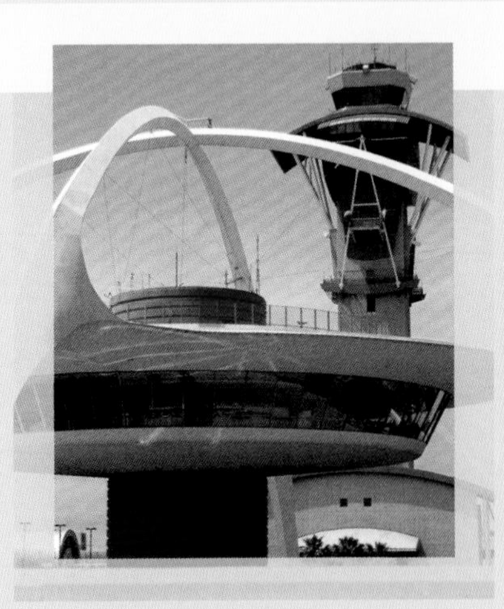

Duties and responsibilities

Air traffic controllers follow the aircraft movement on a radar screen as well as with visual reference points, provide landing and take-off instructions, and transmit weather information by radio. The figure below shows an atmospheric disturbance moving from the Great Lakes region toward Québec. Distances are in kilometres.

Figure 1: Weather radar image

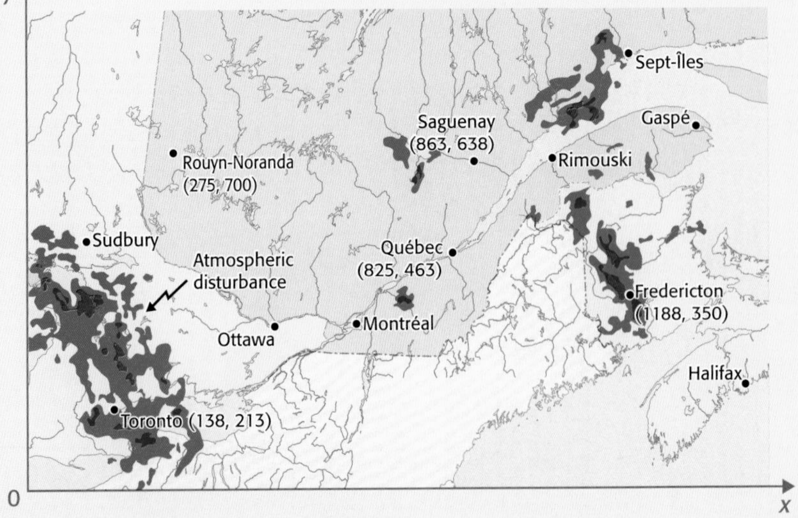

y
Sept-Îles
Gaspé
Saguenay (863, 638)
Rimouski
Rouyn-Noranda (275, 700)
Sudbury
Atmospheric disturbance
Québec (825, 463)
Fredericton (1188, 350)
Ottawa
Montréal
Halifax
Toronto (138, 213)
0
x

An air traffic controller must call for help or alert both the Security and the Search and Rescue Teams when a plane is experiencing difficulties. The adjacent figure shows a plane in distress near a coastal airport. The scale is in kilometres.

Figure 2: Airplane in distress

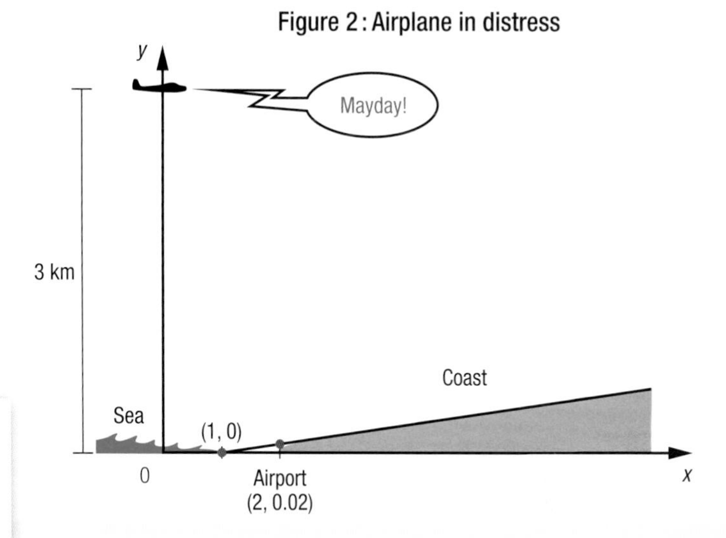

In 1927, the *International Radio Telegraph Convention* adopted the word "Mayday" as the universal distress call. It is an English modification of the French expression *venez m'aider*.

Work on the ground

Air traffic controllers must also direct traffic on the ground in order to avoid collisions or unnecessary movements. The aerial photo below shows the path taken by two planes on different runways. The scale is in metres.

Figure 3: Aerial photo of an airport

1. The two airplanes in Figure **3** leave their departure points at the same time and move toward their respective hangars. Airplane **A** moves at a velocity of 3 m/s and airplane **B** at 4 m/s. Is there a risk of collision?

2. An airplane flying over Québec City in the direction of Toronto flies at a rate of 225 km/h. According to the information in Figure **1**, that airplane will face an atmospheric disturbance in mid-flight between Québec and Toronto. At what time will it experience the disturbance?

3. An airplane in distress drops 10 m per 100 m horizontal flight and makes an emergency landing in a field. According to the information given in Figure **2**, at what distance from the airport should Search and Rescue Teams focus their search?

overview

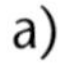

1 For each of the following segments find:

1) the length 2) the slope 3) the coordinates of the midpoint

a)

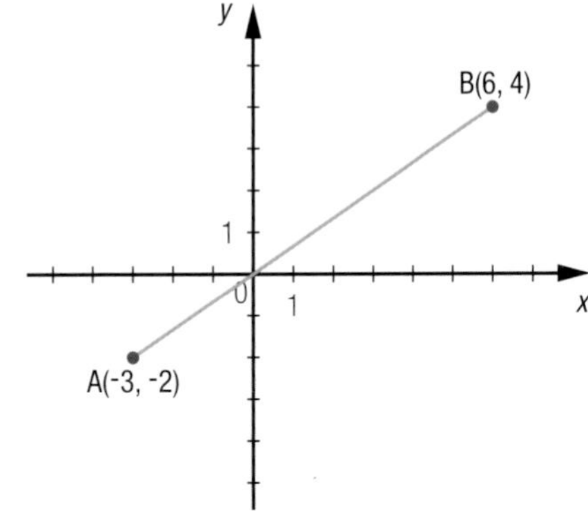

b)

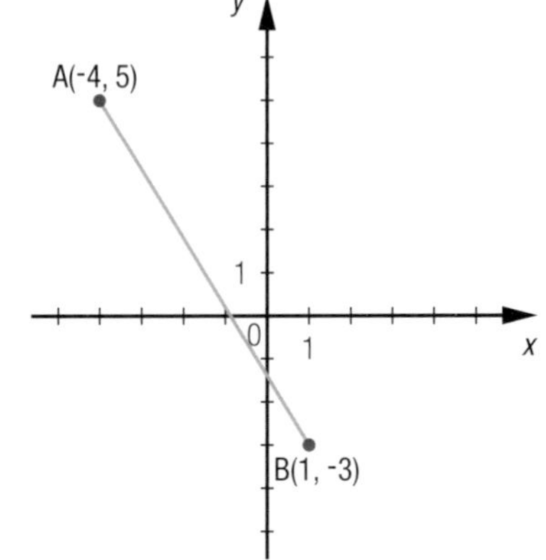

2 Determine the coordinates of the point:

a) situated $\frac{3}{5}$ along segment CD with endpoints C(3, 10) and D(18, 20)

b) situated $\frac{2}{3}$ along segment FE with endpoints E(-3, 5) and F(15, 26)

c) that divides segment AB with endpoints A(8, 7) and B(-16, 2) in a ratio of 2:2

d) that divides segment CB with endpoints B(-1, -5) and C(83, 58) in a ratio of 2:5

3 Prove that the opposite sides of the rhombus below are parallel.

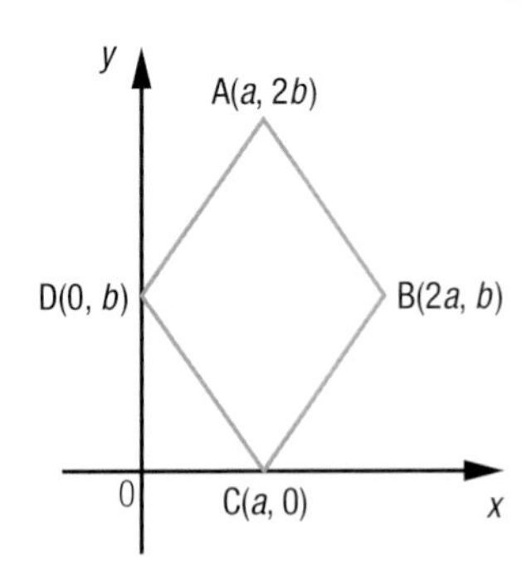

4 Solve the following systems of equations.

a) $y = -x - 5$
$y = 3x - 21$

b) $-3x - y + 9 = 0$
$y = -x$

c) $y = 3.5x$
$y = 5x - 1$

d) $-3x + 2y = 4$
$4x - 18y = 10$

e) $8x + y = -100$
$x = 3y$

f) $-x + 0.1y = 2.8$
$-x - 10.7y = -25.28$

5 Determine the inequality represented by each of the solution sets shown below.

a)

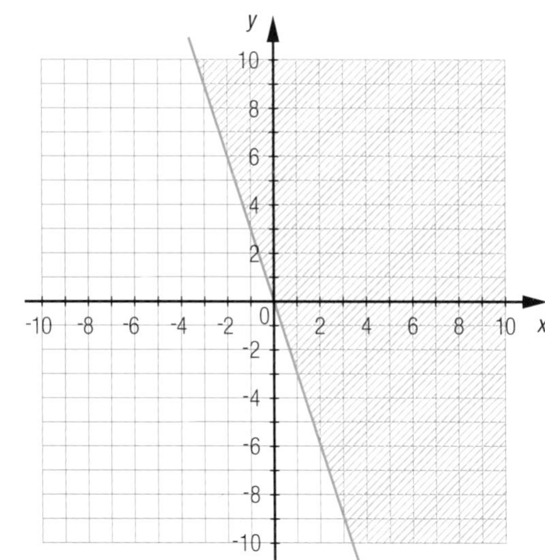

b)

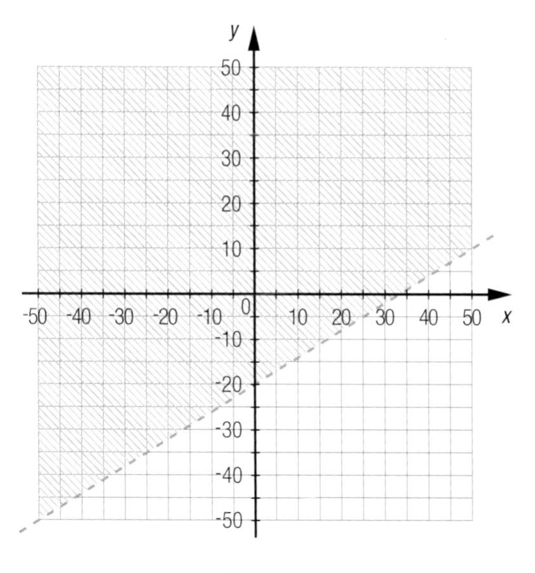

c)

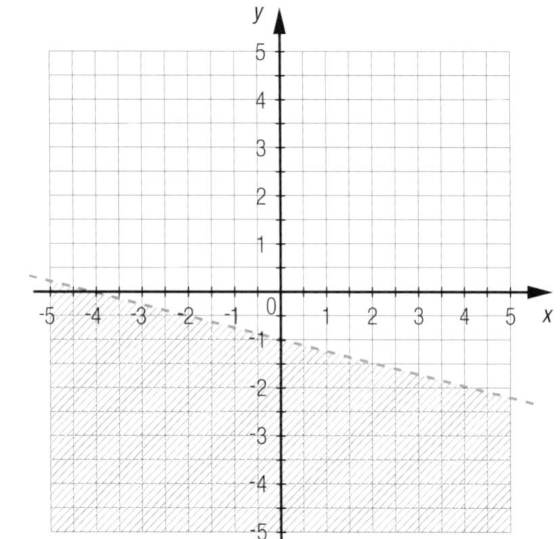

d)

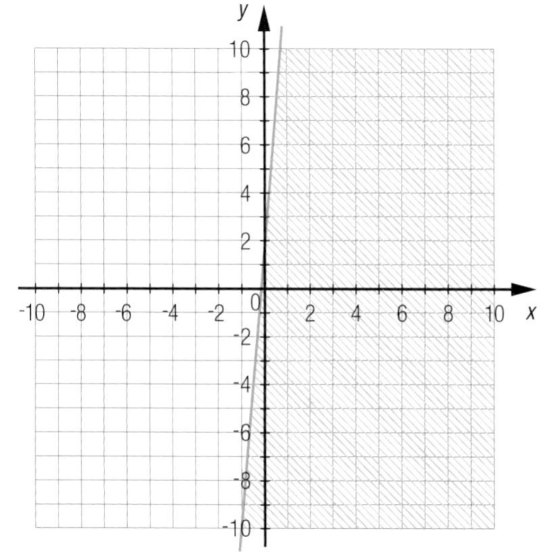

6 Two positive whole numbers are chosen at random. Twice the first number added to the second number is less than 80.

a) If x represents the first number and y the second, determine an inequality that represents this situation.

b) In each case, determine whether the following coordinates are solutions to the inequality defined above.

1) (0, 80) 2) (20, 20) 3) (30, 30)

4) (0, 0) 5) (40, 0) 6) (35, 11)

7 From the information given, find the position of the two lines in relation to each other.

a) Lines l_1 and l_2 have the same slope and different y-intercepts.

b) Lines l_3 and l_4 have the same y-intercept. The slope of l_3 is 2 and that of l_4 is $-\frac{1}{2}$.

c) The slope of l_5 is 3. The slope of l_6 is -3 and both lines pass through the point P(4, 6).

8 For each of the situations below, do the following:

1) Identify the unknowns and represent them using different variables.
2) Describe the situation as an inequality.

a) Mount Everest is the world's tallest peak. Its height is at least 300 m more than 4 times that of Mount Iberville, the highest peak in Québec.

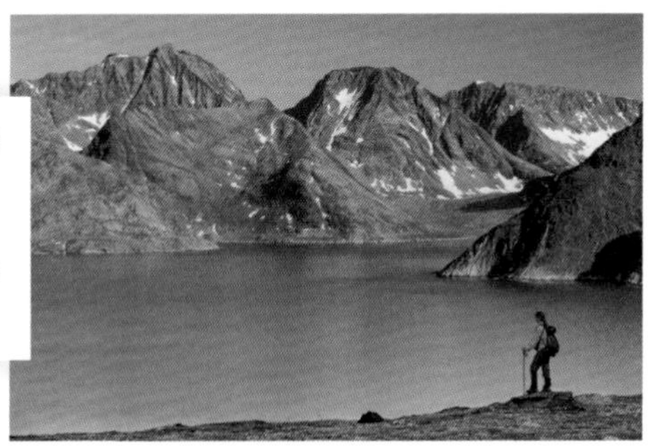

The Torngat Mountains, east of Ungava Bay, are the highest mountains in Québec: Mount Iberville is the highest peak at 1600 m. The landscape is composed of glaciated terrain found in alpine areas such as the Rocky Mountains.

b) At the 2004 Olympics in Athens, China won more than three times the number of medals won by the host, Greece.

c) In 2007, the population of the United States was at least 9 times that of Canada.

d) In Québec, the maximum load allowed for a 10-wheel vehicule and its cargo is 25 250 kg.

9 Find the equation of the line:

a) parallel to $y = 3x - 12$ and that passes through (2, 1)

b) perpendicular to $2x + 2y - 6 = 0$, whose y-intercept is 8

10 Find the area of the quadrilateral below.

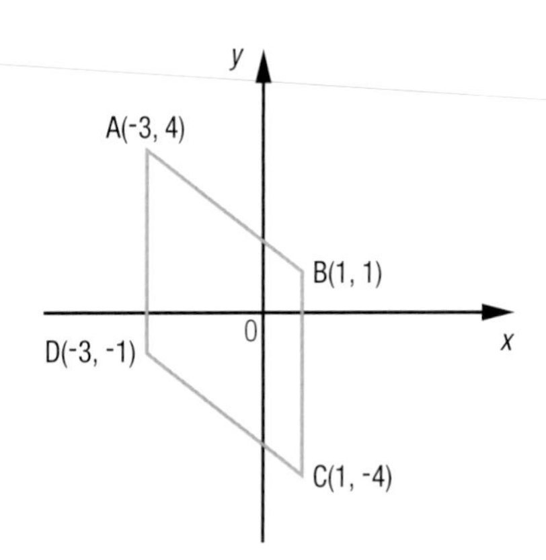

11 Graphically represent the solution set for the following inequalities.

a) $x \geq 4$

b) $y < 9$

c) $y > -x + 5$

d) $-15x + 3y \leq -120$

e) $\frac{x - y}{2} \leq 5$

f) $\frac{x}{3} + \frac{y}{5} - 2 > 0$

12 Philip is trying to decide between two jobs he was offered. Both involve selling new cars. Dealership **A** offers him a salary of $200 plus 4% of his weekly sales. Dealership **B** offers him a salary of 5% of his weekly sales.

a) Describe the situation using a system of equations.

b) Solve the system of equations.

c) Which is the best offer for Philip? Explain your answer.

Battery-operated electric cars are a solution to the pollution problem caused by gasoline-powered vehicles. Under optimal conditions, battery-powered cars can travel an average of about 80 to 200 km depending on to the type of battery used. These cars are more environmentally friendly, but the purchase price is higher especially because the batteries, which are made of toxic heavy metal, must be recycled. The electric car is mainly used in urban settings.

13 The length of a rectangular lot is 4 times its width. Its perimeter is 140 m. What is its area?

14 In preparing 930 mL of fruit juice, the babysitter used 5 times as much orange juice as banana juice. How much banana juice did he use in preparing this mixture?

15 Below is the street-level floor plan of a house. The scale is in metres. What is the price per square metre to cover the floor if the total cost is $4,796?

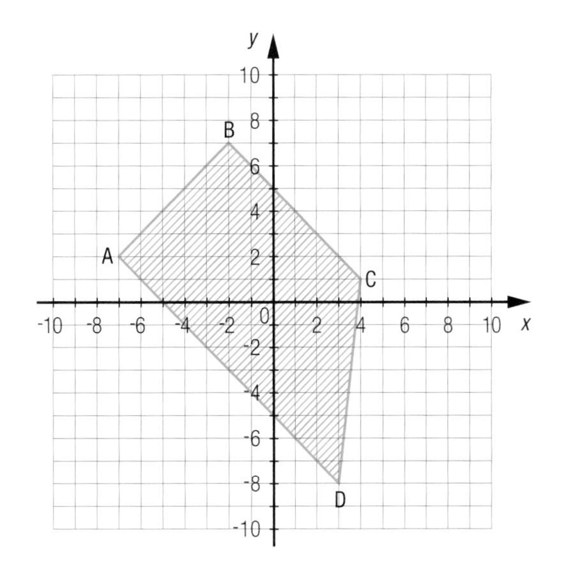

16 Below are the the flight plans of two commercial aircrafts:

Airplane A	Airplane B
• Departure point coordinates: (-60, -650). • Arrival point coordinates: (270, 2650). • Flight altitude: 10 000 m.	• The flight plan is represented by the line $16x + 2y - 332 = 0$. • Flight altitutde: 12 000 m.

a) What distance must Airplane **A** cover if the scale of the Cartesian plane is in km?

b) Airplane **A** must stop at an airport situated 1/3 along its route. What are the coordinates of this airport?

c) At a certain point, Airplane **B** passes directly over Airplane **A**. What are the coordinates of this point?

d) The flight plan of Airplane **C** is parallel to that of Airplane **B**. What is the equation associated with this flight plan if Airplane **C** passes over the point (16, 38)?

The Airbus A380 is the largest passenger plane ever built. The fuselage is mainly oval, and not round, and this design allows the plane to have two complete decks the full length of the aircraft. With three different classes of service, the airplane can accommodate 555 passengers. When configured for chartered or economy services, the airplane holds 853 passengers. The Airbus A380 made its inaugural flight to Canada on November 12, 2007 arriving at Montréal Trudeau Airport.

17 Line l_1 passes through points A(-12, 10) and B(4, 6). Line l_2 is perpendicular to l_1 and passes through points B(4, 6) and C(-1, y). Find the value of y.

18 Following is information about Mika's next hiking trip:

- Her mean walking speed between the departure point and Camp **A** is 3.5 km/h.
- Her mean walking speed between Camp **A** and Camp **B** is 4.5 km/h.
- Her mean walking speed between Camp **B** and the destination point is 2.5 km/h.
- Her time of departure is 8 a.m.

Considering that all distances are in kilometres, determine to the nearest minute:

a) Mika's arrival time at Camp **A**

b) Mika's arrival time at Camp **B**

c) Mika's arrival time at the final destination

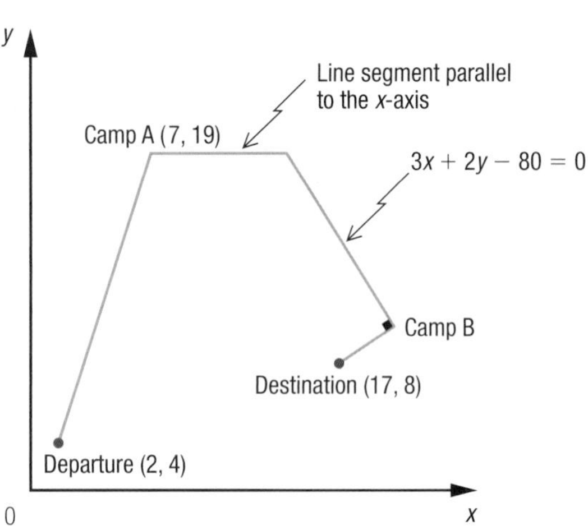

19 A jar contains a total of 25 marbles; some are coloured blue, some yellow and some green. There is 1 more yellow than blue; there are 4 fewer green than yellow. How many marbles are there of each colour?

20 Find the coordinates of the centre of gravity for each of the steel plates shown. The centre of gravity is the point where the plate can be placed in balance on the end of a pin.

a)

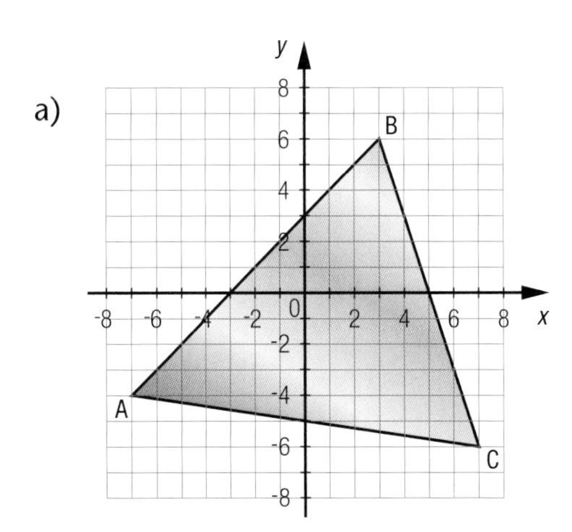

b)

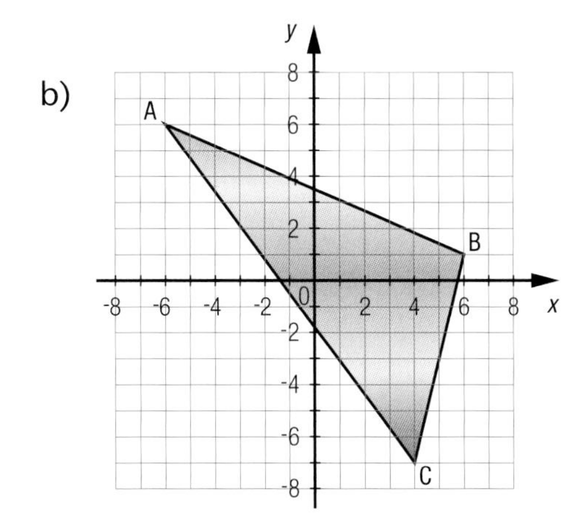

c)

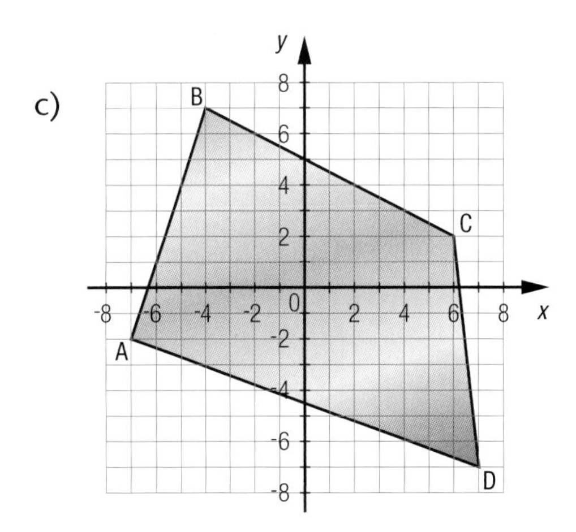

d)

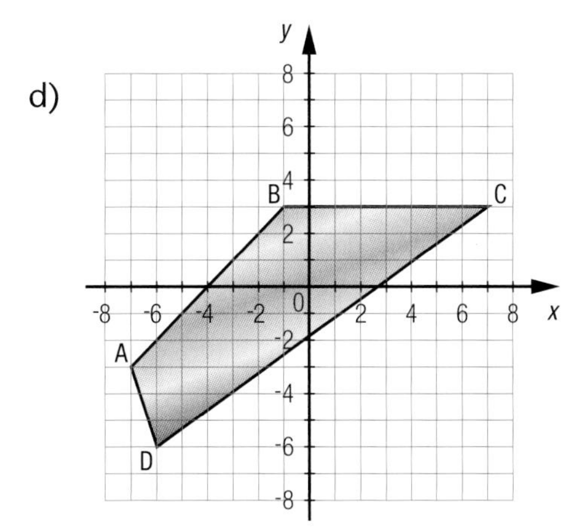

21 **HYDROCHLORIC ACID** Concentrated hydrochloric acid is a clear or slightly yellow vaporous liquid used in stripping paint, polishing metals, treating ore and manufacturing pharmaceutical, photographic and food products. Determine how much 5% hydrochloric acid and 20% hydrochloric acid must be mixed to obtain 10 mL of 12.5% hydrochloric acid.

22 The adjacent diagram illustrates a stage where a singer is performing. At a specific moment during the show, the singer stands at the intersection of the median at A and the altitude C. What are the coordinates of the singer at this exact moment?

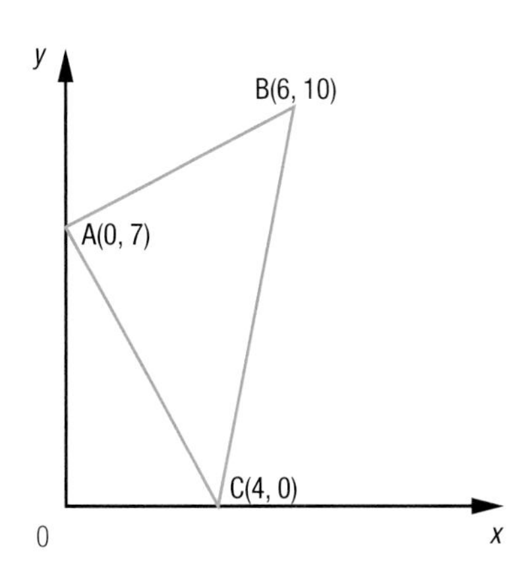

bank of problems

23 A construction company needs to build the section of highway represented in the Cartesian plan below.

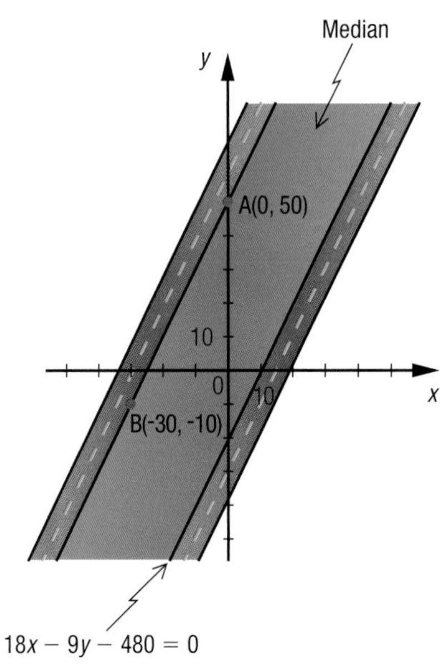

Considering that the scale is in metres, find the width of the green space situated between the two lanes of the highway.

24 A natural gas leak forces the municipal authorities to evacuate an area of the city as a precautionary measure. This area is represented in a Cartesian plane by the inequalities below. In this Cartesian plane, each point with integer coordinates represents a residence.

$y \geq -x + 12$

$-2x + y \leq 8$

$y < 16$

$2x - y - 20 < 0$

How many residences need to be evacuated by the authorities?

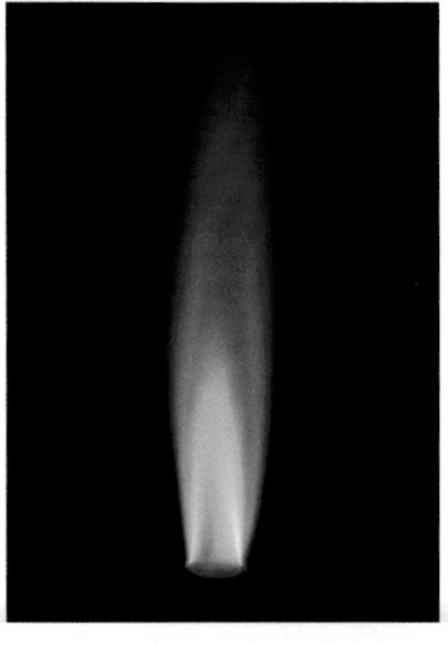

Natural gas is a gaseous fossil fuel found in porous rock. It is one of the least polluting energy sources used to produce heat and electricity. Natural gas is also used as fuel for cars, particularly, in Italy and Argentina. Underground distribution networks of natural gas preserve the landscape.

25 Two drivers followed the strategies below during a 850 km race.

Driver A

- She maintains a speed of 270 km/h for the entire race.

Driver B

- She maintains a mean speed of 280 km/h for the first 90 minutes of the race.
- For the remainder of the race, she maintains a mean speed of 260 km/h until the finish line.

Which strategy will have the best results?

The Gazelle Aicha Rally is the only real rally for women in the world. Since 1990, it has showcased amateur females aged 25 to 65 from all nationalities and different socio-cultural backgrounds. It is not a race, and speed is not a factor. For 8 days, the participants travel off-road for about 2500 km. Equipped with old topographic maps and a compass (no GPS!) the "gazelles" have to track down the beacons that line the route.

26 Find the possible dimensions of the rectangular garden below given the following:

- The area is less than or equal to 1400 m^2.
- The perimeter is greater than 78 m.
- The length and width of the garden are integers.

$(5x + 4)$ m

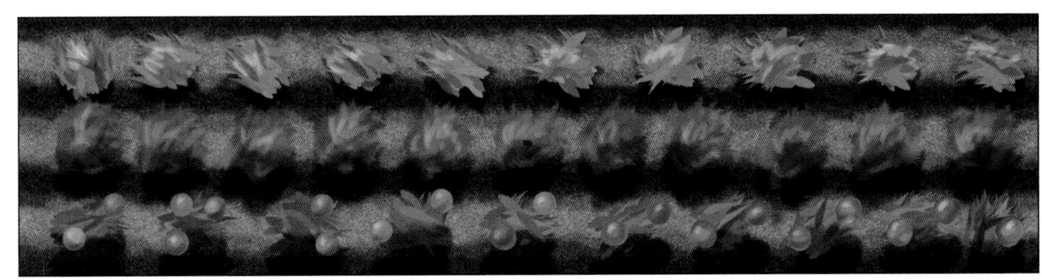

$(4x - 1)$ m

Some cities such as Montréal and Québec give citizens access to plots of land for gardening. Certain small gardens are adapted for people with reduced mobility. For a very small fee, from May 1 to November 1, the gardeners opt for ecological practices: organic fertilizers, composting, water conservation reuse, and recycling.

VISION 2

Statistical measures and linear correlation

How thick will the ozone layer be in 2025? How can you compare the populations of two species of fish? How can you analyze the evolution of the height of human beings over time? How does the result of one examination compare to all the other results of a group? In "Vision 2" you will learn to use a contingency table and a stem-and-leaf plot to group and display data. You will study how to describe and quantify the relationship between two variables in a distribution, and you will use some statistical measurements to compare a data value to the rest of the distribution.

Arithmetic and algebra | Geometry | **Statistics** | Probability

- One-variable distribution
- Percentile
- Mean deviation
- Stem-and-leaf plot
- Two-variable distribution
- Contingency table
- Linear correlation
- Correlation coefficient
- Regression line

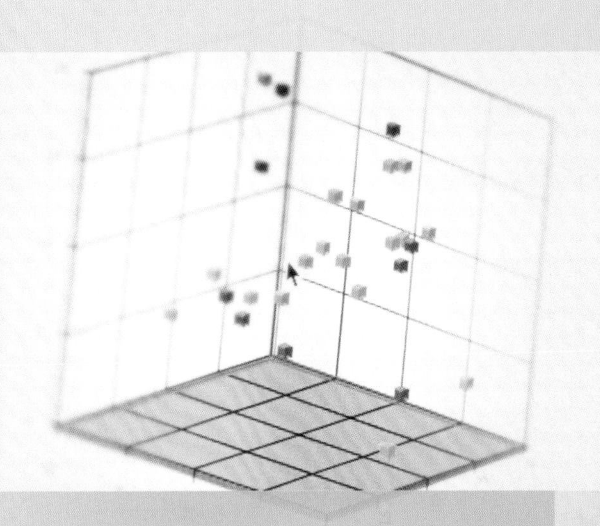

REVISION 2

PRIOR LEARNING 1 Profits and productivity

An insurance company offers a productivity bonus to its representatives when the mean profit in a month is greater than the mean profit of the previous month.

Profits and productivity bonuses

Mean profit in January ($)	17,000	53,000	120,000
Mean profit in February ($)	20,000	70,000	167,000
Productivity bonus ($)	750	4,250	11,750

Below are the profits (in $) reported in March and April by this company:

March

1,000	1,100	1,400	1,600	2,010	2,200	3,400
3,800	4,005	4,500	4,500	4,870	5,110	5,505
5,600	5,780	5,990	6,400	8,000	8,700	8,800
9,000	9,100	9,320	9,440	9,900	10,030	11,000
11,230	11,300	11,500				

April

Profit ($)	Frequency
[10,000, 11,000[	1
[11,000, 12,000[	1
[12,000, 13,000[	2
[13,000, 14,000[	6
[14,000, 15,000[	8
[15,000, 16,000[	5
[16,000, 17,000[	7

a. Draw the box-and-whisker plot representing the profits reported in March.

b. For profits reported in April, calculate:

1) the mode
2) the median
3) the range

c. By how much did the productivity bonus increase for the representatives in April?

Towards the end of the 17th century, London was an important centre of maritime commerce. Edward Lloyd ran a coffee shop which became the meeting point for seamen and shipowners. Maritime companies wishing to insure their vessels arranged meetings there with people ready to assume such a risk. Real insurance contracts were drawn up. Even today Lloyd's of London is the pinnacle of maritime insurance.

MEASURES OF CENTRAL TENDENCY

Measures of central tendency are used to describe the centre of an ordered distribution and the position of the distribution's data values relative to this centre. The **mode**, the **median** and the **mean** are measures of central tendency.

Measure of central tendency	Frequency distribution	Distribution of data grouped into classes
The **mode** indicates the data value that appears the most often in a distribution.	The mode is the value with the highest frequency.	The class with the highest frequency is called the **modal class.** The midpoint of the modal class provides an estimate of the mode.
The **median** indicates the middle position distribution.	In an ordered distribution: • If the number of data values is odd, the **median** is the middle value. • If the number of data values is even, the **median** is the mean of the two middle values.	The class that contains the median is called the **median class.** The middle value of the **median** class provides an estimate of the value. of the median.
The **mean** indicates the centre of equilibrium of a distribution.	$\text{Mean} = \dfrac{\left(\text{sum of the products of the data values and their respective frequency}\right)}{\text{number of data values}}$	$\text{Mean} \approx \dfrac{\left(\text{sum of the products of the midpoints of the data and their respective frequency}\right)}{\text{the number of data}}$

If the importance of each data value is not the same, the mean is called the **weighted mean.**

E.g. A geography course has three terms. Each of the terms has a different importance:

$$\text{Final grade} = 75 \times 0.2 + 72 \times 0.3 + 88 \times 0.5$$
$$= 80.6\%$$

Geography course

Term	Grade (%)	Weight (%)
1	75	20
2	72	30
3	88	50

In a distribution of condensed or grouped data, the frequencies indicate the importance of different values or classes. The mean of these distributions corresponds to a weighted mean.

MEASURES OF DISPERSION

Measures of dispersion are used to describe the concentration or the dispersion of the data values in a distribution. The **range** is a measure of dispersion.

Measure of dispersion	Frequency distribution	Distribution of data grouped in classes
The **range** is a measure that indicates how concentrated or dispersed the data values are in a distribution.	The range is the difference between the highest data value and the lowest data value.	The range is the difference between the upper limit of the highest class interval and the lower limit of the lowest class interval.

E.g. 1) Consider an ordered distribution that consists of 15 data entries:

2, 2, 2, 3, 3, 4, 5, 6, 7, 8, 8, 8, 8, 10, 11.

Mode = 8

Median = 6 (8th data value)

$$\text{Mean} = \frac{2 + 2 + 2 + 3 + 3 + 4 + 5 + 6 + 7 + 8 + 8 + 8 + 8 + 10 + 11}{15} = 5.8$$

Range = 11 − 2 = 9

2) Below is a frequency distribution table:

Families

Number of children	1	2	3	4	5	Total
Frequency	6	16	12	10	9	53

Mode = 2 children since the highest frequency is 16.

Median = 3 children (27th data value)

$$\text{Mean} = \frac{1 \times 6 + 2 \times 16 + 3 \times 12 + 4 \times 10 + 5 \times 9}{53} = 3 \text{ children}$$

Range = 5 − 1 = 4 children

3) Below is a frequency distribution of data grouped into classes:

Kennel

Height of dogs (cm)	[20, 40[	[40, 60[	[60, 80[	[80, 100[	[100, 120[	Total
Frequency	18	19	13	20	10	80

Modal class = [80, 100[cm

Mode ≈ 90 cm

Median class = [60, 80[cm

Median ≈ 70 cm

$$\text{Mean} \approx \frac{30 \times 18 + 50 \times 19 + 70 \times 13 + 90 \times 20 + 110 \times 10}{80} = 66.25 \text{ cm}$$

Range = 120 − 20 = 100 cm

BOX-AND-WHISKER PLOTS

Quartiles are values that divide a distribution into four subsets, called quaters, each containing an equal amount of data. The first quartile is identified as "Q_1," the second as "Q_2," and the third as "Q_3."

Box-and-whisker plots allow you to analyze the dispersion or concentration of the data or to compare two distributions of similar populations. In box-and-whisker plots, each quarter contains the same number of data values.

E.g. The adjacent illustration represents the box-and-whisker plot for given data.

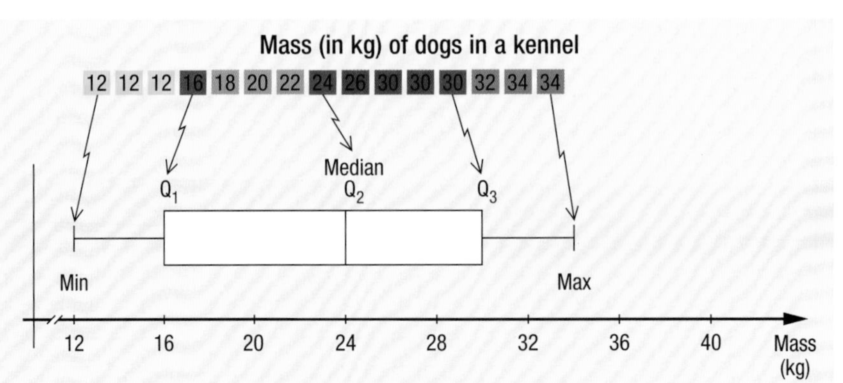

knowledge in action

1 The distribution below describes the wing span (in mm) of 16 monarch butterflies.

93	93	94	96	96	96	98	100
100	101	101	101	101	101	103	105

a) What is the mode of this distribution?

b) What is the range of this distribution?

The monarch butterfly is without a doubt the best known of North American butterflies. It migrates twice a year from Mexico to Canada and back, a feat probably not exceeded by any other insect on earth.

2 Among the following measurements, which is not a measurement of central tendency?

A Median **B** Range **C** Mode **D** Mean

3 Complete this table.

Distribution	Mode	Median	Mean	Range
3, 6, 7, 7, 8, 8, 8, 12, 14, 15, 17, 18, 21, 23, 28, 30, 30				
5, 5, 5, 5, 5, 5, 6, 6, 6, 7, 8, 9, 9, 9, 9, 9, 9, 10				
12, 14, 15, 23, 24, 25, 33, 34, 35, 44, 44, 44, 44, 44				
6, 7, 12, 14, 16, 18, 20, 22, 25, 27, 29, 34, 37				

4 The adjacent table represents the ages of a university's graduates:

a) What is the total number of graduates?

b) What is the modal class?

c) What is the range?

d) What is the median class?

e) What is the mean age of all graduates?

Graduation 2008

Age	Frequency
[20, 25[	87
[25, 30[	231
[30, 35[	145
[35, 40[	78
[40, 45[	63
[45, 50[	32

5 Represent the data below in a table of data grouped into classes. The first class is [10, 20[.

12, 14, 15, 16, 20, 21, 23, 24, 24, 24, 24, 30, 35, 35, 36, 37, 40, 41, 42, 42, 45, 46, 47, 50, 55, 61, 61, 61, 67, 67, 68, 69, 70, 75, 75, 84, 85, 89, 89, 90, 92, 94, 97, 97

6 In each case find:

1) the mode 2) the median 3) the mean

a) **Children**

Number of children	Frequency
0	4
1	6
2	7
3	3

b) **Number of cups of coffee**

Number of cups	Frequency
0	10
1	18
2	14
3	7

c) **Height of horses**

Height (cm)	Frequency
[100, 140[	5
[140, 180[	9
[180, 220[	13
[220, 260[	1

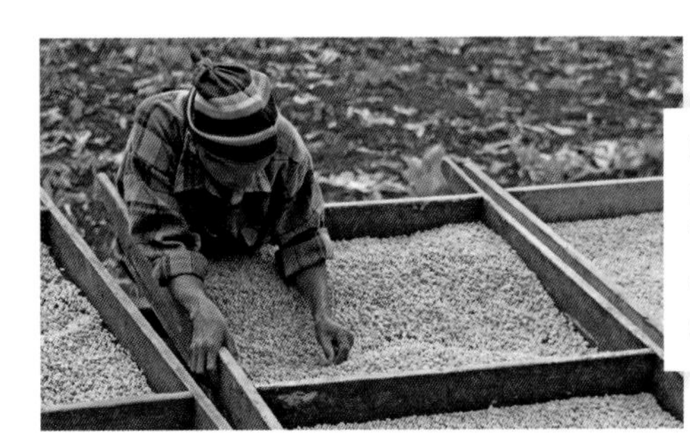

Fair trade uses the purchasing power of consumers to eliminate the exploitation of small producers of coffee. This assures these producers of a comfortable existence while sustaining agricultural methods favourable to the environment.

7 Below are Julie's grades. What is her weighted mean?

Marks

Subject	Grade (%)	Credits
French	95	6
Mathematics	92	6
History	88	4
English	88	4
Physical Education	86	2
Science	85	2

8 Following are 4 distributions:

A 5, 5, 5, 5, 5, 6, 7, 10

B 5, 6, 6, 7, 8, 8, 10

C 1, 2, 3, 4, 5, 5, 5, 5, 6

D 1, 3, 4, 5, 5, 5, 5, 5, 6

Indicate which have the following characteristics:

- the mode and range are the same
- the median is less than the mean

9 Construct a box-and-whisker plot to represent each of the following distributions.

a) 2, 3, 3, 4, 5, 6, 6, 6, 6, 7, 8, 9

b) 12, 14, 14, 16, 20, 22, 22, 23, 24, 25

c) 56, 57, 57, 60, 64, 64, 64, 65, 66, 70, 76

d)

Age	13	14	15	16	17
Frequency	3	4	12	6	2

10 This box-and-whisker plot represents the times of 15 swimmers who participated in a competition.

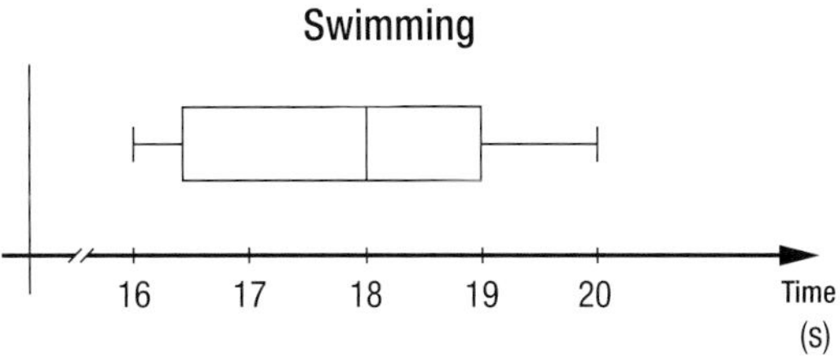

a) What is the variable being studied?

b) Is the variable being studied qualitative, discrete quantitative or continuous quantitative?

c) What percentage of swimmers finished in less than 18 s?

d) In which quarter are the results most condensed?

e) What is the range of this distribution?

11 Each of the following distributions displays the grade out of 50 given to each member of two groups in a drama competition.

Group ①

38, 40, 40, 40,
41, 42, 43, 45,
47, 47, 48, 50

Group ②

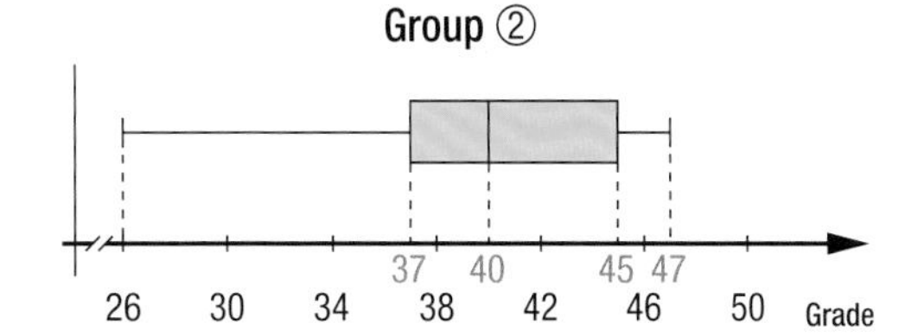

a) Complete the following table.

Measure	Group ①	Group ②
Median		
Range		
Interquartile range		

b) Explain why it is not possible to find the mode of Group ②.

c) Which group has the highest mean? Explain your answer.

12 Four students obtained a grade of 75% on an exam. Which student has the best result relative to his or her own group?

Student group Ⓐ

50, 56, 60, 60, 60,
60, 60, 60, 60, 65,
65, 67, 70, 72, 73,
75, 76, 78, 80, 83,
83, 84, 86, 87, 87

Student group Ⓑ

Grade	Frequency
[50, 60[	1
[60, 70[	5
[70, 80[	14
[80, 90[	5
[90, 100[	2

Student group Ⓒ

Grade	Frequency
55	4
65	6
75	5
85	12
95	2

Student group Ⓓ

65 75 85 95

60 70 80 90 100

SECTION 2.1 Diagrams and statistics

This section is related to LES 3.

PROBLEM Selection

During car races, race controllers on the side of the track inform the race car drivers of events such as accidents, debris on the track or the end of a race.

In order to select controllers for an upcoming car race, the Organizing Committee evaluated applicants based on their knowledge of the rules, their judgement, and their overall efficiency by administering a 100 point test.

Below are the test results in numerical order for the 234 people who wrote the test:

$$16, 18, 30, 42, 44, 44, \underbrace{\ldots}_{\text{35 results}}, 56, 58, 58, 58, 60, 60, 60, \underbrace{\ldots}_{\text{130 results}}, 80, 80, 80, 80, 80, 80,$$

$$82, 84, 86, 86, 88, 88, 88, 90, \underbrace{\ldots}_{\text{33 results}}, 96, 96, 96, 96, 96, 98, 98, 100, 100$$

The Organizing Committee divided these results into 100 groups of an almost equal amount and then selected the best candidates to be part of the top 20 groups.

Race controllers use flags of different colours and designs to communicate important information to the race car drivers. Each flag has a specific meaning. Below are a few samples:

Anthony obtained 86 points. Will he be selected?

The checkered flag indicates the end of a race.

The green flag indicates the track is open and clear of danger.

The white flag with a red cross indicates the track is wet.

The solid yellow flag indicates danger.

ACTIVITY 1 Diving competition

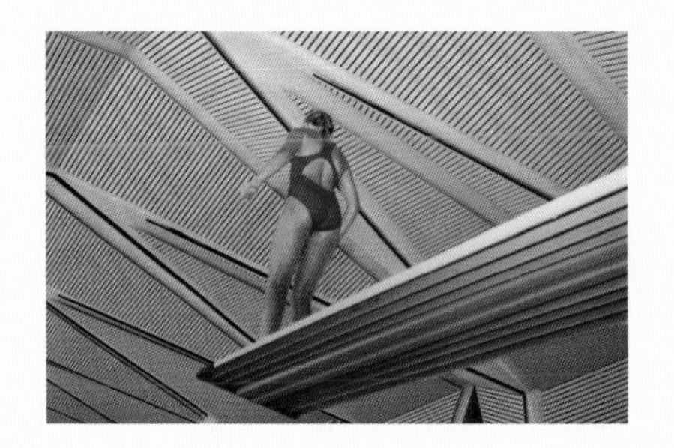

Diving is the sport of plunging into water, usually headfirst, with the addition of gymnastic and acrobatic stunts. In open competition, divers use the 1 m and 3 m springboards as well as the 10 m platform.

In July 2007, the Pan American Games were held in Rio De Janeiro, Brazil. Below are the final scores for the Women's 10 m Platform Competition:

Women's 10 m diving

Name	Country	Score (total points)	Deviation from the score from the mean of the score
Espinosa		380.95	
Ishimatsu		364.60	
Veloso		356.25	
Heymans		348.20	
Ortiz Galicia		340.90	
Marleau		326.60	
Dunnichay		309.60	
Mena Yaima		285.75	
Pineda Zuleta		270.10	
Sae		267.40	
Ortiz		263.15	
Buelvas		261.80	

a. What is the mean of the women's diving scores?

b. Which diver had the score:

1) closest to the mean?

2) furthest from the mean?

c. Fill in the last column of the table above.

d. What is the mean deviation of the women's diving scores?

The Pan American Games are held every four years and are open to athletes from all countries in the Americas to compete in various sports.

The Cuban, Jose Guerra, won the Men's 10 m platform with a final score of 527.40 points. During this final competition, the mean score was 443.24 points while the mean deviation was 44.17 points.

e. Of the two competitions, the Men's and the Women's, which was the tightest?

f. For their respective groups, which diver stood out more in their respective group, Espinosa or Guerra?

In 2003, Canadian Émilie Heymans won the 10 m platform world championship title as well as three gold medals at the Pan American Games in Santo Domingo.

ACTIVITY 2 The red oak

In southern Québec, a forestry technician has measured the diameter (in cm) of all the red oak trees within a forested area of 100 m^2. Below is the data collected:

40	45	53	55	55	55	55	60
60	63	64	64	65	67	67	68
70	70	70	71	72	75	75	79
82	83	83	91	97	102	113	

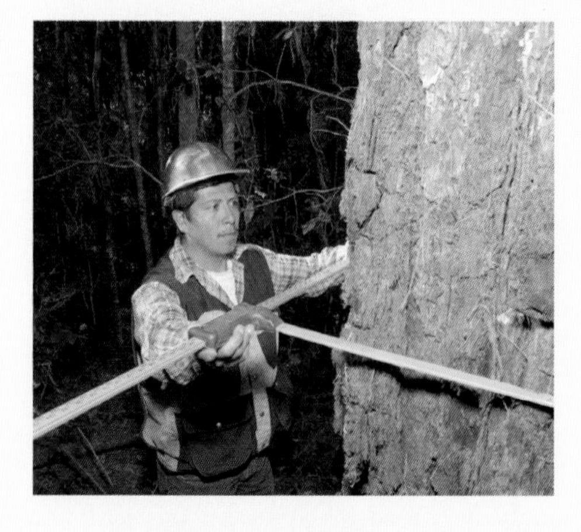

In order to analyze the data, the technician makes a stem-and-leaf plot:

Red oak trunk diameters (in cm)

4	0	5							
5	3	5	5	5	5				
6	0	0	3	4	4	5	7	7	8
7	0	0	0	1	2	5	5	9	
8	2	3	3						
9	1	7							
10	2								
11	3								

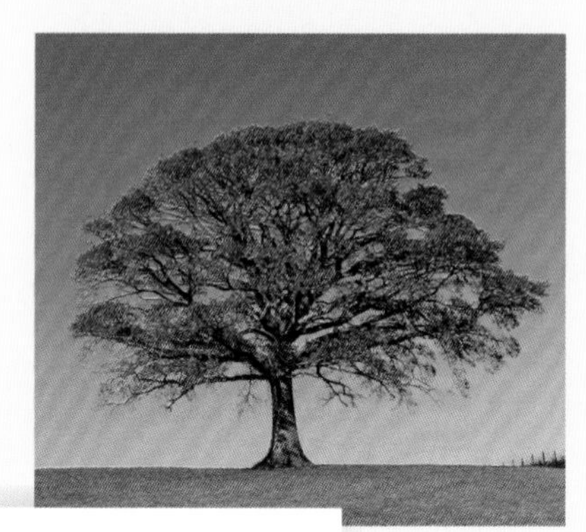

Oak is a hardwood resistant to insects and mould. Wood from this tree is used in woodworking and its chips for smoking fish.

a. Note this distribution:

1) What do the numbers in the left column represent?
2) How many data values are on the 5th line?
3) On which line is the data value of 71 cm?
4) Where is the data value of 97 cm?
5) What do the data on the same line have in common?

b. In this distribution, what diameter corresponds to:

1) the mode?
2) the median?

c. What are some of the advantages of a stem-and-leaf plot over a list?

d. In this distribution, what is the percentage of red oaks with a trunk diameter:

1) less than 65 cm?
2) equal to 70 cm?
3) less than or equal to 75 cm?
4) greater than 83 cm?

e. Add the following values to the stem-and-leaf plot: 59 cm, 81 cm, 96 cm, 96 cm, 109 cm.

Techno math

A spreadsheet allows you to perform statistical calculations on numbers entered into cells. For example, it is possible to determine the measures of dispersion and the measures of position in a distribution.

Each of the data values in a distribution is entered into a cell.

Formulas are used to automatically make calculations using the data values in a distribution.

Screen 1

	A	B	C	D	E	F	G	H	I	J	K	
1	4	5	6	7	7	7	8	9	12	12		
2	12	15	18	18	19	20	20	21	22	25		
3	26	28	30	30	31	32	33	35	40	41		
4	41	42	46	46	46	46	46	47	47	49		
5	49	51	53	54	54	55	57	58	58	58		
6												
7	**Distribution**											
8	Total number of values										50	→ =NB(A1:J5)
9	Mean deviation										15.44	→ =MEAN.DEVIATION(A1:J5)
10												
11	**Calculation based on a data value:**										**30**	
12	Number of values less than										22	→ =NB.SI(A1:J5;"<"&K11)
13	Number of values equal to										2	→ =NB.SI(A1:J5;"="&K11)
14	Percentile rank										46	→ =ROUNDED(100*(K12+K13/2)/K8;0)

By changing the data value in cell K11, the values of other cells are automatically updated.

Screen 2

	A	B	C	D	E	F	G	H	I	J	K
11	**Calculation based on a data value:**										**7**
12	Number of values less than										3
13	Number of values equal to										3
14	Percentile rank										9

Screen 3

	A	B	C	D	E	F	G	H	I	J	K
11	**Calculation based on a data value:**										**51**
12	Number of values less than										41
13	Number of values equal to										1
14	Percentile rank										83

a. In Screen **1**:

1) What data value appears in cell D2?

2) In which cell can you find the data value 28?

b. In the formulas in Screen **1**, what is meant by the expression A1:J5?

c. Using a spreadsheet and the distribution in Screen **1**, determine the percentile of:

1) 12
2) 26
3) 53

d. Using a spreadsheet and the distribution in Screen **1**, determine the results of the formulas:

1) =MODE(A1:J5)
2) =MEDIAN(A1:J5)
3) =AVERAGE(A1:J5)

e. Using a spreadsheet find:

1) the standard deviation of the adjacent distribution
2) the percentile of 127

100, 101, 103, 105, 105, 107, 108, 110, 110, 112, 113, 115, 116, 118, 120, 124, 124, 126, 127, 127, 128, 129, 130, 130, 131, 134, 134, 135, 136, 137, 138, 140, 140, 141, 142, 144, 145, 145

knowledge 2.1

ONE-VARIABLE DISTRIBUTIONS

A **one-variable distribution** is a distribution whereby the set of values collected during a statistical analysis focuses solely on one variable.

STEM-AND-LEAF PLOT

Stem-and-leaf plots are used to represent data values for one or two distributions that are placed on one or both sides of a central column (stem). In this type of plot, note the following:

- Each line is associated with one class of values.
- Each data value is split into two parts (stem and leaf): the column of the first digit(s) represents the stem and the last digit represents the leaves.

E.g. 1) The data values below correspond to the heart beats of 30 people following a stress test. Complete a stem-and-leaf plot for the following list of heart beats per minute.

70, 73, 73, 76, 78, 81, 82, 85, 85, 87, 88, 88, 89, 90, 92, 92, 96, 97, 99, 101, 101, 101, 104, 106, 106, 107, 112, 114, 115, 118.

Below is the display of this distribution using a stem-and-leaf plot:

Cardiac rates (number of beats/min)

Stem								
7	0	3	3	6	8			
8	1	2	5	5	7	8	8	9
9	0	2	2	6	7	9		
10		1	1	1	4	6	6	7
11	2	4	5	8				

2) The two sets of data values below correspond to the length (mm) of 59 salmon.

Males: 115, 148, 154, 180, 190, 195, 218, 225, 232, 232, 258, 259, 273, 273, 306, 311, 315, 342, 343, 369, 385, 390, 435, 444, 445, 446, 450, 462, 473, 538, 555, 566, 567, 570.

Females: 105, 112, 136, 147, 152, 180, 181, 258, 259, 271, 271, 276, 332, 341, 362, 363, 369, 370, 399, 407, 419, 489, 500, 525, 531.

Below is the display of this distribution using the stem-and-leaf plot:

Lengths of 59 salmon (mm)

Male								Stem	Female						
		95	90	80	54	48	15	**1**	05	12	36	47	52	80	81
73	73	59	58	32	32	25	18	**2**	58	59	71	71	76		
90	85	69	43	42	15	11	06	**3**	32	41	62	63	69	70	99
	73	62	50	46	45	44	35	**4**	07	19	89				
			70	67	66	55	38	**5**	00	25	31				

MEASURE OF DISPERSION: MEAN DEVIATION

A measure of dispersion describes the spread of the concentration of the data values in a distribution. The **mean deviation** is a measure of dispersion that indicates the mean of the deviations of the data values from the mean of the distribution.

$$\text{Mean deviation} = \frac{\text{Sum of deviations from the mean}}{\text{Total number of values}}$$

E.g. Below is a data set of eight values: 1, 4, 5, 6, 8, 8, 9, 11. The mean of this distribution is 6.5.

Calculation of mean deviation

Value	Mean of the distribution	Deviation from the mean
1	6.5	$\|1 - 6.5\| = 5.5$
4	6.5	$\|4 - 6.5\| = 2.5$
5	6.5	$\|5 - 6.5\| = 1.5$
6	6.5	$\|6 - 6.5\| = 0.5$
8	6.5	$\|8 - 6.5\| = 1.5$
8	6.5	$\|8 - 6.5\| = 1.5$
9	6.5	$\|9 - 6.5\| = 2.5$
11	6.5	$\|11 - 6.5\| = 4.5$

The deviation equals the absolute value of the difference of the two values.

Mean deviation: $\frac{5.5 + 2.5 + 1.5 + 0.5 + 1.5 + 1.5 + 2.5 + 4.5}{8} = 2.5$

MEASURE OF POSITION: PERCENTILE

A measure of position describes the location of one value in relation to the others in the distribution. **Percentiles** are a measure of position that indicate the percentage of values below or equal to the given value.

With the help of percentile, a distribution can be divided into one hundred groups each containing 1% of the data values. The rank of each sub-group is the percentile of the data values it contains.

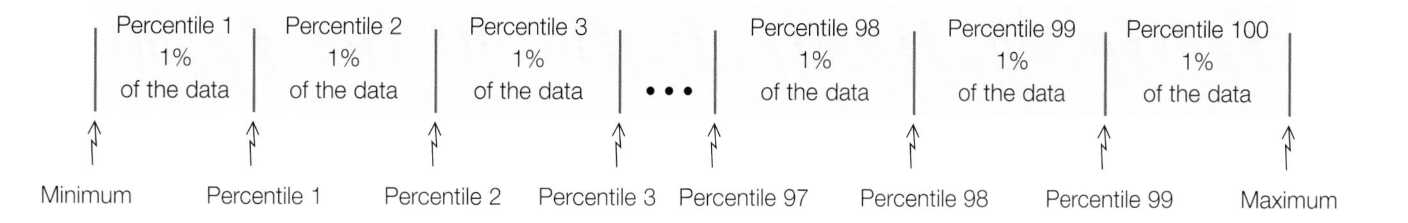

The formula below shows how to calculate the percentile of a data value within a distribution. If the result is not an integer, then it is **rounded upwards to the nearest integer.**

$$\text{Percentile of } x = \left(\frac{\text{number of data values less than } x + \dfrac{\text{number of data values equal to } x}{2}}{\text{total number of values}} \right) \times 100$$

To identify the data value whose percentile is known, do the following:

1. Determine the position or the total number of data values less than or equal to the value by doing the calculation below. If the result is not an integer, it is rounded **downwards to the nearest integer.**

$$\frac{\text{percentile}}{100} \times \text{total number of values}$$

2. Find the position of the data value in the ordered distribution for the rank calculated.

E.g. Below is a distribution containing 158 values:

6, 7, 8, ..., 19, 21, 21, 21, 23, 24, ..., 50, 51, 52, 55, 56, 56, 57, 58, ..., 89, 89, 90

(61 values between 8 and 19; 41 values between 24 and 50; 36 values between 58 and 89)

Calculation	Interpretation
1) Percentile of 21: $\left(\frac{65 + \frac{3}{2}}{158}\right) \times 100 \approx 42.09$	The percentile of the number 21 is 43. Or, 43% of the numbers in the distribution are less than or equal to 21.
2) Percentile of 52: $\left(\frac{113 + \frac{1}{2}}{158}\right) \times 100 \approx 71.84$	The percentile of the number 52 is 72. Or, 72% of the numbers in the distribution are less than or equal to 52.
3) 75th percentile: $\frac{75}{100} \times 158 = 118.5$	The data value with a 75th percentile is in the 118th position of the ordered distribution. The value is therefore 57.
4) 71st percentile: $\frac{71}{100} \times 158 = 112.18$	The data value with a 71st percentile is in the 112th position of the distribution. The value is therefore 50.

practice 2.1

1 In each case, calculate the deviation between each of the following values.

a) 12 and 23 b) 0 and 142 c) 5 and 5

d) -8 and -13 e) -10 and 6.4 f) -12.39 and 12.39

2 Calculate the mean deviation for each of the following distributions.

a) 12, 14, 15, 17, 19, 22, 25 b) 1, 25, 26, 27, 28, 29, 60

c) 15, 15, 15, 16, 16, 16 d) 20, 20, 20, 20, 20, 20

3 **2007 CANADIAN CHAMPIONSHIP** Below are the scores from the women's gymnastics competition for the vault and the uneven bars:

Vault: 13.3 14.7 13.95 14 13.9 14.35 14.4 14.4 13.75 13.95 14.1 12.8 14.2 13.8 13.75 13.25 14.5 13.4 13.65 13.9 13.5 14 13.1 13.7 12.95 12.4 12.65 14.4 14 12.5 13.5

Bars: 14.55 13.35 14.4 14.45 12.65 13.05 12.3 12.7 11.5 12.4 12.1 12.65 11.45 12.8 12.95 12.6 11.6 10.85 12.6 10.2 11.45 11.4 10.8 10.65 11.9 10.55 9.8 10.95 11.85 12 8.8 12.85

The scores in red are for gymnast Hélène Dellio. Compared to the other gymnasts, for which of the two events did Hélène place better? Justify your answer.

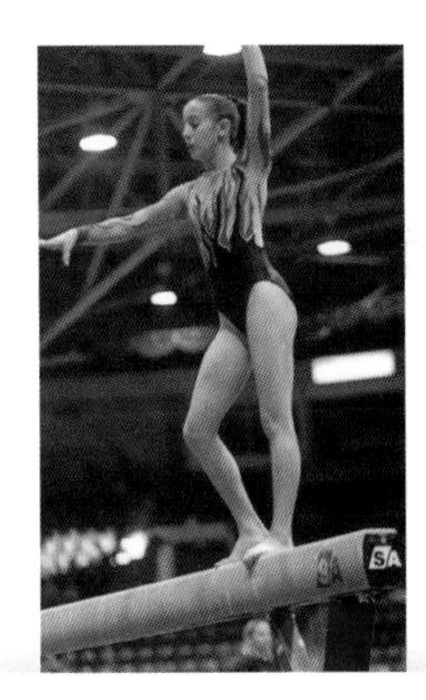

Hélène Dellio, a Québec gymnast, was part of the Québec National Gymnastics Senior Team at the Canada Games in 2007.

4 **CHILD DEVELOPMENT** In daycare, children are tested using various specialized tools and methods to verify their aptitude for starting elementary school. Each child under five years old is evaluated on a 50 point scale in five different developmental categories. The interpretive chart for the results is as follows.

Children at daycare

Test results (percentile)	Interpretation
Less than the 10th percentile	Vulnerable
Between 10th and 25th percentile	At risk
Between 25th and 75th percentile	Almost ready for school
75th percentile or above	Ready for school

Following the test results for all children under five years of age at Daycare **A**:
22, 28, 32, 34, 34, 35, 35, 35, 38, 39, 40, 41, 43, 45, 46, 47, 47, 48, 49, 50

How many children at Daycare **A** are:

a) ready for school? b) at risk?

5 At a chess tournament, Julio ranked 10th out of a total of 94 participants. What was the percentile of the person just ahead of him and of the person just after him?

Throughout the centuries, many artists have painted scenes of people playing chess. This oil canvas, a work done by the Italian Caravaggio, was painted in Venice in 1610.

6 For each of the following distributions, determine the percentile for the data value identified by a red box.

a) 23, 24, 25, 26, 27, 28, 31, 34, 36, 38, 38, 38, 38, 42, 42, [45], 46, 47, 47, 47, 48, 50, 51, 53, 54, 57, 58, 67, 67, 68, 69, 70, 71, 71, 73, 74, 79, 80, 80

b) 120, 120, 123, 130, 135, 135, 138, 141, 142, 144, 145, 146, 147, 147, 153, 153, [153], 155, 160, 172, 172, 173, 174, 175, 176, 177, 178, 180, 185, 187

c) 12, 15, 18, 18, ..., 321, [322], 328, ..., 841, 844, 849
(first "..." : 230 values; second "..." : 431 values)

7 Below are two stem-and-leaf plots:

Table 1

6	3	5	8	8	9	
7	0					
8	1	4	7	7	8	8
9	1	2	5	8	9	
10	0	8				
11	1	5	5			

Table 2

5	5	7	9				
6							
7	2	5	6	8			
8	1	2	3	4	6	6	7
9	0	0	0	2	5		
10	1						

For each of the plots shown above, find:

a) the mean deviation

b) the 87th percentile

c) the data value that is nearest to the 80th percentile

8 The data values below are the results from a statistical study.

34, 35, 36, 37, 38, 45, 46, 47, 48, 49, 49, 49, 49, 51, 52, 53, 54, 55, 56, 57, 58, 59, 59, 59, 65, 66, 67, 68, 69, 70, 72, 76

a) What data value corresponds to the 86th percentile?

b) What is the percentile of the data value 49?

c) What data value has 78% of the other values lower than it?

d) What data value has 70% of the other values that are equal or lower than it?

9 The table below contains data on the mean mass of male adult dogs of various breeds.

Adult dogs

Breed	Mean mass (kg)	Breed	Mean mass (kg)
Chihuahua	2	Irish Setter	26
Yorkshire Terrier	3	Belgian Shepherd Dog (Malinois)	27
Dwarf Spitz	4	German Hound	29
Italian Greyhound	4	French Spaniel	29
Shih Tzu	6	Braque Weimar	34
Miniature Poodle	6	Golden Retriever	34
West Highland White Terrier	8	Boxer	34
Cairn Terrier	8	Labrador	36
Cavalier King Charles	9	German Shepherd	36
Dachshund (Teckel)	10	Doberman	39
Pyrenean Shepherd	13	Rottweiler	47
French Bulldog	13	Leonberg	57
English Cocker Spaniel	13	Dogue de Bordeaux	59
Whippet	14	Bullmastiff	59
Breton Spaniel	18	Bernese Mountain Dog	60
Staffordshire Bull Terrier	24	Irish Wolfhound	63
Collie	24	Newfoundland	64
Siberian Husky	24	Great Dane	71
Sharpeï	25	Saint-Bernard	82
English Bulldog	26	Mastiff	87

a) Present the above data in the form of a stem-and-leaf plot.

b) What is the percentile (of the mass) of:

1) the Labrador? 2) the Shih Tzu? 3) the Chihuahua?

c) Which dog's mass corresponds to the 84th percentile?

d) What would the percentile of a Great Dane be if its mass was 90 kg instead of 71 kg?

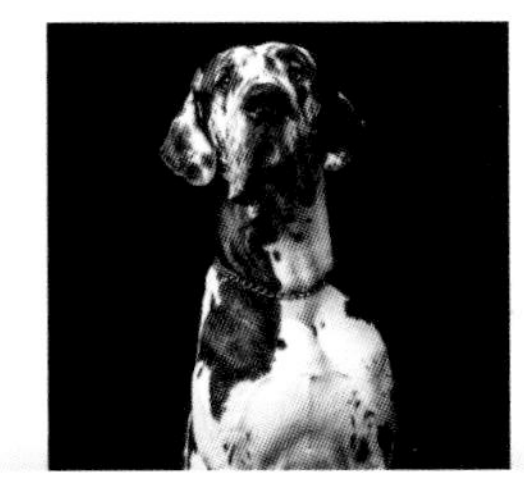

The Great Dane also known as the German Mastiff is a giant among dog breeds. A good nature and imposing stature are qualities that make the Dane an excellent guard dog.

The Shih Tzu, literally "Lion Dog," is an ancient breed which originated in Tibet. It was the favoured breed of the Chinese empresses.

Originating in Newfoundland and Labrador, the Labrador Retriever was so popular in the 1700s that a law was passed limiting the number of Labs to one per household.

10 a) Find two numbers for which the mean is equal to 50 and the mean deviation is 10.

b) Find two numbers for which the mean is equal to 30 and the mean deviation is 10.

c) Find two numbers for which the mean is equal to 50 and the mean deviation is 20.

d) Create a distribution of six different numbers for which the mean is equal to 50 and the mean deviation is 10.

11 At a basketball training camp, the percentage of successful shots is calculated and the best shooters are chosen for the team. Which is a better position: a 77% and a 90th percentile or 90% and a 77th percentile? Justify your answer.

12 In which quarter does a data value fall if it is in the 78th percentile?

13 In a distribution of 15 different values, what is the percentile of the median?

14 At a dance audition, dancers are separated into three groups. Then each dancer is compared to the others in their respective groups. In which group would a dancer most likely be accepted if they obtained a score of 86%? Explain your answer.

Group ①: 86, 86, 86, 86, 86, 86, 86, 86, 86, 86, 86, 86, 86, 86, 86, 86, 86, 86, 86, 86, 86, 86, 86

Group ②: 20, 23, 26, 30, 45, 56, 67, 67, 67, 68, 71, 72, 74, 76, 80, 86, 86, 87, 87, 87, 89, 90, 93

Group ③: 80, 80, 81, 81, 82, 82, 82, 82, 83, 83, 83, 84, 85, 85, 85, 86, 86, 87, 88, 88, 89, 90, 90

15 For which of these two distribution lists is the 89th percentile the highest?

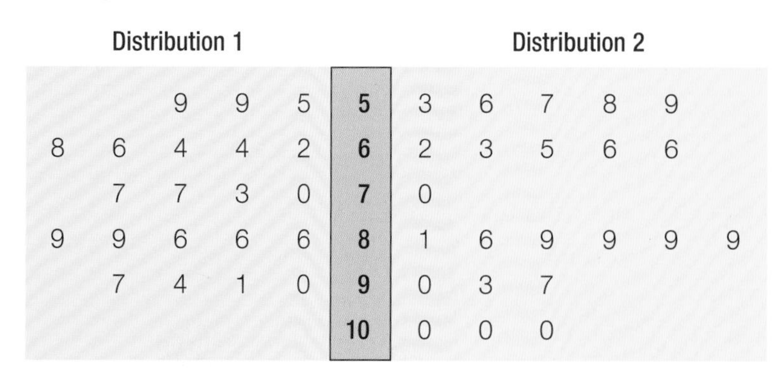

Distribution 1						Distribution 2					
		9	9	5	**5**	3	6	7	8	9	
8	6	4	4	2	**6**	2	3	5	6	6	
	7	7	3	0	**7**	0					
9	9	6	6	6	**8**	1	6	9	9	9	9
	7	4	1	0	**9**	0	3	7			
					10	0	0	0			

16 One way to quantify the level of poverty in Québec is to calculate the deviation between a family's income and the poverty level in a given region. The following is the total family income (in $) for the twelve families in Region **A**:

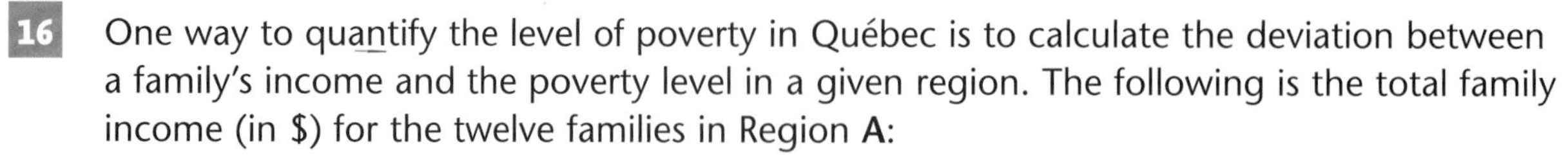

12,000 12,345 13,000 13,560 14,210 14,670 15,050 15,500 15,530 15,700 16,210 18,900

Considering that the poverty level in this region is $15,000, what is the mean deviation for these families?

17 In a laboratory, mice are put through a maze. The time it takes for each mouse to complete the maze is recorded. A piece of cheese is then put at the exit of the maze and the trial times for each mouse are once again recorded. Below are the results:

Course time without cheese (s)		Course time with cheese (s)
	1	8
1	**2**	3 4 6 8 9
9	**3**	0 0 0 0 3 5 5 7 7
9 7 5 3 2 2	**4**	2 6 8 9 9 9 9
8 8 6 6 4 4 1	**5**	
6 4 4 2 0 0 0	**6**	

Compare the results and indicate:

a) the distribution where the race times are the closest

b) the distribution where the mode is the highest

c) the percentage of improvement in trial times when cheese is put at the exit

18 In order to become future air traffic controllers, candidates are subjected to a three step hiring process. After each step, a certain number of candidates are chosen to move on to the next step. Following are the test results, in percentages:

Step 1: The interview

70, 74, 76, 76, 80, 80, 80, 81, 82, 82, 84, 84, 84, 86, 87, 88, 88, 88, 88, 88, 89, 89, 90, 90, 91, 92, 92, 93, 94, 94, 97, 100

Step 2: The written exam

60, 60, 65, 68, 71, 72, 72, 74, 75, 76, 78, 79, 80, 82, 83, 83, 83, 84, 84, 84, 85, 88, 89, 90

Step 3: The practical exam

65, 65, 65, 65, 66, 67, 69, 70, 70, 72, 77, 78, 80, 80, 80, 82, 85, 87

a) What percentage of candidates are retained for:

1) the 2nd step? 2) the 3rd step?

b) Compare the dispersion of the results from one step to another.

c) What is the percentile associated with the test score 82% in the:

1) 1st step? 2) 2nd step? 3) 3rd step?

d) If the candidates who reached the 90th percentile after the 3rd step are hired, how many people would be chosen in this sample?

Being an air traffic controller is a career loaded with considerable responsibility and stress. These professionals need to be in excellent physical condition and have nerves of steel because they are responsible for the lives of all the occupants on airplanes in the local airspace that they control.

19 **CAR RACING** Below is a sample of the data taken from a car race held in the Netherlands in 2007:

Maximum speed reached during qualification (km/h)	180.859	179.072	178.492	178.851	178.411	177.745
	177.555	177.689	177.048	176.220	175.735	174.964
	174.942	174.855	174.172	170.596		
Maximum speed reached during the actual race (km/h)	177.819	177.617	177.584	177.566	177.199	177.186
	177.179	177.167	177.152	176.996	176.738	176.574
	176.140	176.124	174.317	173.858		

a) If the stem represents the whole part of the number and the leaves represent the number after the decimal point, draw a stem-and-leaf plot that represents the above data.

b) In which distribution are the speeds closest together? Justify your answer using a statistical measure.

c) Considering that the speeds in red are those of Alexandre Tagliani, determine the distribution where his speed is closest to the mean maximum speed of all the other race car drivers?

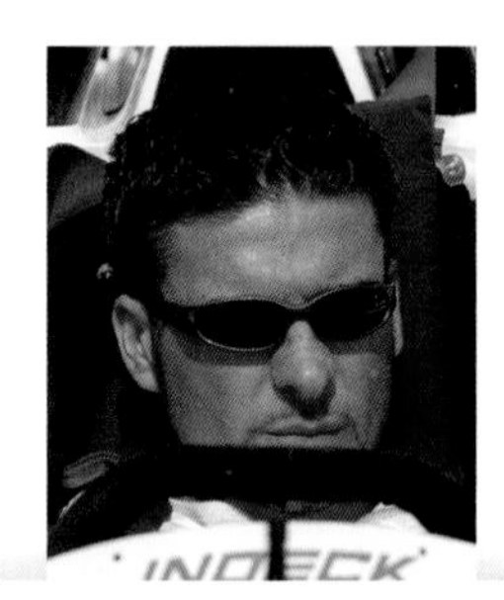

Alexandre Tagliani is a Québec race car driver who races in the Champ Car Series. With Italian roots, it was while visitng Italy with his grandfather that Alexandre discovered his passion for automobile sports.

20 Over a one year period, is the mean deviation of the daily temperatures in Mexico greater or less than that in Canada? Justify your answer.

The climate in Mexico varies according to altitude. On the coasts it is warm and humid whereas the higher altitude areas have a much cooler temperature. The rainy season is from May to October.

21 **THE GRAND TOUR OF THE TRAILS** In 2006, Guy Leclerc finished 117th out of 141 participants in the Saint Bruno 5 km race entitled "The Grand Tour of the Trails." In 2007, he improved his percentile rank by five spots amongst the 126 participants. What was his ranking in 2007?

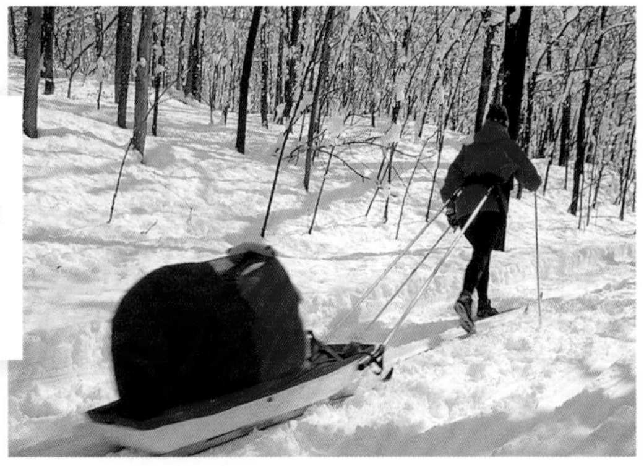

The race, the Grande Virée des Sentiers, takes place in Mount Saint Bruno National Park. A network of 27 km of trails coils around the five lakes in this park which is located 20 km southeast of Montréal. In the winter, the trails are used for cross-country skiing.

SECTION 2.2 Qualitative interpretation of correlation

This section is related to LES 4.

PROBLEM Education

The table below is an excerpt of data from a random survey regarding the level of education and annual income of 27 respondants from the same country.

Education and income

Number of years of schooling	Annual income ($ in thousands)	Number of years of schooling	Annual income ($ in thousands)	Number of years of schooling	Annual income ($ in thousands)
8	20	12	38	17	65
17	45	15	44	17	55
13	34	17	60	12	40
12	40	12	100	15	34
12	23	9	22	16	48
9	30	10	27	16	52
10	30	22	230	20	80
15	45	24	180	22	130
16	46	17	60	17	100

Analysts have confirmed the existence of a relationship between people's level of education and their annual income.

Analyst A

Analyst B

There is no correlation whatsoever! It's purely random. There are many entrepreneurs who have had little or no education who have become millionaires!

Analyst C

Analyst D

There is a direct link between education and income, the more years of scolarity, the higher the annual income. It's completely logical!

Of these four analysts, who best describes the data found in the survey?

ACTIVITY 1 An experimental drug

Epilepsy is a neurological disorder characterized by periodic epileptic seizures of varying degrees of intensity ranging from loss of consciousness for a few seconds to full body convulsions. The attacks or episodes are brought on by uncontrollable electrical short-circuits in the brain.

A research team has been testing a new anti-seizure drug. They would like to establish the minimum dosage of the medication required to reduce the number of epileptic seizures. Below are some of their research findings:

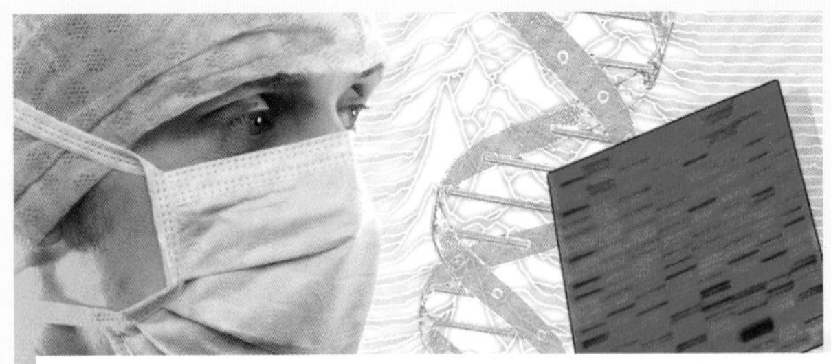

The causes of epilepsy and epileptic seizures are still somewhat of a medical mystery today. Since the time of Hippocrates, the ancient Greek doctor, it has been believed there is a certain genetic component to this illness. Current progress in the field of molecular biology should enable researchers to better pinpoint the genes responsible for causing epilepsy.

Anti-seizure medication

Dose (mg) / Drop in attacks (%)	[0, 100[	[100, 200[	[200, 300[	[300, 400[	[400, 500[	[500, 600[	Total
[0, 14[	7	2	0	0	0	0	
[14, 28[	22	8	0	0	0	0	
[28, 42[	3	12	3	0	0	0	
[42, 56[	0	18	7	5	0	0	
[56, 70[	0	4	9	8	0	1	
[70, 84[	0	0	0	12	8	4	
[84, 98[	0	0	0	0	0	7	
Total							

a. How many epileptics:

1) participated in the study?

2) tested doses of [0, 300[mg?

3) experienced a decrease of 70% or more in the number of their attacks?

4) tried doses of [400, 500[mg and experienced a decrease in the number of attacks by [70, 84[%?

5) tested the medication and showed no improvement?

b. What percentage of epileptics whose seizures decreased by at least 56% were taking a dosage of 300 mg or more?

c. Can the research team conclude the following:

1) The higher the dose of medication, the more the physical condition of the patient improves. Explain your answer.

2) There is a correlation between the amount of medication absorbed by a patient and the decrease of the frequency of the seizures. Explain your answer.

Fatigue, stress, background sounds, trauma, and even fasting are all factors that can trigger an epileptic seizure.

ACTIVITY 2 Sibling rivalry

Few athletes achieve top world-ranking within any given sport. Therefore it is even more surprising and exceptional that two members of the same family should both achieve such status. This is the case for sisters Venus and Serena Williams, two of the world's finest tennis players for the past decade. Below are their rankings from 1999-2007:

World tennis rank

Year	1999	2000	2001	2002	2003	2004	2005	2006	2007
Serena	4	6	6	1	3	7	11	95	7
Venus	3	3	3	2	11	9	10	48	8

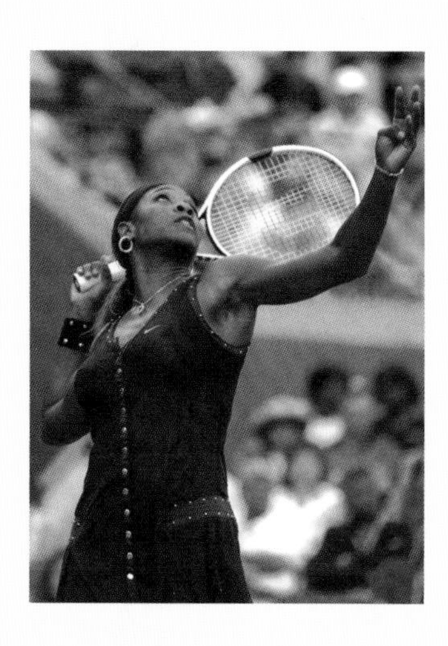

Although younger than Venus, Serena has outplayed her sister in most of their matches.

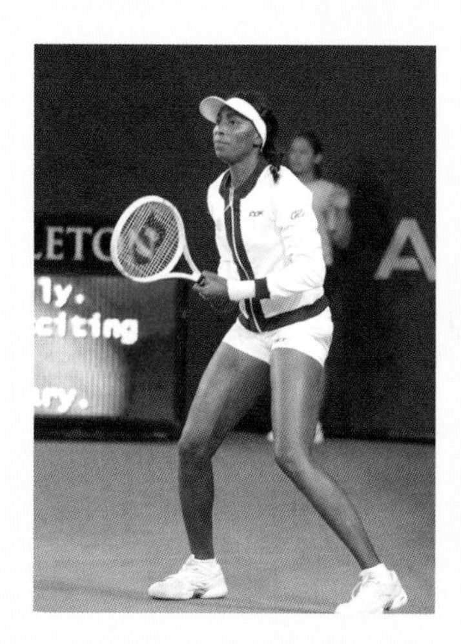

a. Which of the two sisters has maintained, on average, the highest ranking overall?

b. Of the two sisters, who is a more consistent player?

c. Is there a correlation between Venus's ranking and Serena's? Explain your answer.

The graph below shows the relationship between Serena's world rank and the percentage of games she won in that same year.

d. Does the scatter plot seem to follow a trend or display a certain inclination on the graph?

e. Does there seem to be a correlation between Serena's rank and the percentage of games she won?

f. Describe the relationship between world-rank and the percentage of games won per year, as either zero, weak, moderate, strong or perfect. Explain your answer.

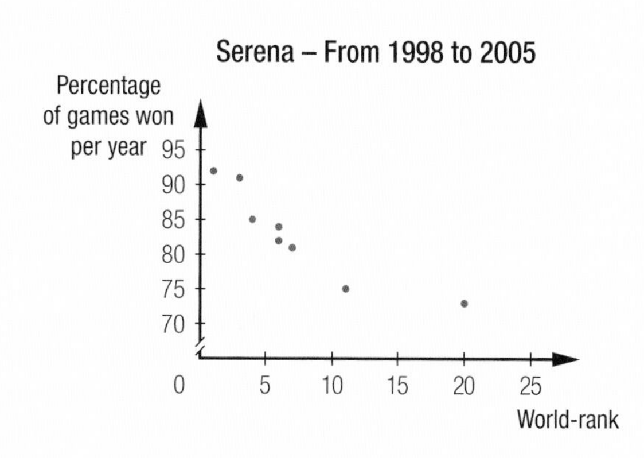

Techno math

A graphing calculator allows you to display different types of graphs. The following explains how you would display a scatter plot:

This table of values presents the data of a two-variable distribution.

x	y
1	29.3
2	37.5
3	41.8
4	50.6
5	51.9
6	54.1
7	53.4
8	59.3
9	62.1

This screen allows you to enter each ordered pair from the table of values.

Screen 1

L1	L2	L3
1	29.3	
2	37.5	
3	41.8	
4	50.6	
5	51.9	
6	54.1	
7	53.4	
8	59.3	
9	62.1	

L3(1)=

This screen allows you to choose the scatter plot as the display mode.

Screen 2

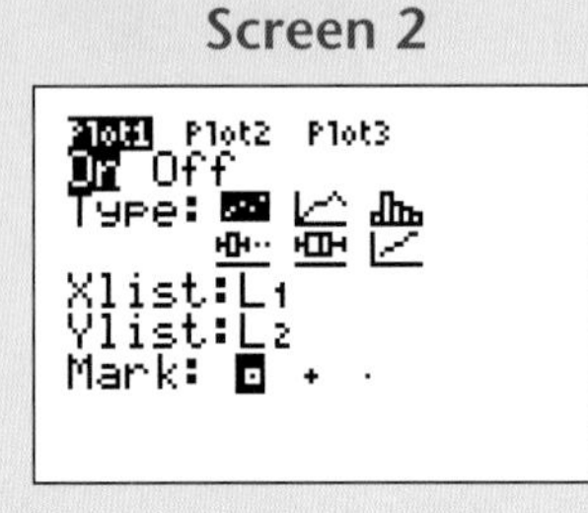

Screen 3

70
0
10
0

By moving the cursor on the graphical display you can select and view the individual coordinates of each point in the scatter plot.

Screen 4

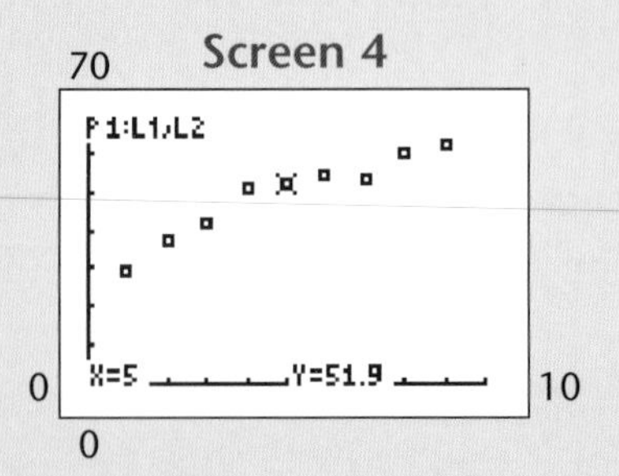

70
0
10
0

a. In Screen **4**, what does the point $x = 5$ and $y = 51.9$ represent?

b. Using a graphing calculator, do the following:

1) Display the scatter plot corresponding to the data table below.

L1	L2	L3
5	82	
10	58	
15	29	
20	22	
25	16	
30	9	

L3(1)=

2) Move the cursor on the scatter plot that corresponds to the table of values below.

L1	L2	L3
0	-14	
20	-6	
40	9	
60	22	
80	19	
100	13	

L3(1)=

TWO-VARIABLE DISTRIBUTION

A **two-variable distribution** corresponds to the set of ordered pairs collected during a statistical study researching two variables from a given situation.

In a statistical study, the term **statistical variable** is used to define the characteristic being researched.

CORRELATION

To study the **correlation** between two statistical variables is to describe the **relation** between quantitative values in a distribution. It is possible to describe the **type**, **direction** and **intensity** of a correlation between two variables.

- The type of correlation reflects the **mathematical model** that best describes the relation between the variables.
- A correlation is said to be **positive** or **negative** depending on the direction of the variation.

Positive: When the values of a variable increase (or decrease), the values of the other variable also increase (or decrease).

Negative: When the values of a variable increase (or decrease), the values of the other variable also decrease (or increase).

- A correlations may be **zero, weak, moderate, strong,** or **perfect** depending on the intensity of the relationship between the variables.

CONTINGENCY TABLE

A **contingency table** allows you to represent a two-variable distribution, and to describe the type and direction of the correlation that may exist between the two variables.

In a contingency table, note the following:

- One of the variables is associated with the first column and the other variable is associated with the first row.
- The data values can be grouped into classes.
- The row and column totals indicate the frequencies of each of the values or classes.
- The lower right cell contains the total number of data points in the distribution.
- The correlation is said to be linear when the majority of the occurences follow one of the two diagonals.

E.g. Each of the following ordered pairs represent the age and number of branches of 20 trees of the same species.

(12, 160) (13, 145) (13, 220) (15, 240) (16, 205) (19, 285) (21, 390) (22, 410) (22, 350) (23, 360) (23, 450) (24, 325) (25, 480) (25, 455) (27, 320) (27, 475) (27, 410) (29, 410) (30, 555) (33, 530)

Trees of the same species

Age (years) \ Number of branches	[100, 200[	[200, 300[	[300, 400[	[400, 500[	[500, 600[	Total
[10, 15[	2	1	0	0	0	3
[15, 20[	0	3	0	0	0	3
[20, 25[	0	0	4	2	0	6
[25, 30[	0	0	1	5	0	6
[30, 35[	0	0	0	0	2	2
Total	2	4	5	7	2	20

This contingency table shows that the correlation between the age and the number of branches is linear and positive.

Various conclusions can be drawn from the data in the table. Following are some examples:

- $\frac{2}{5}$ of the trees in this sample are 25 years or older
- 80% of the trees containing [300, 400[branches are less than 25 years old
- $\frac{5}{6}$ of the trees in the [25, 30[age group have 400 or fewer branches
- 45% of the trees have [200, 400[branches

SCATTER PLOT

A **scatter plot** allows you to represent a two-variable distribution and to describe the type, the direction and the intensity of the correlation that may exist between the two variables.

In a scatter plot, note the following:

- One of the variables is associated with the *x*-axis, and the other variable is associated with the *y*-axis.
- Each of the ordered pairs in the distribution is represented with a point.
- The correlation is said to be linear when the points tend to form a straight line. The correlation is said to be zero if the points are distributed randomly and becomes stronger as the points come closer to forming a straight line. The scatter plots on the next page describe the different characteristics of linear correlation.

Zero Correlation

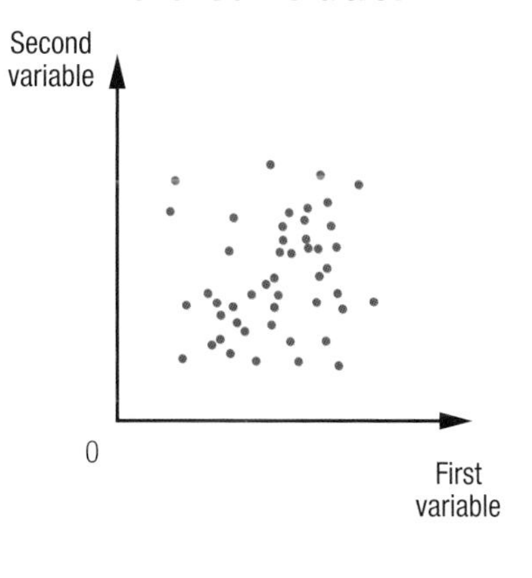

Negative Correlation

Weak Correlation

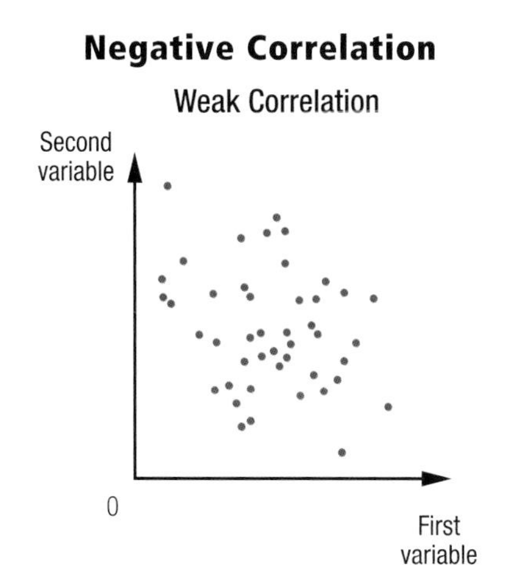

Moderate Correlation

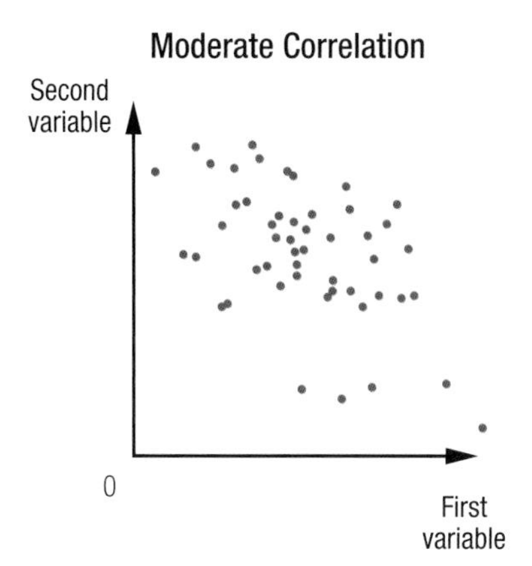

Strong Correlation

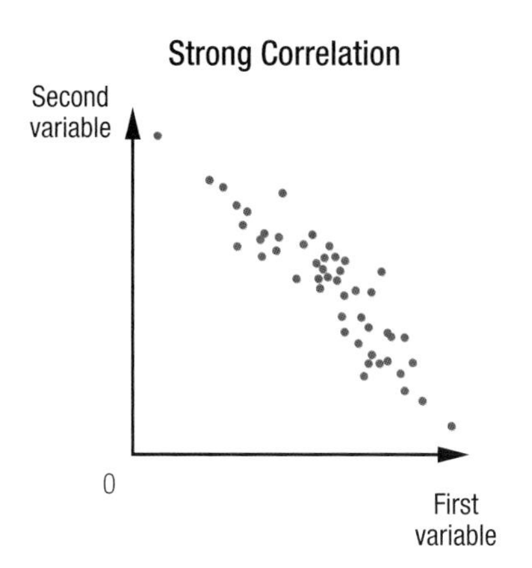

Perfect Correlation

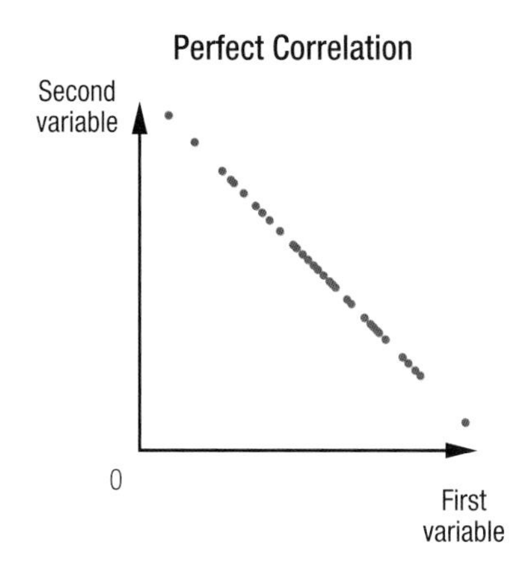

Positive Correlation

Weak Correlation

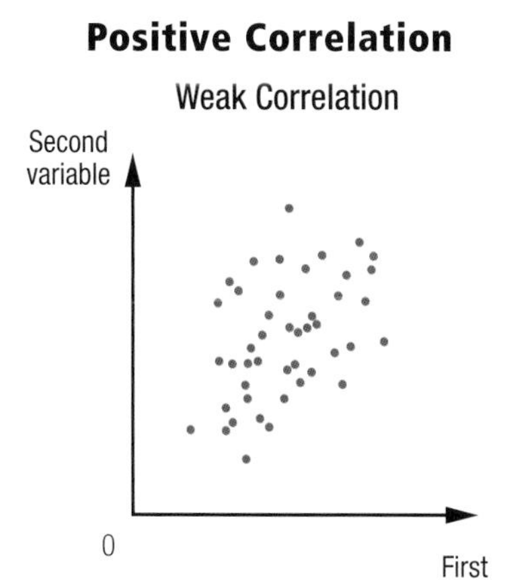

Moderate Correlation

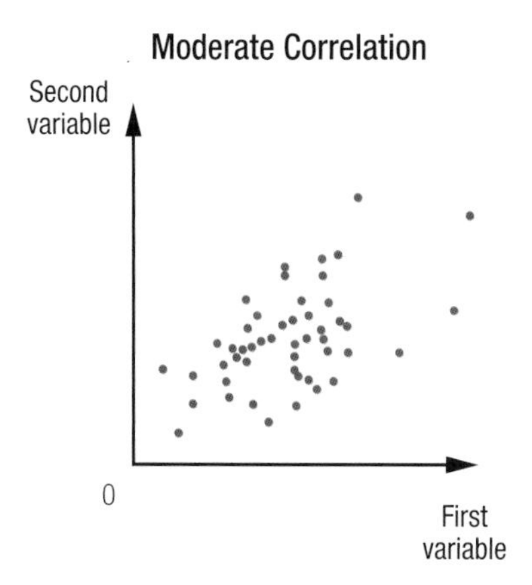

Strong Correlation

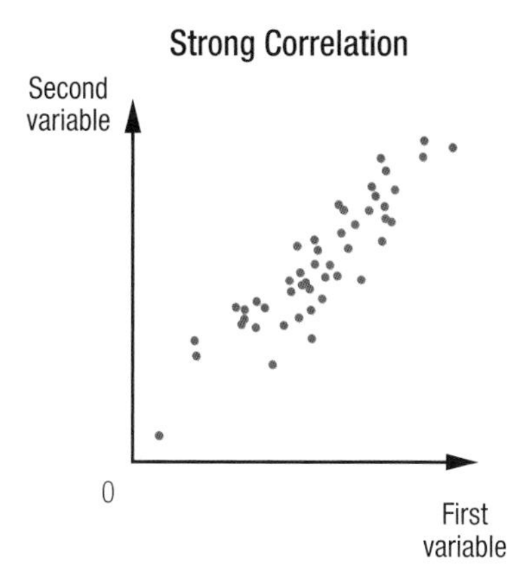

Perfect Correlation

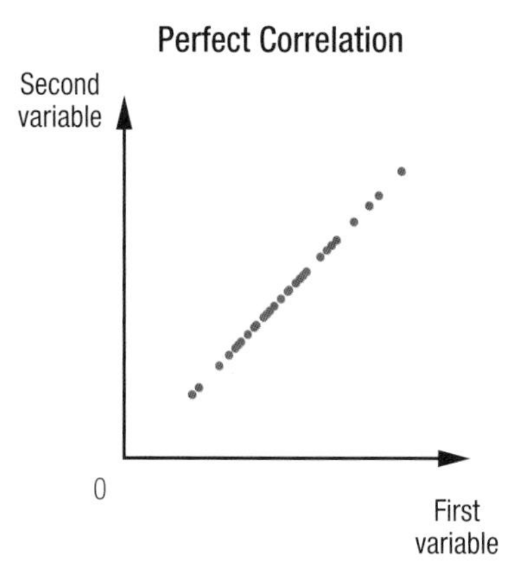

practice 2.2

1 Complete the contingency table below, showing the legal minimum age for marriage in different countries.

Legal age of marriage

Woman's age \ Man's age	18	19	20	21	Total
16		0	2	3	17
17	7	0	5	6	
18	15	0		15	39
19		0	8	13	21
Total	34	0			95

In Québec the legal age for marriage is 16. However, people under 18 years old must obtain their parent's permission first.

2 The contingency table below shows the mother tongue of three groups of high school students.

High school

Group \ Mother tongue	French	English	Spanish	Creole	Other	Total
①	15	6	3	6	4	34
②	12	10	8	2	1	33
③	10	13	2	4	3	32
Total	37	29	13	12	8	99

Given this data, find:

a) the size of the sample

b) the percentage of Francophones in this sample

c) the percentage of Anglophones in Group ②

d) the percentage of people in the sample who are neither Francophones nor Anglophones

3 Consider the relationship between the price of a sweater and the distance between where the sweater was made and where it was sold. Would you think the correlation is:

a) strong and positive? b) weak and negative? c) zero?

4 Below are data concerning the circumference of the head and waist size of 15 women.

Measurements of 15 women

Circumference of head (cm)	Waist size (cm)	Circumference of head (cm)	Waist size (cm)	Circumference of head (cm)	Waist size (cm)
57	86	59	96	55.4	81
56.2	90	58.7	94	55	78
58.1	93	57.6	95	56.3	83
58.7	92	56	80	57.1	85
57.4	87	56.3	84	58.2	88

a) Construct a scatter plot to represent this situation.

b) Describe the correlation between the two variables.

5 For each of the four sets of data below, indicate whether there is positive, negative or zero correlation between the variables.

a)

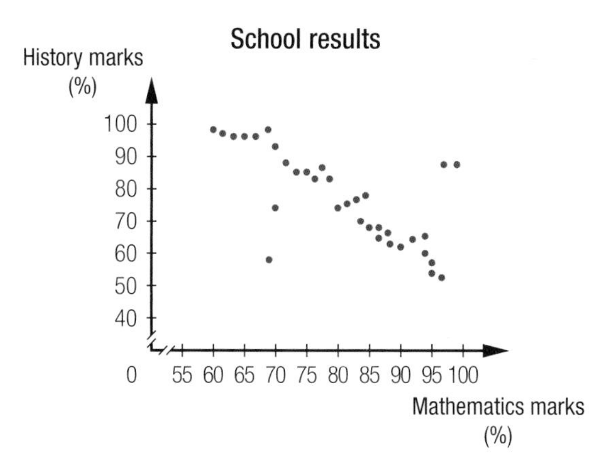

b)

Accord Inc.

Number of years of seniority	Annual revenue ($ in thousands)	Number of years of seniority	Annual revenue ($ in thousands)
1	30	6	50
2	32	6	56
2	33	6	56
3	35	7	57
4	36	8	60
4	40	8	63
4	40	8	68
5	41	9	74
5	46	10	74

c) The number of hours of sleep before an exam and the final mark (in %) on the exam:

(2, 50) (3, 50) (4, 45) (4, 35) (5, 50)
(5, 53) (6, 60) (6, 45) (7, 70) (7, 74)
(8, 80) (8, 75) (8, 78) (8, 83) (8, 85)
(9, 79) (9, 84) (9, 90) (9, 89) (9, 92)
(9, 93) (9, 87) (10, 85) (10, 87)
(10, 92) (10, 90) (10, 93)

d) Number of jobs and annual income of the same person.

Annual income ($ in thousands) \ Number of jobs	1	2	3	4
[0, 30[	1	2	2	1
[30, 60[	8	2	6	2
[60, 90[	2	6	1	1
[90, 120[	6	3	4	7

6 For each of the scatter plots below describe the linear correlation.

a) Screen 1

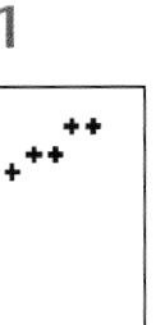

b) Screen 2

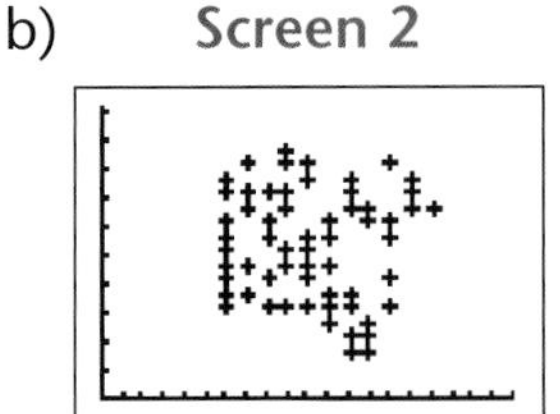

c) Screen 3

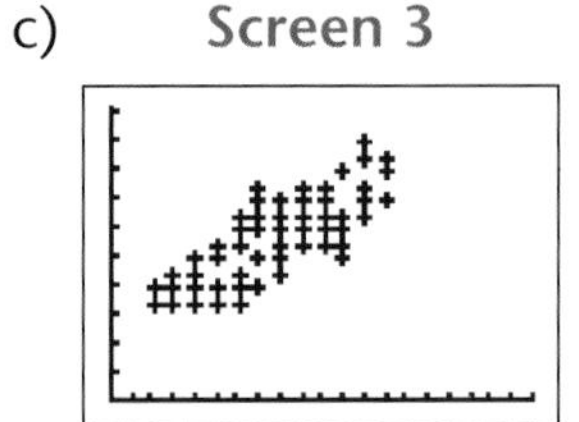

d) Screen 4

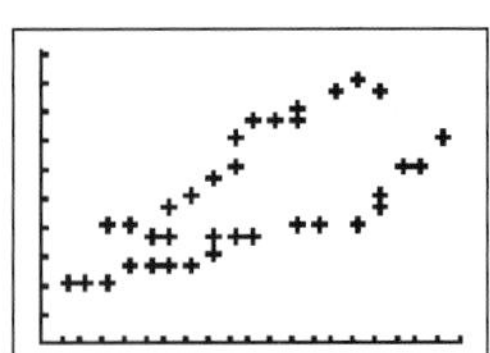

7 HOCKEY The adjacent table displays information on players for the Montréal Canadiens at the start of the 2007–2008 season.

a) Complete the contingency table below.

Start of 2007-2008 season

Games played \ Points scored	[0, 10[	[10, 20[	[20, 30[	[30, 40[	[40, 50[	Total
[0, 10[						
[10, 20[						
[20, 30[						
[30, 40[						
[40, 50[						
Total						

b) How many players who played 20 games or more scored less than 30 points.

c) How many players played 30 games or more and scored 40 points or more?

d) How many players played fewer than 20 games or scored less than 10 points?

e) What percentage of players scored fewer than 30 points and played at least 40 games?

f) Does a correlation exist between the number of games played and the points scored? Explain your answer.

Start of the 2007-2008 season

Player	Games played	Points scored
Alex Kovalev	44	41
Tomas Plekanec	44	38
Saku Koivu	43	33
Christopher Higgins	44	31
Andrei Markov	44	31
Mark Streit	44	28
Andrei Kostitsyn	40	25
Roman Hamrlik	44	18
Guillaume Latendresse	40	16
Michael Ryder	39	15
Mike Komisarek	44	12
Bryan Smolinski	30	12
Mathieu Dandenault	43	11
Kyle Chipchura	36	11
Sergei Kostitsyn	14	8
Patrice Brisebois	29	8
Tom Kostopoulos	39	7
Maxim Lapierre	17	6
Steve Bégin	25	4
Francis Bouillon	39	3
Josh Gorges	24	3
Mikhail Grabovski	12	2
Ryan O'Byrne	11	2
Corey Locke	1	0

In 2000, the Montréal Canadiens Hockey Club created the Montréal Canadiens Children's Foundation, an organization to help sick and disabled children. Since its creation, this foundation has given nearly $6 million to different charities throughout the Province of Québec.

8 Of the scatter plots below, which shows the strongest linear correlation?

A

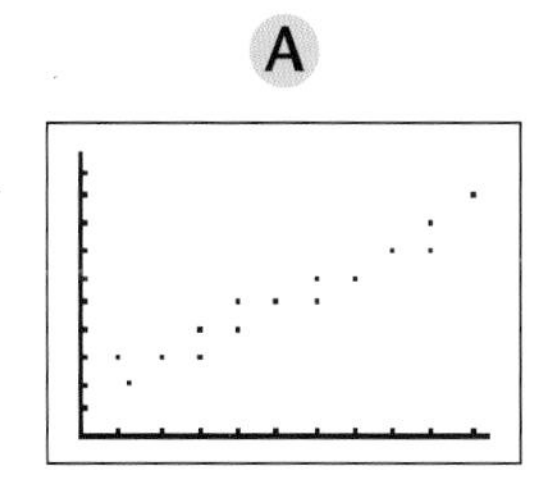

B

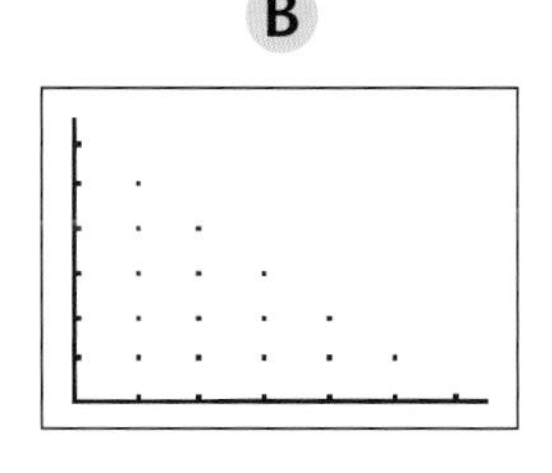

C

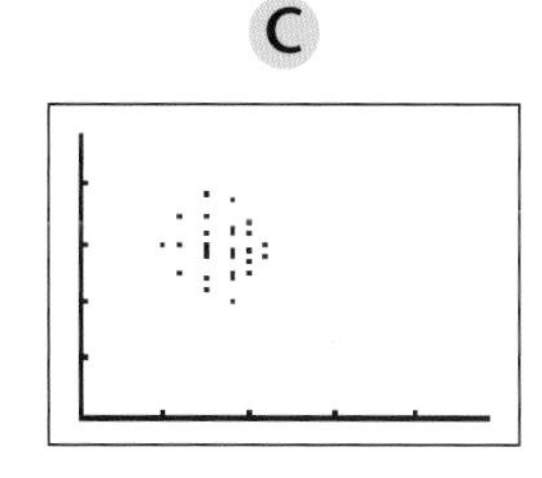

D

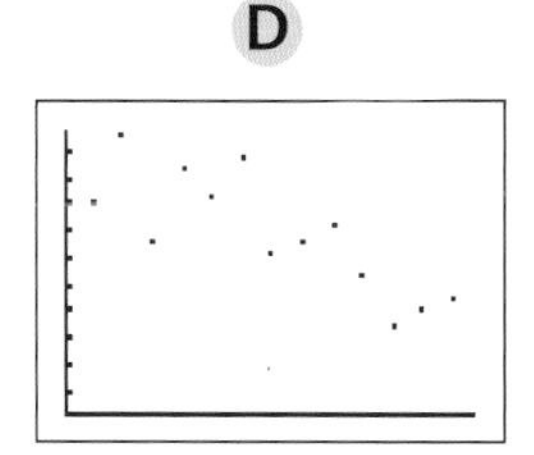

9 For each of the contingency tables below, state if the linear correlation is:

1) positive or negative
2) zero, weak, moderate, strong or perfect

a) **Players on a football team**

Mass (kg) \ Age	15	16	17	18	19	Total
[40, 50[	6	1	0	0	0	7
[50, 60[	2	3	2	0	0	7
[60, 70[	1	5	6	1	1	14
[70, 80[	0	0	2	8	2	12
[80, 90[	0	0	1	1	3	5
Total	9	9	11	10	6	45

b) **Swimmers on a swim team**

Mass (kg) \ Age	15	16	17	18	19	Total
[40, 50[	6	5	3	2	1	17
[50, 60[	4	4	3	2	2	15
[60, 70[	2	7	3	3	3	18
[70, 80[	0	6	2	2	2	12
[80, 90[	0	0	0	1	1	2
Total	12	22	11	10	9	64

c) **Secondary cycle 2**

Length of foot (cm) \ Height (cm)	[120, 140[	[140, 160[	[160, 180[	[180, 200[	Total
[15, 20[	12	3	0	0	15
[20, 25[	5	12	3	0	20
[25, 30[	0	6	10	4	20
[30, 35[	0	2	7	6	15
Total	17	23	20	10	70

10 In a random survey of 50 adults, the following two questions were asked in the following order:

- How many years of schooling do you have?
- How many children do you have?

Following are the results:

(12, 0) (12, 4) (13, 2) (13, 3) (13, 1) (14, 1) (14, 0) (9, 0) (9, 6) (10, 1) (11, 3) (11, 5) (11, 3)
(7, 7) (7, 2) (7, 0) (15, 1) (15, 2) (15, 2) (15, 3) (15, 4) (16, 5) (22, 0) (21, 1) (17, 3) (18, 2)
(18, 1) (16, 2) (16, 5) (16, 3) (17, 2) (17, 0) (17, 1) (17, 2) (17, 3) (19, 1) (19, 5) (20, 3) (17, 2)
(18, 0) (20, 0) (12, 3) (16, 3) (15, 1) (16, 1) (13, 2) (12, 3) (20, 4) (19, 2) (18, 1)

Based on this sample, is there a correlation between these two variables? Justify your answer.

11 DIET Using the table below, indicate if there is a correlation between the amount of protein and the amount of fat contained in the various food items.

100 gram portions of food items

Food item	Protein (g)	Fat (g)	Food item	Protein (g)	Fat (g)
Pears	0.4	0.4	Salmon	16	8
Pineapple	0.5	0.1	Chicken	21	8
Carrots	0.6	0.5	Ham	22	22
Potato	2	0.1	Raspberries	1	0.6
Eggs	13	12	Corn	10	5
Milk	3.5	3.9	White rice	7.4	0.8
Cheese	7.9	3.3	Sardines	20	12
Chick peas	18	5	Tuna	27	13

Fats provide much of the energy for the human body: 1 g of fat provides 9 kcal. Fats, stored in the fat tissue, serve as a reserve energy supply. Even though fats are indispensable for the healthy working of the body, a diet too rich in animal fats can lead to various illnesses, including heart disease.

12 a) Complete the following table:

Students in your class

Student	Distance between the elbow and the middle finger (cm)	Length of the student's foot (cm)
1		
2		
3		
...	...	...

The distance between the elbow and the tip of the middle finger, called the "cubit," is an ancient unit of measurement.

b) Is there a linear correlation between the two measurements above? Justify your answer.

c) For each student, calculate the ratio $\frac{\text{distance from the elbow to the tip of the middle finger}}{\text{length of foot}}$. What do you notice?

13 Discuss the correlation between the variables in each of the following cases:

a) The points on a scatter plot are clustered along a line that has a negative slope.

b) The scatter plot resembles a circular shape.

c) A scatter plot is scattered along a line with a positive slope.

14 The table below shows the number of evening and weekend hours worked by 30 teenagers along with their academic average.

Hours worked by teenagers

Number of hours worked	Academic average (%)	Number of hours worked	Academic average (%)	Number of hours worked	Academic average (%)
0	85	7	78	14	65
0	78	8	80	14	63
0	89	8	76	15	70
0	83	8	73	15	68
0	82	8	75	15	70
3	85	8	76	16	62
3	78	8	70	18	58
3	76	9	69	20	56
4	73	10	67	22	55
6	73	12	67	24	54

Which of the following statements is true? Justify your answer.

A) The correlation between the two variables is strong and positive.

B) The correlation between the two variables is strong and negative.

C) The correlation between the two variables is weak and positive.

D) The correlation between the two variables is weak and negative.

15 GASOLINE To counter the unpredictable fluctuations in the price of gasoline in the Montréal area, a website shows the best places to buy gasoline at any given time of day. The adjacent scatter plot shows the price of gasoline over a 60 hour period from November 5, 2007 to November 7, 2007. Is there a correlation between the elapsed time and the price of gasoline? Justify your answer.

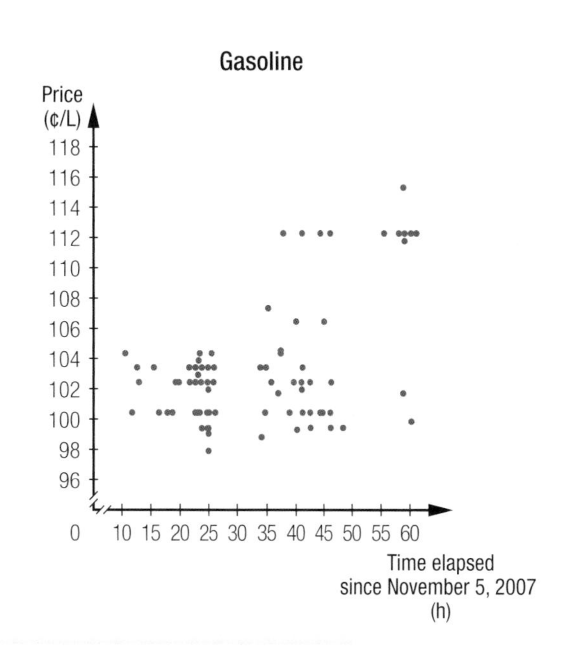

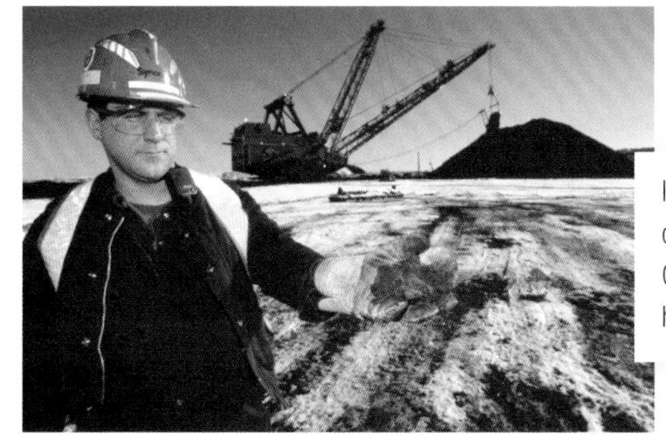

In 2004 the Alberta tar sands represented 40% of Canada's crude oil production. This is an important resource for Canada. However, the open-pit mining of these tar sands has had a negative impact on the environment.

16 The adjacent scatter plot shows the relationship between the total amount of class time in CEGEP courses and the corresponding amount of study time.

a) Construct a contingency table from this data.

b) Describe the correlation between these two variables.

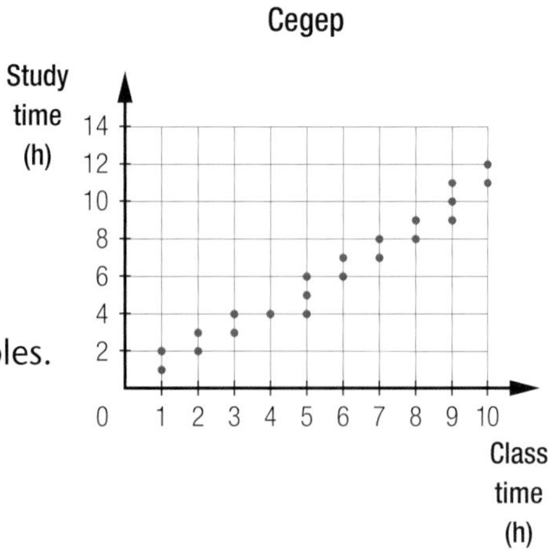

17 WEATHER On March 4, 1971, there was a blizzard that paralyzed much of the province of Québec. Most schools were closed for several days. The table below shows the weather conditions at Pierre Elliott Trueau Airport for that day.

Meteorological conditions

Time	Temperature (°C)	Wind speed (km/h)	Visibility (km)	Barometric pressure (kPa)	Wind chill factor (°C)
0:00	-5.4	31	1.6	100.33	-13
1:00	-5	31	1.6	100.17	-13
2:00	-5	35	1.2	100.01	-14
3:00	-5	40	0.8	99.88	-14
4:00	-5	40	0.6	99.66	-14
5:00	-5	40	0.6	99.4	-14
6:00	-5	45	0.6	99.2	-15
7:00	-5	40	0.6	99.09	-14
8:00	-4.4	47	0.2	98.71	-14
9:00	-4.4	53	0.2	98.5	-14
10:00	-4.4	58	0	98.2	-15
11:00	-4.4	42	0.2	97.97	-13
12:00	-4.4	74	0	97.69	-16
13:00	-5	74	0	97.49	-17
14:00	-5	71	0	97.36	-16
15:00	-4.4	61	0	97.29	-15
16:00	-4.4	39	0	97.4	-13
17:00	-4.4	29	0.2	97.51	-12
18:00	-5	32	0.2	97.61	-13
19:00	-5.6	35	0.6	97.71	-14
20:00	-5.6	39	2.4	97.85	-15
21:00	-5.6	27	12.9	97.99	-13
22:00	-5	27	24.1	98.12	-13
23:00	-5	29	24.1	98.22	-13

From the variables shown in the table above, choose two variables for which the linear correlation can be described as strong and negative. Justify your choice by drawing the corresponding scatter plot.

18 The adjacent table shows the age and favourite colour of some students.

Is there a relationship between these two variables? Explain your answer.

High school students

Favourite colour \ Age	14	15	16	17	18	19	Total
Blue	8	0	0	0	0	0	8
Red	0	8	3	0	0	0	11
Yellow	0	3	8	3	0	0	14
Black	0	0	3	8	3	0	14
Green	0	0	0	3	8	3	14
Other colours	0	0	0	0	3	8	11
Total	8	11	14	14	14	11	72

19 Forty-five adult males had the volume of their brain and their Intelligence Quotient (IQ) measured. Below are the data that were collect:

Brain

Volume (cm^3)	IQ	Volume (cm^3)	IQ	Volume (cm^3)	IQ	Volume (cm^3)	IQ	Volume (cm^3)	IQ
1000	85	1025	100	1189	78	1111	113	1010	100
1030	115	1080	110	1045	90	1220	117	1245	78
1100	100	1105	155	1176	116	1209	140	1202	89
1080	145	1110	136	1101	123	1089	99	1056	98
1113	120	1009	97	1209	132	1028	94	1105	110
1200	100	999	102	1210	150	1100	107	1221	109
1167	75	1103	110	1006	120	1099	112	1048	112
1150	89	1125	111	1090	100	1109	129	1130	123
1203	80	1150	130	1200	89	1121	112	1178	120

Is there a linear correlation between the volume of a person's brain and his or her IQ? Explain your answer.

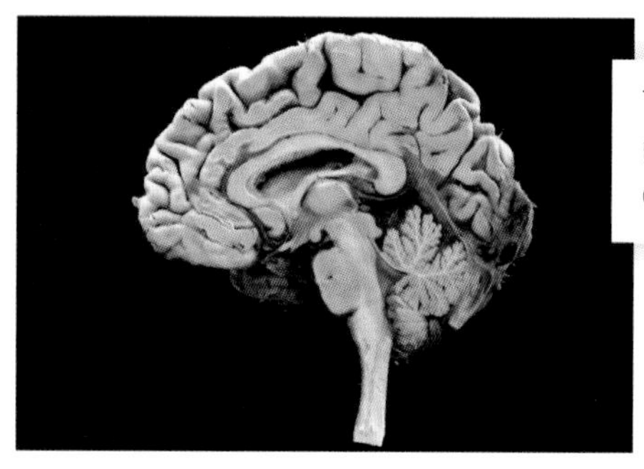

The brain, with its 95 billion neurons, is the most complex organism in the world.

To determine the intelligence quotient of children, they are submitted to a series of tests in order to determine their mental age. The IQ corresponds to the ratio of their mental age to their actual age. For example, an eight-year-old child with a mental age of 10 has an IQ = $\frac{10}{8} \times 100 = 125$.

SECTION 2.3 Quantitative interpretation of correlation

This section is related to LES 4.

PROBLEM Kilocalories and lipids

Some foods contain fats known as lipids. Often maligned, lipids are nevertheless a necessary source of energy required to keep our body healthy, and our diet must include them. There is, however, "good" and "bad" fat, and overconsumption of the latter can lead to health problems.

The table below provides information on the quantities of kilocalories and lipids in a 100-gram serving of various foods.

Energy intake

Food (100 g portion)	Number of kilocalories	Quantity of lipids (g)	Food (100 g portion)	Number of kilocalories	Quantity of lipids (g)
Bacon	130	4.5	Sesame seeds	560	50
Butter	752	83	Peanut oil	900	100
Lean ground beef	204	14.7	Ham	229	15
Dark chocolate	514	30	Cooked ham	135	6.5
Chips	557	31	Mayonnaise	357	39
Chicken leg	226	13.5	Chocolate mousse	445	27
Turkey	109	2.4	Nuts	670	64
Snails	82	1	Raisin bread	325	12
Fries	274	15	Salami	460	42
Cheddar chesse	407	33	Sardines	163	9
Feta cheese	267	21	Lemon pie	189	10
Gouda cheese	347	29	Apple pie	302	15.2
Parmesan cheese	380	27	Veal	170	11

A lot of importance is attributed to the quality of fatty acids present in edible fats. Saturated fatty acids (butter, deli meats, meat, manufactured products such as biscuits, cakes, and various prepared meals) that are overconsumed may lead to cardiovascular disease. On the other hand, unsaturated fatty acids (olive oil, almonds, avocado, fatty fish, duck) can reduce bad cholesterol risks and contain omega-6 which plays a protective role for arteries, and omega-3 an indispensable component for nerve and brain cells.

Considering that a 100 g portion of chocolate muffin contains 16 g of lipids, how many kilocalories will it produce?

ACTIVITY 1 Gestation

A gestation period corresponds to the number of days that a mother carries her baby in her womb. Animals' gestation periods vary from one species to another. The following table lists some data on this subject.

Animals

Animal species	Female mass (kg)	Gestation period (days)
Rabbit	1	35
Cat	9	65
Wolf	40	64
Leopard	50	94
Pig	80	114
Lion	150	108
Bear	295	200
Moose	550	230
Bison	600	250
Cow	730	264
Giraffe	1200	410
Hippopotamus	1500	310

At birth, and from its very first day alive, the fawn follows its mother who breast feeds it. At just a few days old, the baby moose can run faster than a human and swim without difficulty.

This scatter plot illustrates the relationship between a female's mass and her gestation period.

a. Associate each of the points on the graph with the corresponding animal species.

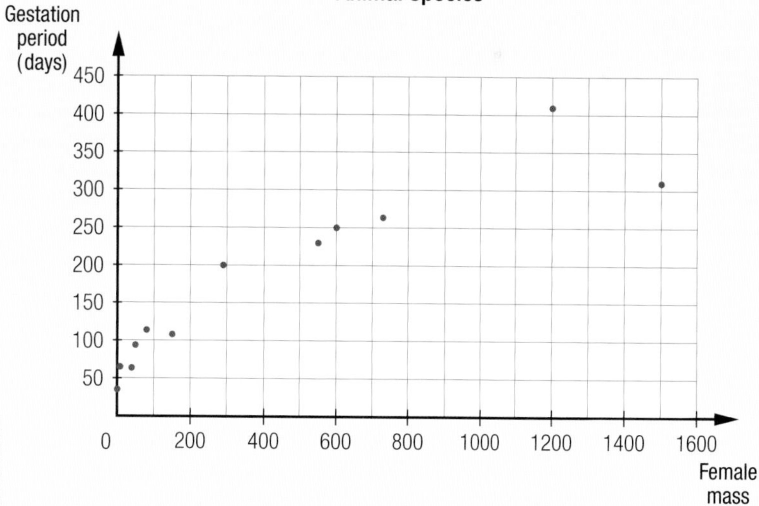

The grizzly bear is the second largest of the bear family after the polar bear. The colour of its fur may vary greatly from one bear to another. These large bears, once found almost everywhere in North America, have now sought refuge in the Western Rockies.

b. Complete the table below.

Animal group	Median mass (kg)	Median gestation period (days)	Ordered pair composed of the median mass and median gestation period
Rabbit, cat, wolf, leopard			M_1(,)
Pig, lion, bear, moose			M_2(,)
Bison, cow, giraffe, hippopotamus			M_3(,)

c. Find the coordinates of point P, so that:

1) the *x*-coordinate corresponds to the mean of *x*-coordinates for points M_1, M_2 and M_3
2) the *y*-coordinate corresponds to the mean of *y*-coordinates for points M_1, M_2 and M_3

d. Determine the slope of the line passing through points M_1 and M_3.

e. In the graph in **a.**, do the following:

1) Plot point P.
2) Draw a line passing through point P having the same slope as the line passing through points M_1 and M_3.

f. Determine the equation of the line drawn in **e.**

g. Given the equation of this line, what would be:

1) the gestation period of a mare with a mass of 1100 kg?
2) the mass of a female tiger whose gestation period is 100 days?

The following table provides data on various monkey species.

Monkeys

Species	Mass of female (in kg)	Gestation period (days)
Chimpanzee	40	227
Gorilla	70	257
Orangutan	40	260
Baboon	20	187
Macaque	7	175
Mandrill	25	176

The mandrill is a primate related to the baboon. It is found in Central Africa. Its main food sources are fruit, nuts and insects. It is primarily a ground dweller, living in groups of approximately 15 members that include a single dominant male and several females with their offspring.

The scatter plot below illustrates the relationship between a female's mass and the gestation period for several monkey species.

h. Associate each of the points on the graph with the corresponding species.

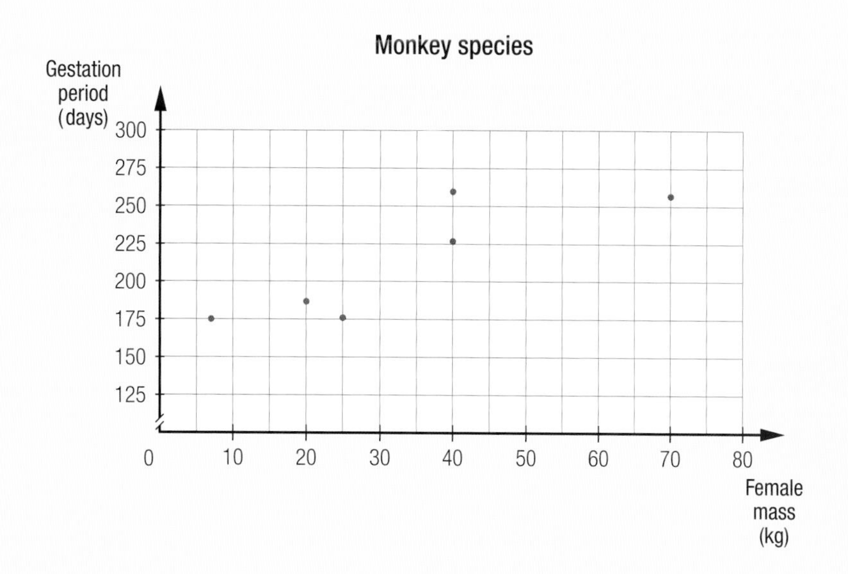

i. Complete the following table.

Monkey group	Mean mass (kg)	Mean gestation period (days)	Ordered pair composed of the mean mass and mean gestation period
Baboon, macaque, mandrill			P_1(,)
Chimpanzee, gorilla, orangutan			P_2(,)

j. In the graph in **h.**, do the following:

1) Plot points P_1 and P_2.
2) Draw a line passing through points P_1 and P_2.

k. Determine the equation of the line passing through points P_1 and P_2.

l. Given the equation of this line, what would be:

1) the gestation period of a red howler monkey that has a mass of 4 kg?
2) the mass of a bonobo that has a gestation period of 230 days?

Human DNA is very similar to that of chimpanzees and macaques. In 2007, a study revealed that humans share 93% of their genes with the rhesus monkey, a monkey used widely in medical research.

ACTIVITY 2 Dreams within reach

The Chavez family are ready to build the house of their dreams. They chose three cities where they would like to live and are analyzing the cost of the land in relation to its area.

In the Cartesian planes below, they have represented the data collected from the three cities. Each scatter plot has been framed in a rectangle and represented by a line.

a. Describe the linear correlation observed in each scatter plot.

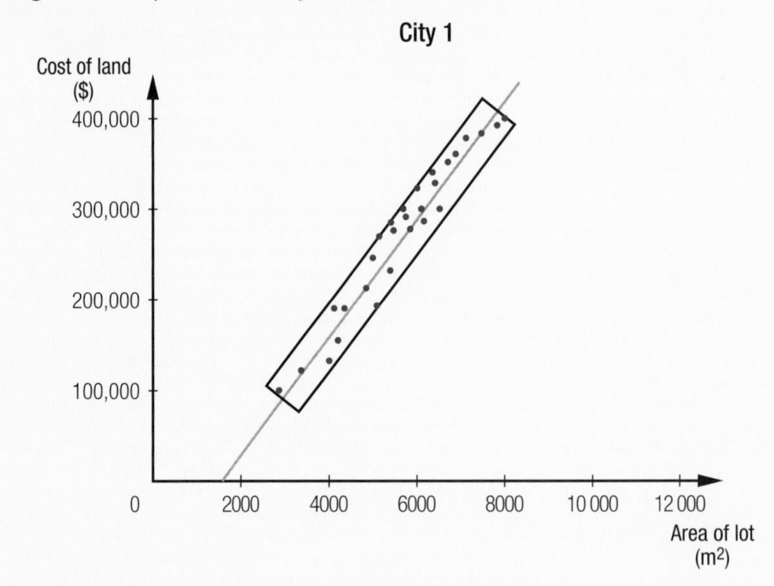

b. In which city does a 6000 m^2 lot cost the least?

c. 1) For which rectangle is the $\frac{\text{length of small side}}{\text{length of large side}}$ ratio the greatest?

2) Describe this rectangle.

d. 1) For which rectangle is the ratio $\frac{\text{length of small side}}{\text{length of large side}}$ the smallest?

2) Describe this rectangle.

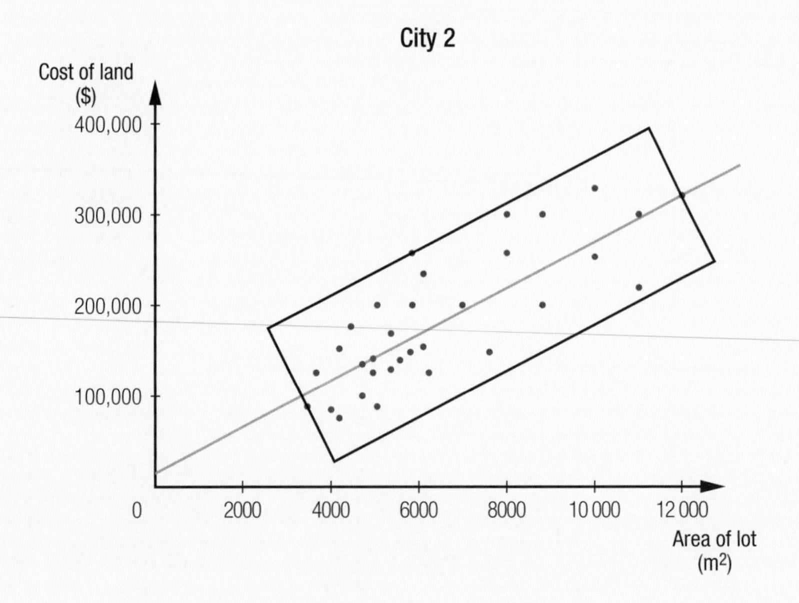

e. When comparing the three scatter plots, what relationship might you establish between:

1) the reliability of a statistical study and the intensity of the correlation?

2) the appearance of the rectangle framing the points and the intensity of the correlation?

f. For which city does the study seem the least reliable? Explain your answer.

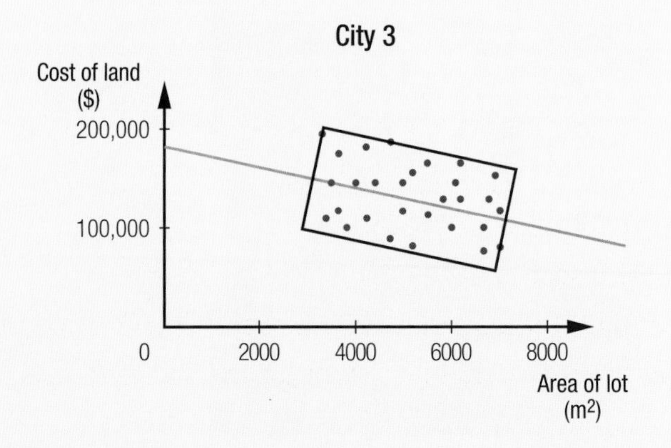

Techno math

A graphing calculator allows you to display a scatter plot and determine the equation of a regression line using the median-median line method.

The table of values below shows the results of an experiment involving two variables.

x	8	15	20	23	26	28	29	37
y	50	41	40	35	38	26	30	11

This screen allows you to enter each ordered pair from the table of values.

Screen 1

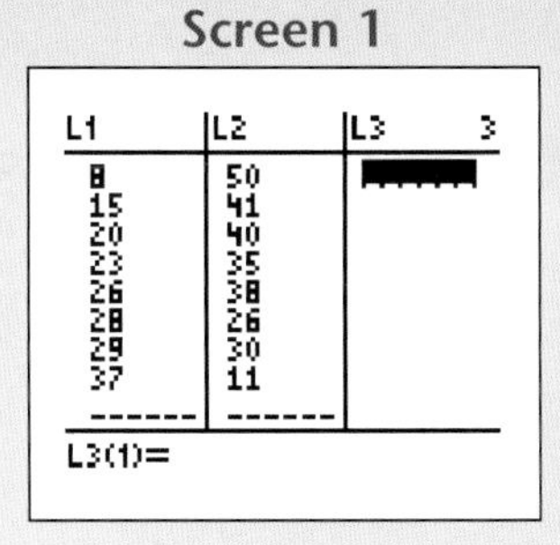

This screen allows you to select the scatter plots as the display mode.

Screen 2

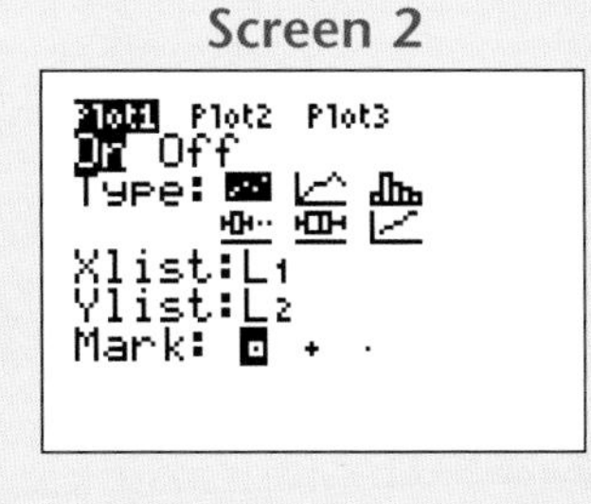

Screen 3

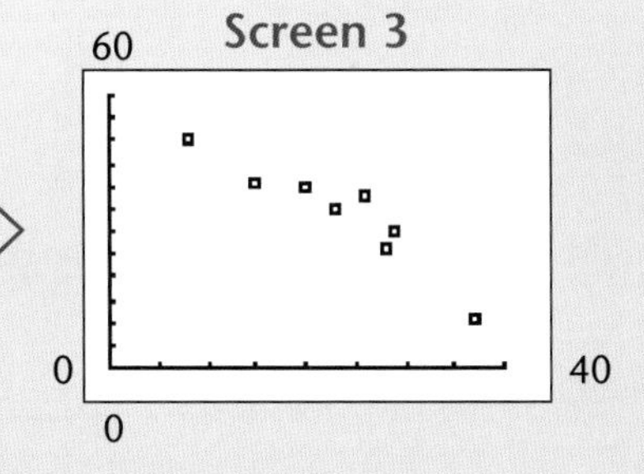

This screen allows you to select the median-median line method to approximate the equation of the regression line.

Screen 4

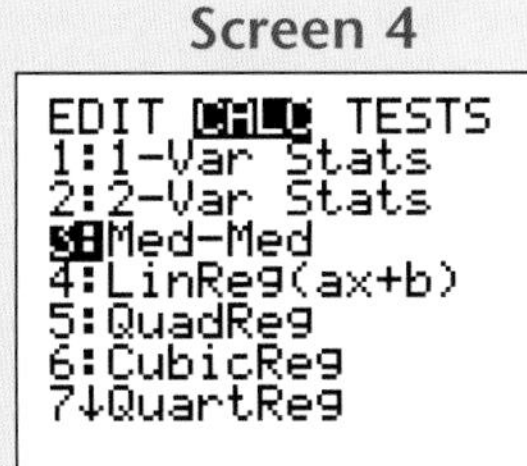

These two screens allow you to determine the equation of the regression line using the median-median line method, and to enter the result in the equation editor.

Screens 5 et 6

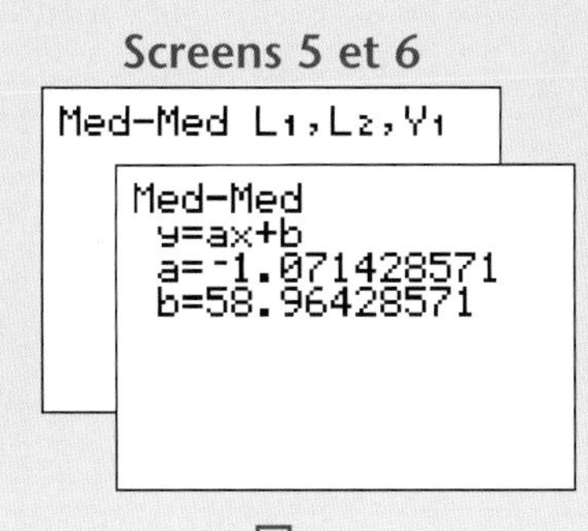

This screen allows you to display the regression line obtained using the median-median line method from the scatter plot.

Screen 7

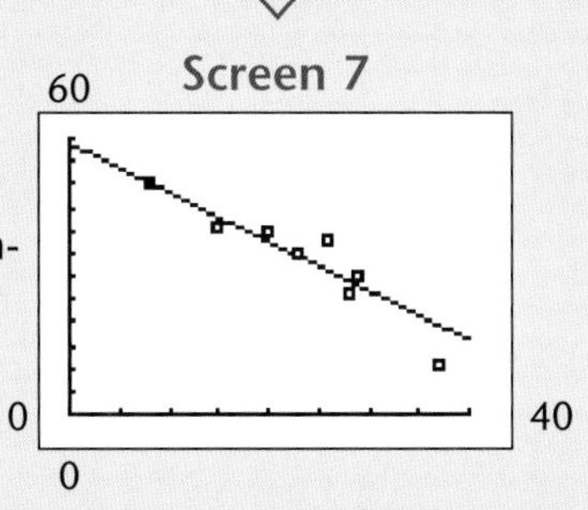

a. In Screen **6**, what do the values of a and b represent?

b. Using a graphing calculator and the adjacent data, do the following:

x	0	10	20	30	40	50	60
y	60.8	65.6	66.6	67.1	70	71.8	74.1

1) Display a scatter plot.
2) Determine the equation of the regression line by using the median-median line method.
3) Draw the regression line through the scatter plot.

CORRELATION COEFFICIENT

It is possible to quantify the intensity of a linear correlation between two statistical variables using a number in the interval [-1, 1]. This number is called the **correlation coefficient** and is designated with the letter ***r***.

Correlation coefficient		Meaning
Negative	Positive	
Near 0	Near 0	Indicates a **zero** linear correlation between the two variables
Near -0.5	Near 0.5	Indicates a **weak** linear correlation between the two variables
Near -0.75	Near 0.75	Indicates a **moderate** linear correlation between the two variables
Near -0.87	Near 0.87	Indicates a **strong** linear correlation between the two variables
Equal to -1	Equal to 1	Indicates a **perfect** linear correlation between the two variables

There are several methods you can use to **approximate the linear correlation coefficient** of a two-variable distribution. One of them is a method of **graphical estimation** involving a **rectangle in a scatter plot**. This method consists of:

1. representing the two-variable distribution with a scatter plot
2. drawing a line that represents most of the points
3. constructing the smallest rectangle framing all the significant points on the scatter plot with two sides parallel to the line
4. approximating the linear correlation coefficient between the two variables using the following formula:

$$r \approx \pm\left(1 - \frac{\text{length of short side}}{\text{length of long side}}\right)$$

E.g.

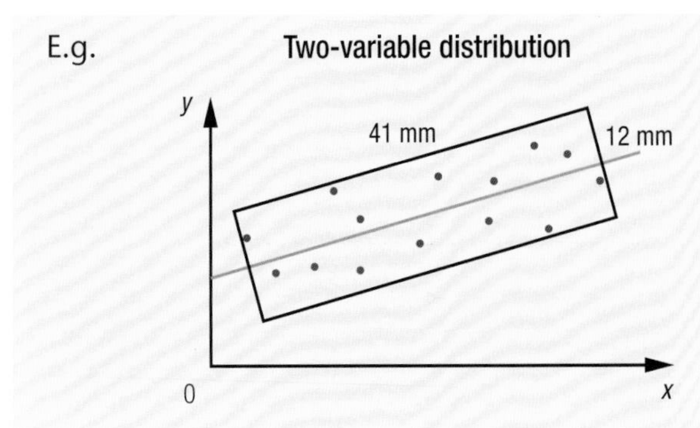

$r \approx 1 - \frac{12}{41} \approx 0.71$

The correlation between the two variables is therefore positive and moderate.

REGRESSION LINE

In a scatter plot relating two statistical variables, the line that best represents all of the points is called a **regression line**. There are various methods to determine the equation of a regression line.

Median-median line method

This is how to determine the equation of a regression line using the **median-median line** method:

1. Order the ordered pairs in the distribution based on their x-coordinates.
2. Divide all of the ordered pairs into three equal groups. If this is not possible, divide them so that the first and last groups each have the same number of ordered pairs.
3. Determine the median x-coordinate and the median y-coordinate in each of the three groups to form the median ordered pairs $M_1(x_1, y_1)$, $M_2(x_2, y_2)$ and $M_3(x_3, y_3)$.
4. Determine the coordinates of point P that correspond respectively to the mean of the x-coordinates and the mean of the y-coordinates of points M_1, M_2 and M_3.
5. Determine the slope of the line that passes through points M_1 and M_3.
6. The regression line is the line that passes through point P and has the same slope as the line that passes through points M_1 and M_3.

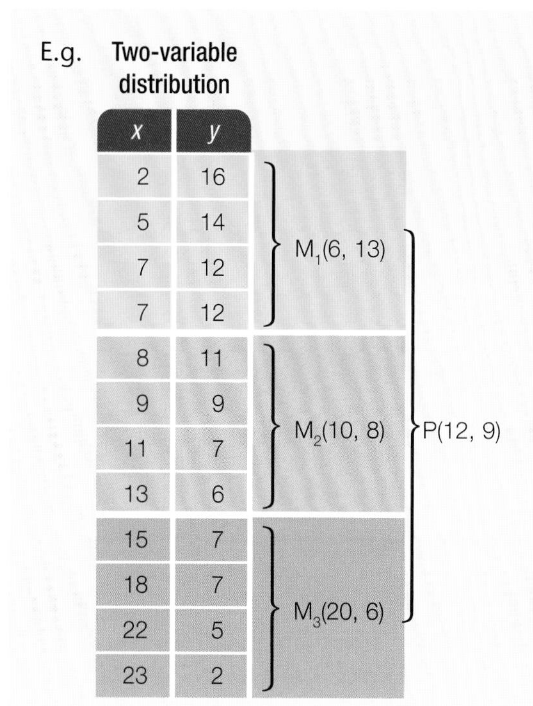

E.g. Two-variable distribution

x	y		
2	16		
5	14		
7	12	$M_1(6, 13)$	
7	12		
8	11		
9	9		
11	7	$M_2(10, 8)$	P(12, 9)
13	6		
15	7		
18	7		
22	5	$M_3(20, 6)$	
23	2		

- The slope of the line passing through points M_1 and M_3 is: $\frac{6 - 13}{20 - 6} = -0.5$.
- The equation of the line with a slope of -0.5 and passing through point P is $y = -0.5x + 15$, which corresponds to the equation of this regression line.

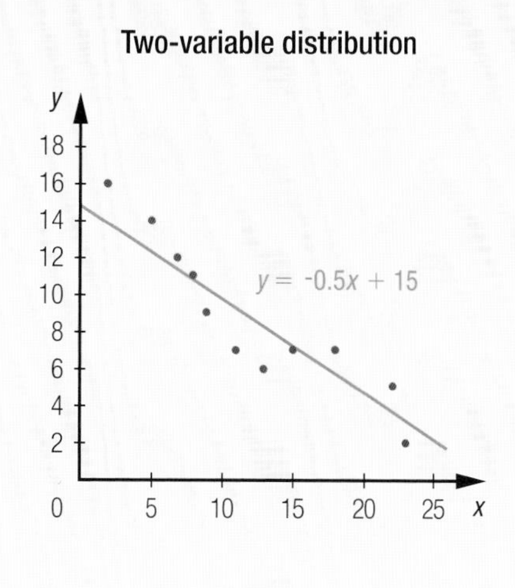

Mayer line method

Following is how to determine the equation of a regression line using the **Mayer line** method:

1. Order the ordered pairs in the distribution according to their *x*-coordinates.
2. Divide all the ordered pairs into two equal groups if possible.
3. Determine the mean of the *x*-coordinates and the mean of the *y*-coordinates in each of the two groups in order to form the mean ordered pairs $P_1(x_1, y_1)$ and $P_2(x_2, y_2)$.
4. The regression line is the line which passes through points P_1 and P_2.

E.g.

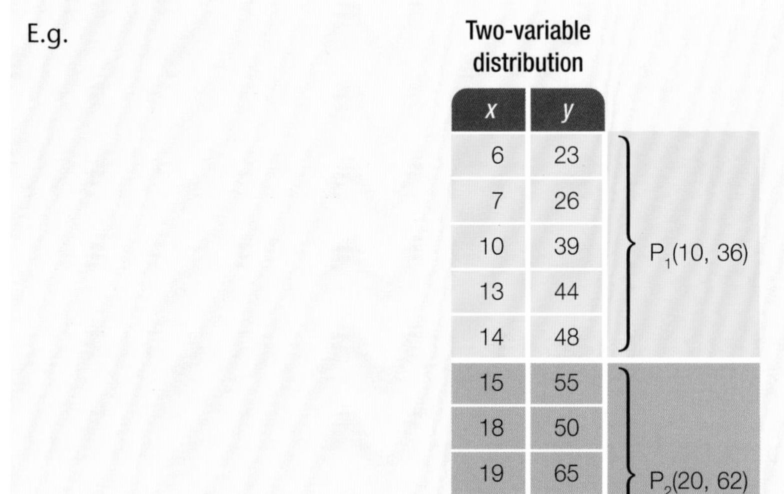

Two-variable distribution

x	*y*	
6	23	
7	26	
10	39	$P_1(10, 36)$
13	44	
14	48	
15	55	
18	50	
19	65	$P_2(20, 62)$
23	68	
25	72	

The equation of the line passing through points P_1 and P_2 is $y = 2.6x + 10$ which corresponds to the equation of the regression line.

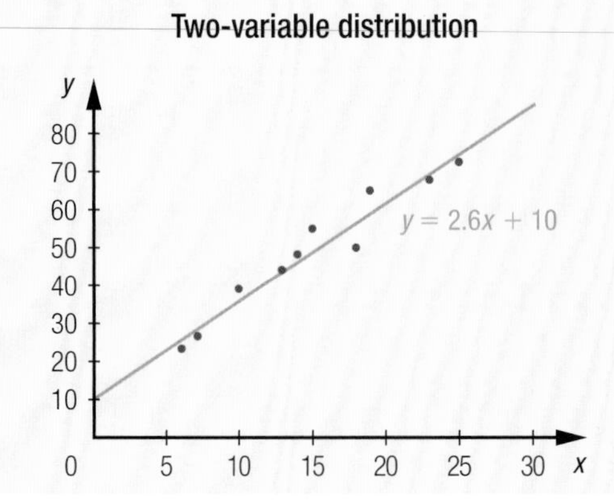

A regression line allows you to predict the value(s) of one of the variables based on the value(s) of the other, and the correlation coefficient allows you to determine the reliability of this prediction.

practice 2.3

1 Associate each of the scatter plots below with one of the linear correlation coefficients.

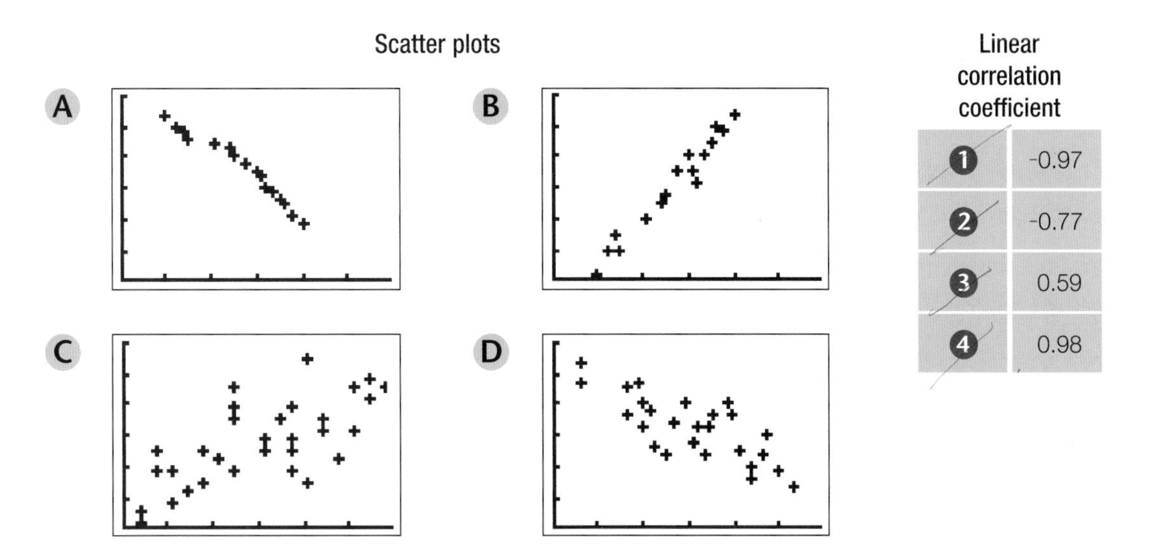

Linear correlation coefficient	
1	-0.97
2	-0.77
3	0.59
4	0.98

2 Below are three scatter plots:

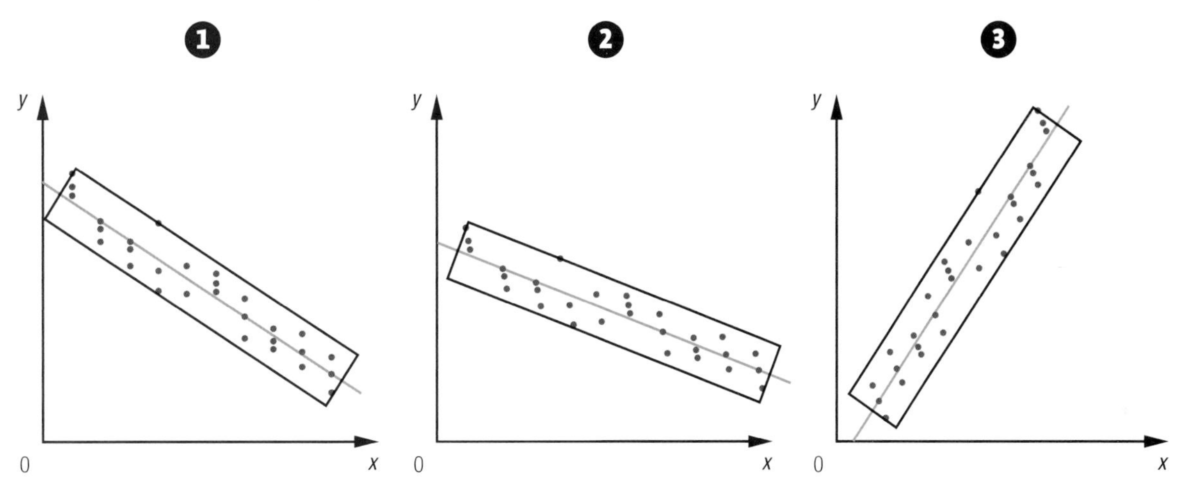

a) Compare:

1) the dimensions of each of the three rectangles
2) the slope of each of the three orange lines
3) the linear correlation coefficient associated with each graph.

b) Graphically estimate the linear correlation coefficient for each of the graphs

3 Which of the linear correlation coefficients below indicates the strongest correlation?

A 0.89 B 0.1 C -0.95 D -0.24

4 Match each description in the left column with a coefficient in the right column.

Description	Linear correlation coefficient
A Zero	1 -1
B Moderate	2 -0.89
C Strong	3 0
D Perfect	4 0.55

5 For each of the adjacent scatter plots, do the following:

1) Graphically estimate the linear correlation coefficient.
2) Determine the equation of the regression line drawn in **a)**.

a) 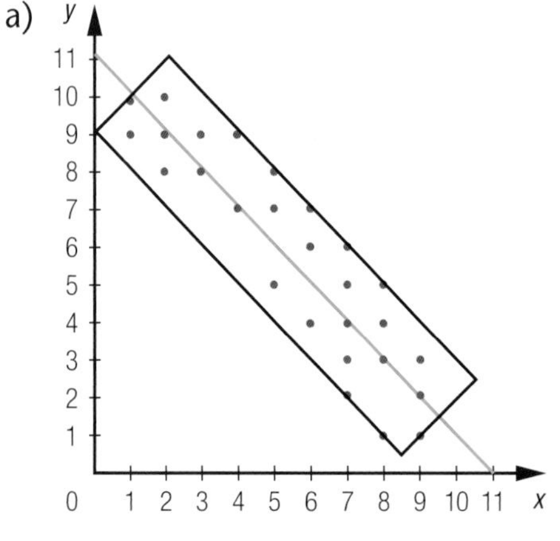

b) 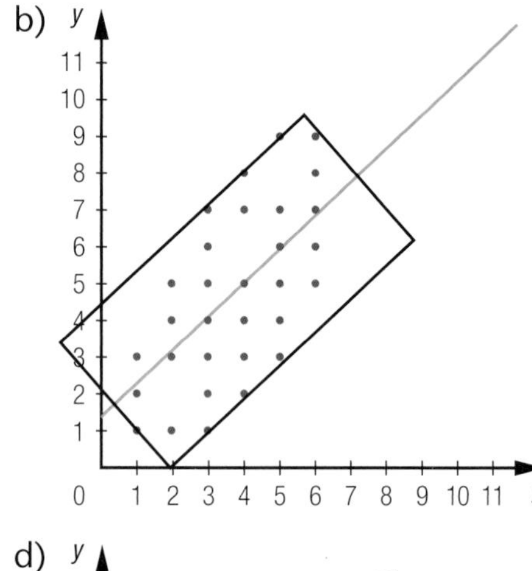

c) 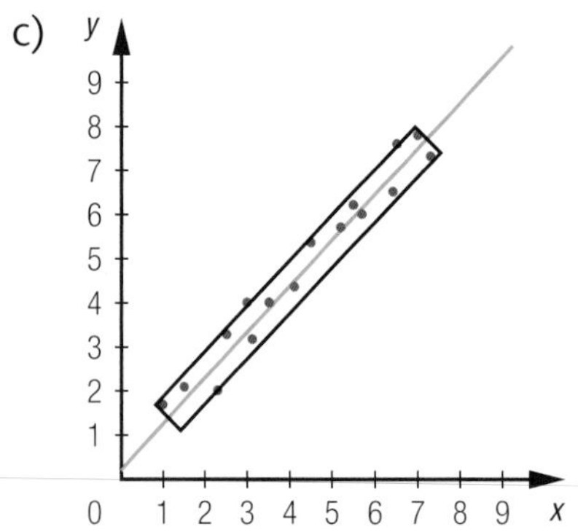

d) 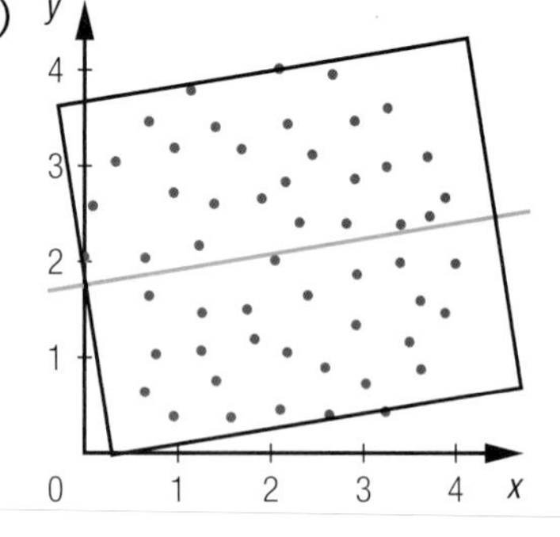

6 Each of the following ordered pairs represents the age of a person and his or her monthly expenditures (in $):

(13, 70) (13, 20) (13, 65) (13, 84) (13, 26) (14, 17) (14, 78) (14, 17) (14, 78) (14, 120) (14, 100) (14, 30) (14, 70) (14, 81) (15, 12) (15, 38) (15, 68) (15, 155) (15, 105) (16, 90) (16, 25) (16, 45) (16, 92) (16, 110) (17, 178) (17, 230) (17, 180) (17, 100) (17, 140) (17, 110) (17, 24) (17, 76)

a) Draw the scatter plot that represents this data.

b) 1) Graphically estimate the linear correlation coefficient between the two variables.
   2) Describe the linear correlation between the two variables.

c) Determine the equation of the regression line.

d) 1) Predict the amount of monthly expenditures of a 23-year-old person.
   2) Is this prediction reliable? Explain your answer.

7 **OZONE** The adjacent graph represents the annual quantity of ozone (in DU) in Canada.

a) Describe the linear correlation between these two variables.

b) Draw a line that represents the majority of the points.

c) Determine the equation of this line representing the regression line for this situation.

d) Calculate the quantity of ozone in 2010.

Ozone in Canada

Quantity (DU): 0, 320, 330, 340, 350, 360, 370, 380

Years: 1950, 1960, 1970, 1980, 1990, 2000, 2010

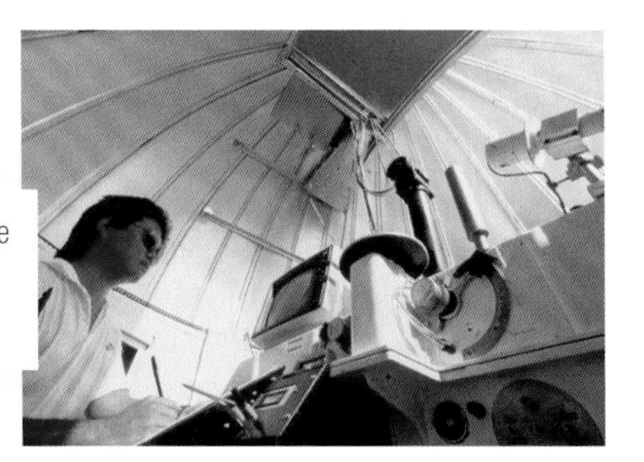

The Dobson spectrophotometer, named after its inventor, John Dobson, is the standard instrument used to measure the ozone layer from the ground. This measure is expressed in Dobson Units (DU).

8 **TRACK AND FIELD** The adjacent table represents the world-record times for the women's 100 m race in track and field.

a) Construct a scatter plot that represents this distribution.

b) Graphically estimate the linear correlation coefficient for this distribution.

c) Describe the linear correlation between the variables.

d) Determine the equation of the regression line.

e) In which year:

1) was the world record 12.5 s?

2) could this record have been below the 10 s barrier?

100 m (women's)

Year	Time (s)
1922	12.8
1926	12.4
1928	12.2
1928	12
1932	11.9
1933	11.8
1934	11.7
1937	11.6
1948	11.5
1952	11.4
1955	11.3
1961	11.2
1965	11.1
1968	11.08
1972	11.07
1976	11.04
1976	11.01
1977	10.88
1983	10.81
1983	10.79
1984	10.76
1988	10.49

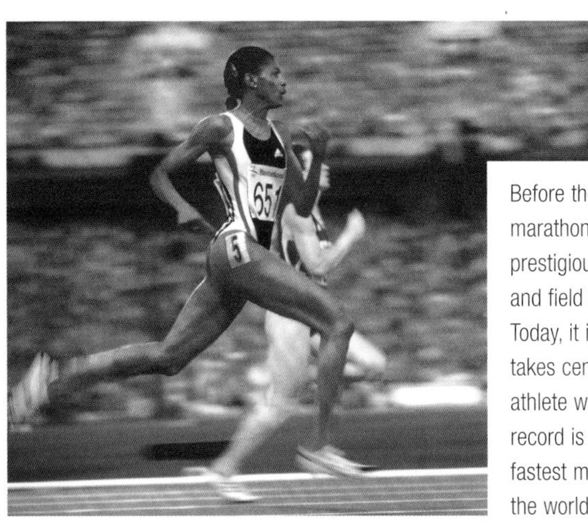

Before the 20th century, the marathon was the most prestigious event of track and field competitions. Today, it is the 100 m that takes centre stage. The athlete who holds the world record is considered the fastest man or woman in the world.

9 Place the scatter plots in increasing order according to the linear correlation coefficient associated with each one.

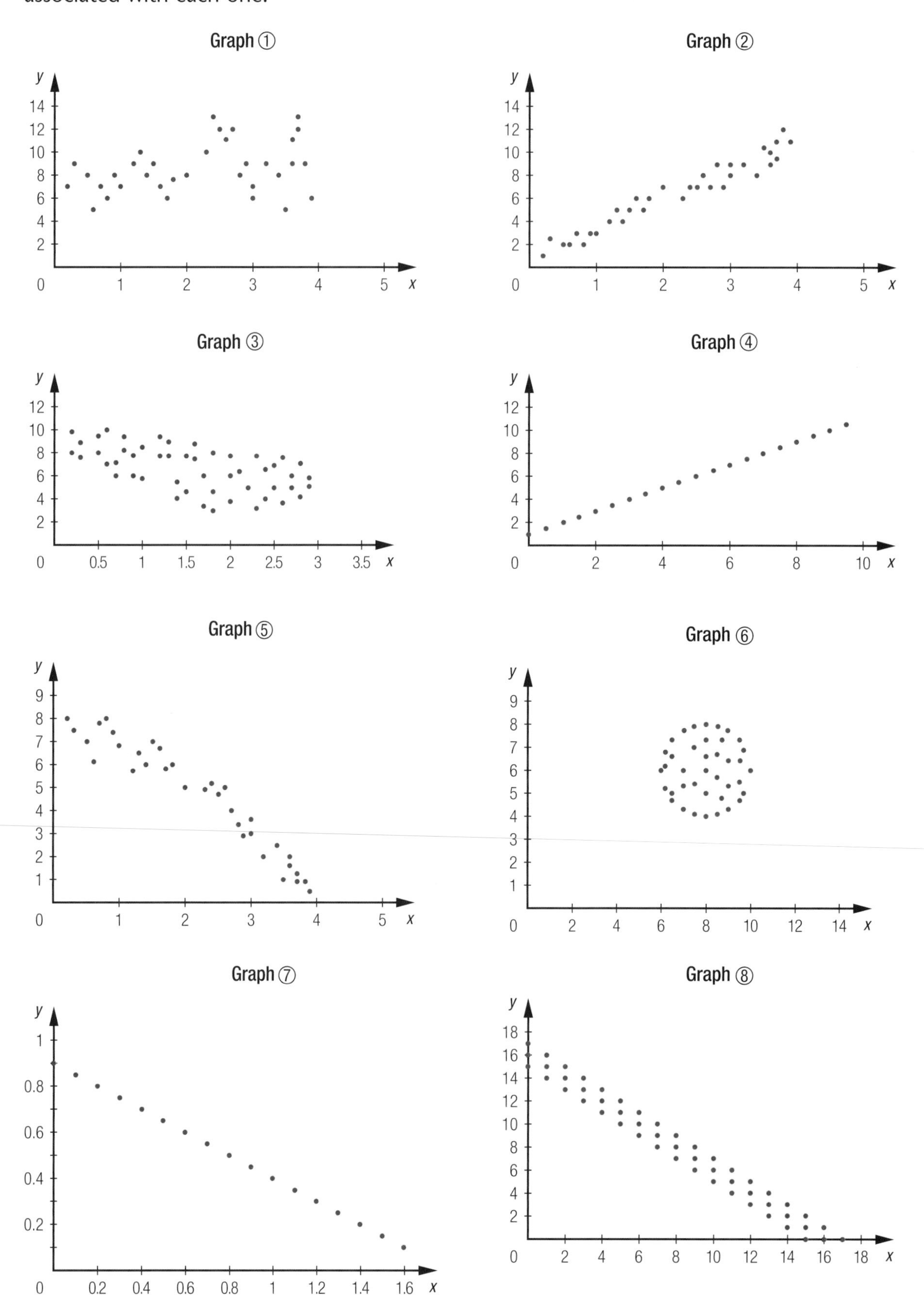

10 Graphically estimate the linear correlation coefficient associated with each of the following scatter plots.

a) Graph 1

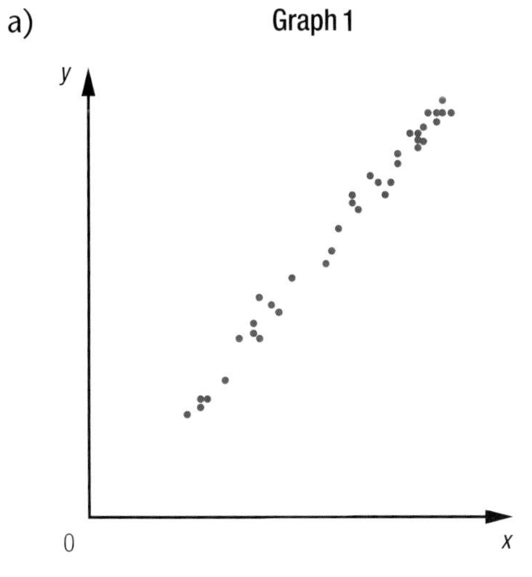

b) Graph 2

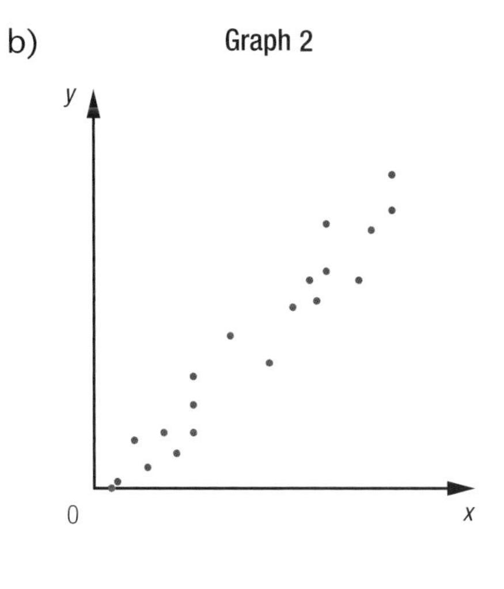

c) Graph 3

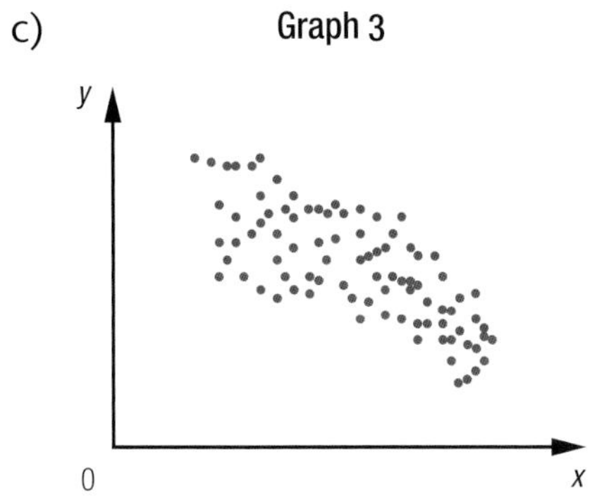

d) Graph 4

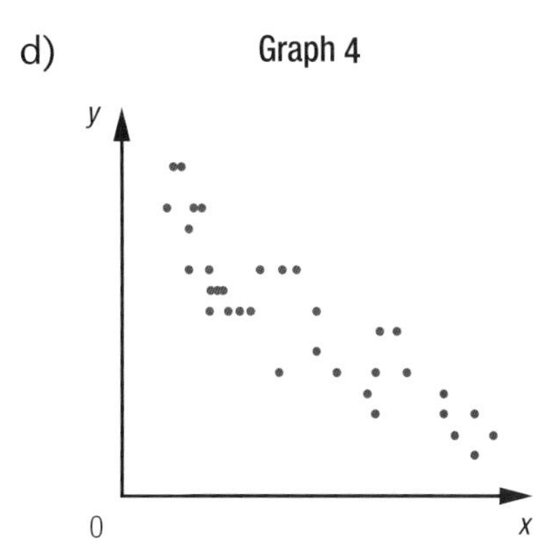

11 a) For each case, determine the equation of the regression line using the median-median line method.

Table of values 1

x	1	1.3	1.7	2	2	2.7	3.1	3.5	3.7	4	4.3	4.7	5.2	5.4	5.7
y	1.3	2.8	1.8	3	2.1	3.4	3.5	3	4.3	3.3	4	5	4.7	5.7	5

Table of values 2

x	12	16	21	24	28	32	34	37	41	44	48	53	55	57	61
y	57	55	50	50	50	47	43	40	35	34	33	30	30	28	21

b) For each case, determine the equation of the regression line using the Mayer line method.

Table of values 3

x	0.8	1.1	1.5	1.8	2	2.2	2.7	2.9	3	3.3	3.6	3.9	4	4.2
y	0	0.5	0.5	1.2	1.3	1	1.5	1.3	2	1.9	2.2	2.5	2.2	2.8

Table of values 4

x	1	1	1	1.5	1.5	1.5	1.5	2	2	2	2.5	2.5	2.5	2.5
y	8	7.5	6.5	7.5	7	8	6	6.5	6	7.5	6	5.5	4	4.5

12 The number of kilocalories consumed by 20 teenagers as they watched television was recorded over a seven-day period. Below are the data that was collected:

Eating habits and television

Time (min)	Kilocalories (kcal)	Time (min)	Kilocalories (kcal)	Time (min)	Kilocalories (kcal)	Time (min)	Kilocalories (kcal)
120	70	350	230	480	580	600	830
160	120	370	270	500	670	640	840
240	140	400	400	520	700	700	870
280	160	420	470	560	780	740	910
310	180	450	540	575	800	770	940

According to the data collected:

a) How long did a teenager who consumed 1000 kilocalories watch television?

b) How many kilocalories did a teenager who watched 2.5 hours of television consume?

c) Is it possible to consume 6000 kilocalories while watching television for a whole day?

13 Below is meteorological data collected in a city over the course of a day:

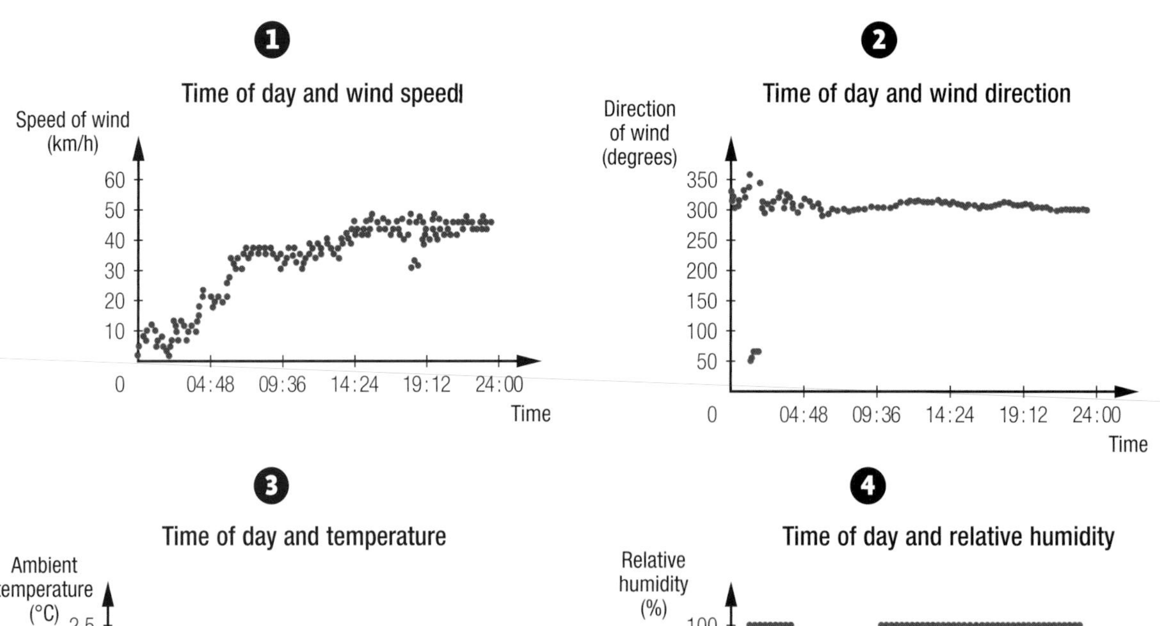

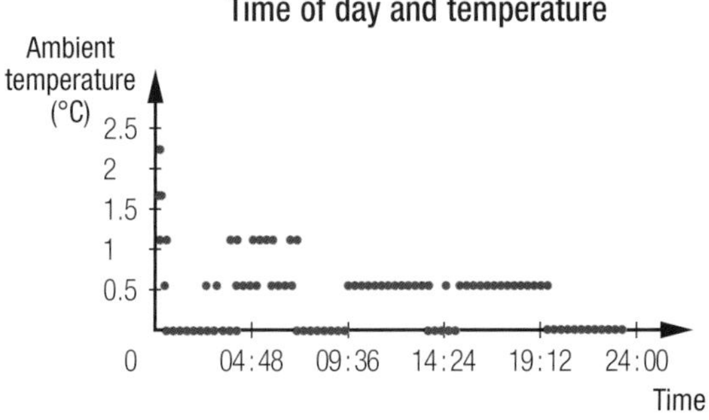

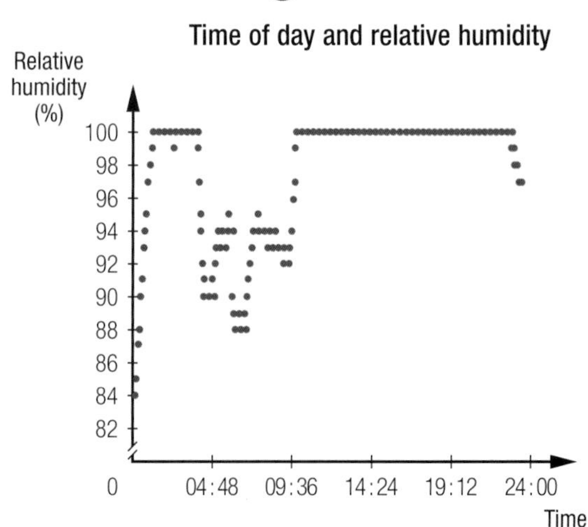

a) Which graph has the strongest linear correlation?

b) Describe the linear correlation between:

1) the speed and direction of the wind

2) the temperature and relative humidity

SECTION 2.4 Interpretation of linear correlation

This section is related to LES 4.

PROBLEM Leonardo da Vinci

Leonardo da Vinci devoted his life to the search for knowledge. He was known for his detailed and elaborate research methods and for his studies of proportions found in the human body.

Leonardo da Vinci (1452-1519)
Italian painter, architect, engineer and sculptor

People were randomly selected to take part in a statistical survey. The graph below represents the relationship between people's height and the length of their outstretched arms.

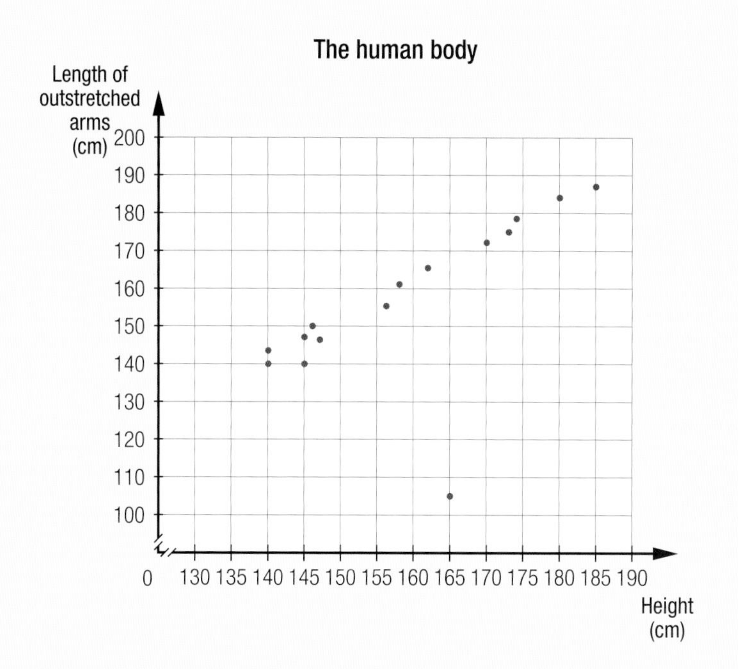

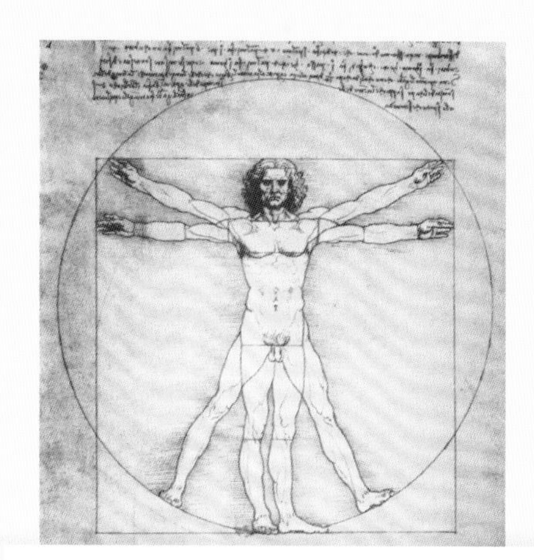

Leonardo da Vinci designed rules for painting a person's body to reflect its size as accurately as possible. These rules establish links among the dimensions of the human body. The *Vitruvian Man* designed by da Vinci depicts a man with outstretched arms standing within a circle and a square simultaneously. The square enables one to see that the length of the outstretched arms corresponds approximately to the figure's height.

What should the length of a 165-cm tall person's outstretched arms be?

ACTIVITY 1 It's amazing!

Fifty objects are placed on a table and 15 volunteers are asked to remember as many of the objects as they can.

Screen **1** displays the number of objects memorized by a person in relation to the time given for observation.

Screen **2** displays the equation of the regression line and the linear correlation coefficient.

Screen **3** displays the line of best fit.

Screen 1

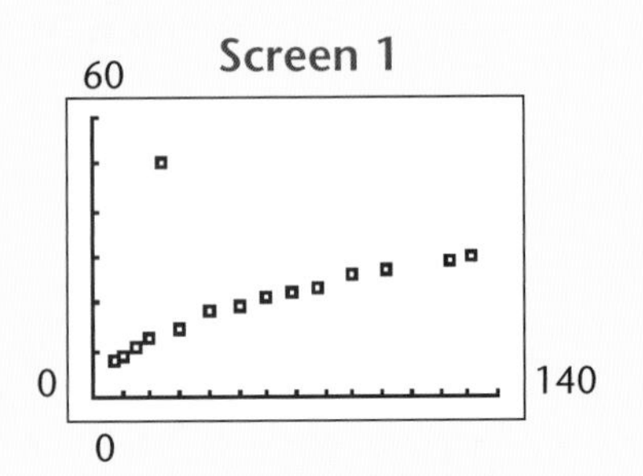

Screen 2

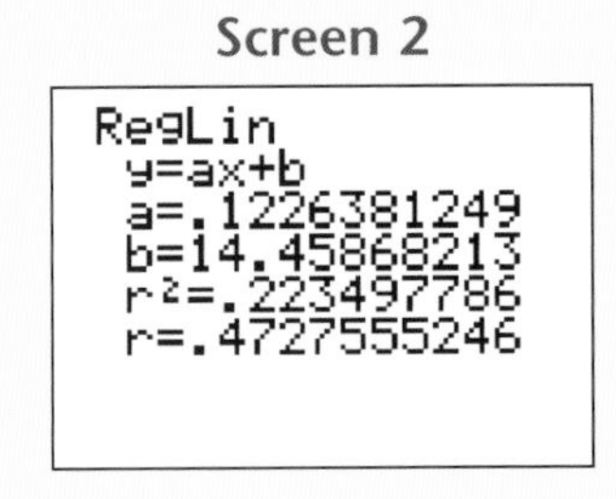

Screen 3

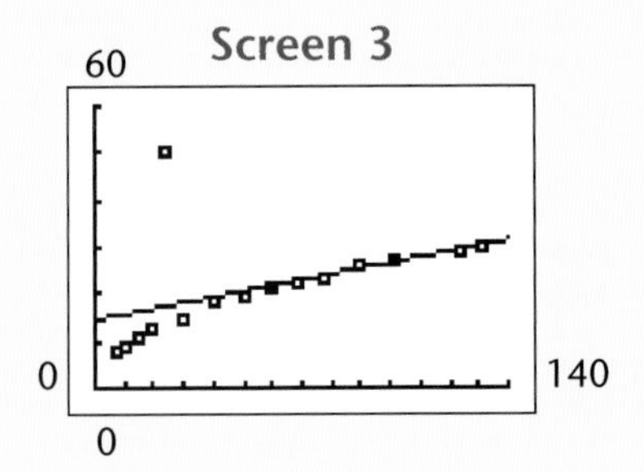

Noticing that one of the results differs greatly from the others, a psychologist decided not to include it. Below is the result:

Screen 4

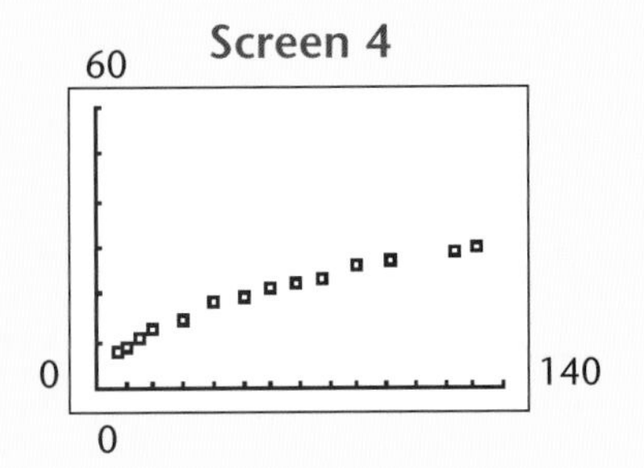

Screen 5

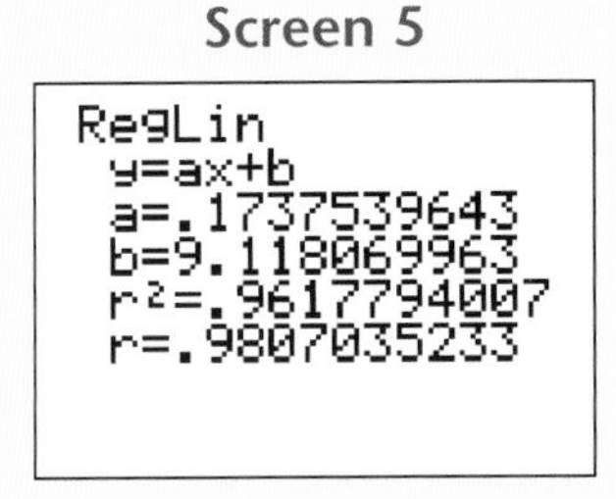

Screen 6

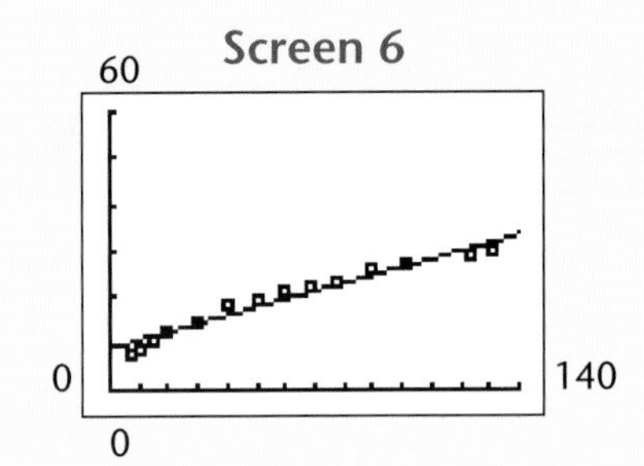

a. Describe the linear correlation observed:
1) in Screen **1** 2) in Screen **4**

b. Compare the linear correlation coefficients in Screens **2** and **5**.

c. A 16th person studies the objects for 32.5 s. According to the results obtained, how many objects should this person remember?

d. In a statistical study, should all data be systematically used to draw conclusions or make predictions? Explain your answer.

In 2007, the German, Gunther Karsten won the world memory championship held in Germany. This championship comprises several disciplines: memorizing numbers, words, cards, etc. After five minutes of study, Gunther Karsten was able to cite 726 numbers from memory. The time allowed to respond was 15 minutes.

ACTIVITY 2 Snow on my report card

This graph represents the relationship between snow accumulation since the beginning of winter and a student's academic results.

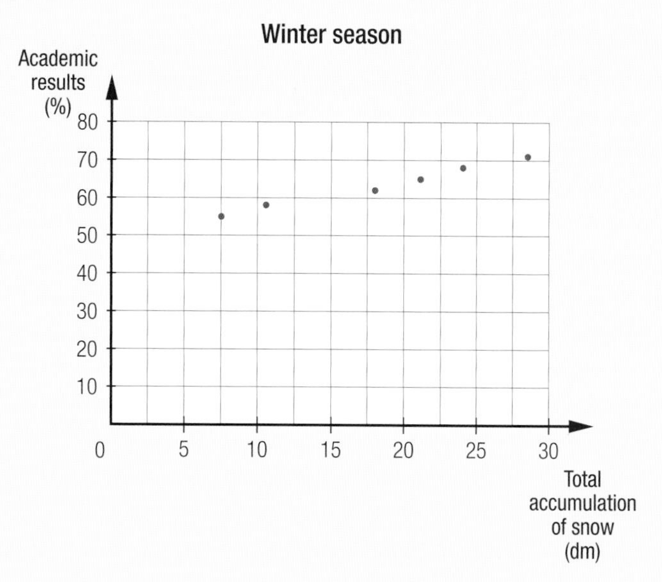

a. Based on the position of the points, answer the following:

1) Can you say that there is a strong linear correlation? Explain your answer.
2) Is it possible to predict the academic results of this student if snow accumulation reaches 60 dm? Explain your answer.
3) Is it possible to predict the snow accumulation if the academic result of this student is 85%? Explain your answer.

b. 1) Does the amount of snow accumulation affect the academic results of a student? Explain your answer.

2) Does a student's academic result affect the total snow accumulation levels? Explain your answer.

c. What can you conclude about the relationship between total snow accumulation and academic results?

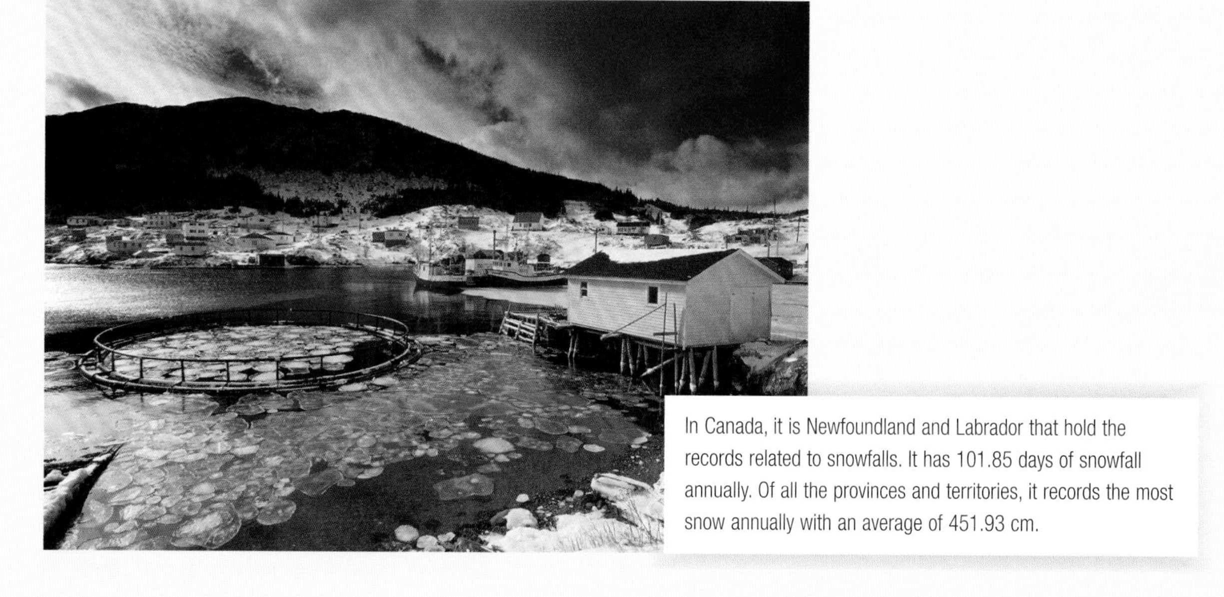

In Canada, it is Newfoundland and Labrador that hold the records related to snowfalls. It has 101.85 days of snowfall annually. Of all the provinces and territories, it records the most snow annually with an average of 451.93 cm.

Techno math

A graphing calculator can be used to define the equation of a regression line and generate the correlation coefficient of a two-variable distribution.

The table of values below provides the ordered pairs of a two-variable distribution.

x	1.6	2.2	3.8	4.1	5.1	6.9	7.3
y	715	720	730	715	760	775	770

This screen allows you to enter each ordered pair from the distribution.

Screen 1

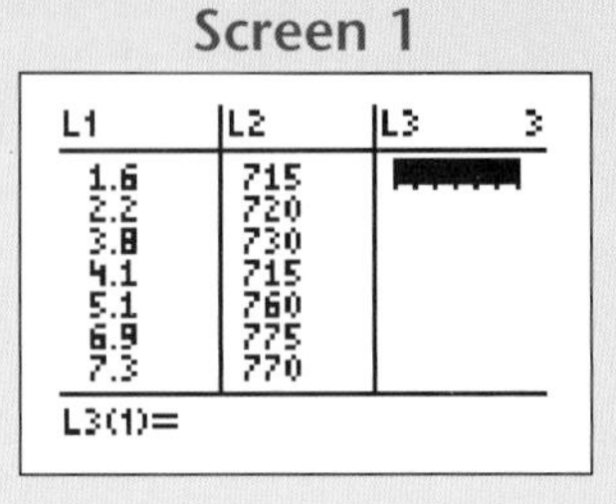

This screen allows you to choose the scatter plot as the display method.

Screen 2

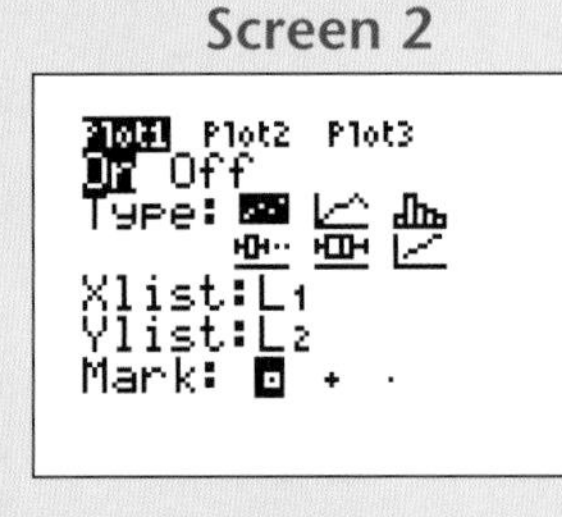

Screen 3

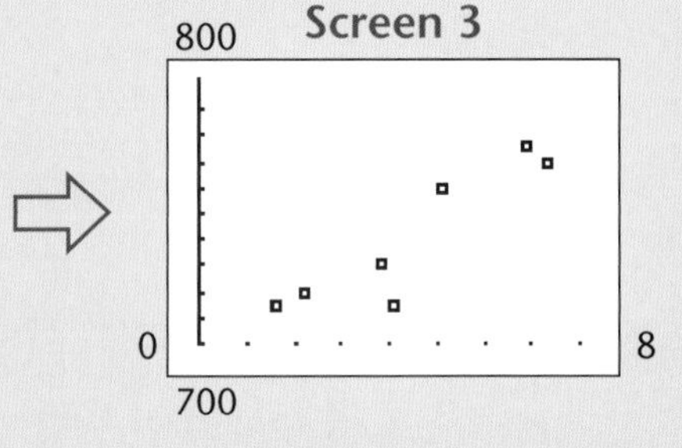

This selection allows you to determine the equation of the regression line.

Screen 4

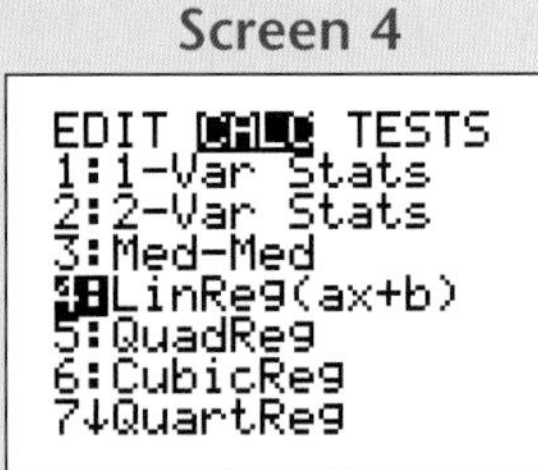

These two screens show you how to determine the equation of the regression line, enter the result in the equation editor and calculate the correlation coefficient.

Screens 5 & 6

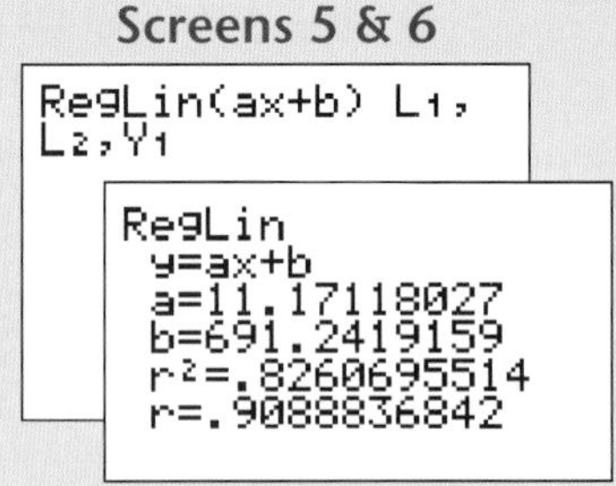

This screen displays the regression line and the scatter plot.

Screen 7

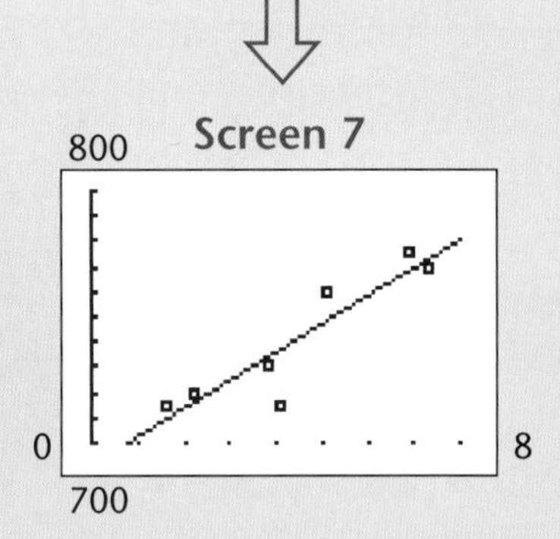

a. In Screen **6**, what do the values a, b and r represent?

b. Using a graphing calculator and the adjacent data, do the following.

x	6	17	25	32	45	52	61
y	99	94	86	60	71	64	59

1) Display a scatter plot.
2) Determine the equation of the regression line.
3) Calculate the correlation coefficient.
4) Plot the regression line and the scatter plot.

knowledge 2.4

LINEAR CORRELATION

Below is how to determine the correlation coefficient between two statistical variables and the equation of a regression line using a **graphing calculator**:

1. Enter the ordered pairs into a table.

E.g.

```
EDIT CALC TESTS
1:Edit...
2:SortA(
3:SortD(
4:ClrList
5:SetUpEditor
```

L1	L2	L3
5	72	
6	57	
10	48	
15	45	
16	33	
18	37	

L3(1)=

2. Display the scatter plot.

E.g.

```
Plot1 Plot2 Plot3
On Off
Type:
Xlist:L1
Ylist:L2
Mark: ▫ + ·
```

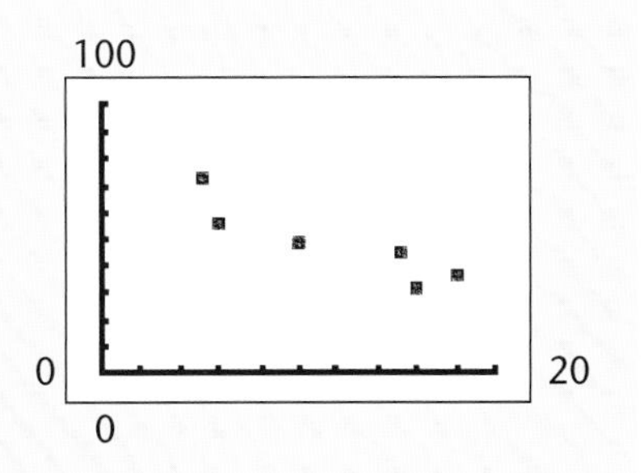

3. Display the equation of the regression line and the correlation coefficient.

E.g.

```
EDIT CALC TESTS
1:1-Var Stats
2:2-Var Stats
3:Med-Med
4:LinReg(ax+b)
5:QuadReg
6:CubicReg
7↓QuartReg
```

```
RegLin(ax+b) L1,
L2,Y1
```

```
RegLin
 y=ax+b
 a=-2.381696429
 b=76.453125
 r²=.8392569648
 r=-.9161096904
```

4. Display the regression line.

E.g.

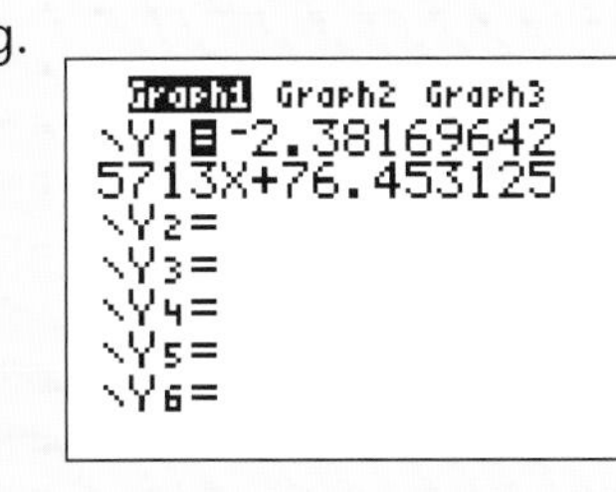

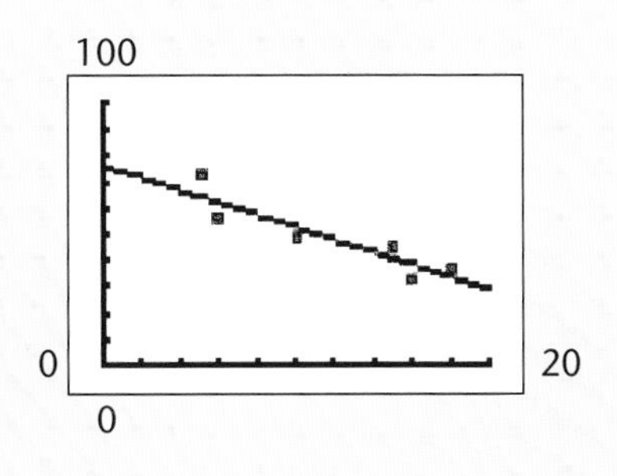

FACTORS TO CONSIDER WHEN INTERPRETING A CORRELATION

Several factors can influence the interpretation of a correlation between two variables. For this reason, one must be vigilant when making predictions and drawing conclusions.

Interpretation	Example
• The link between two variables can be one of cause and effect: that is when with one of the variables has a direct effect on the other. In such cases, the correlation is perfect and the relation between the two variables is defined by a rule.	The correlation between altitude and temperature is perfect since the temperature varies in direct relation to altitude.
• The correlation between two variables can be significant without the two variables being directly linked to each other. They can both depend on a third variable which, as it varies, generates variations for the first two variables.	In the summer, it may seem that there is a strong correlation between the number of ice cream cones sold and the number of air conditioning units sold in a given city while in fact these two variables depend on another variable, is, the temperature.
• Considering a correlation as being linear while another model would be more appropriate.	The population growth of a major city can be studied according to a linear correlation. However, using an exponential model would be more appropriate.
• It sometimes may happen that there is a correlation between two variables only over a given interval.	Over the interval [5, 10] years, the correlation between a person's age and his or her height is linear. However, before and after this interval, the linear model is not the best fit.
• A two-variable distribution may include outlier data, notably due to manipulation or measurement errors.	The degree of precision of the instrument used during data collection is poor.

A strong correlation indicates the existence of a statistical link. However, it does not explain the reason and nature of the link. Subsequently, one will try to qualitatively and quantitatively assess this link and make predictions, keeping in mind the limitations of these predictions.

practice 2.4

1 Indicate which of the scatter plots below contain at least one outlier.

A

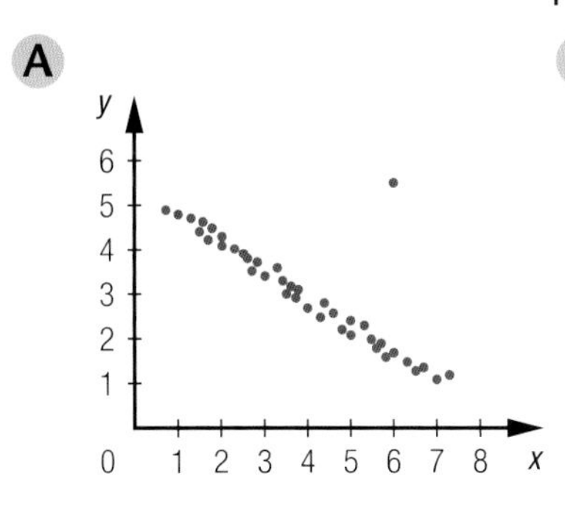

B

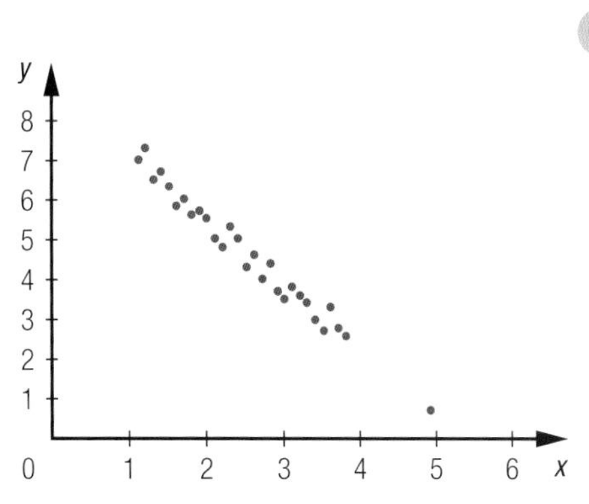

C

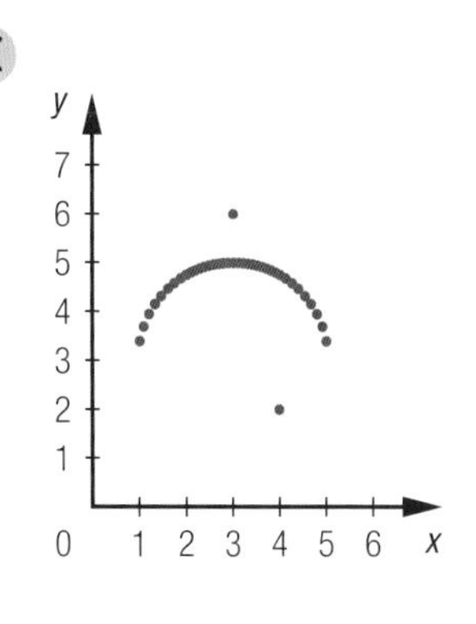

2 For each of the table of values below, do the following.

1) Calculate the linear correlation coefficient.
2) Indicate the equation that defines the regression line.

a)

x	1	1.1	1.2	1.3	1.4	1.5	1.6	1.7	1.8	1.9	2	2.1	2.2	2.3	2.4	2.5	2.6	2.7
y	0	0.8	1	0.5	1.5	2	1.3	1	2	1.4	2.5	4	3.4	3	5	4.3	4	4.7

b)

x	1.1	1.2	1.3	1.4	1.5	1.6	1.7	1.8	1.9	2	2.1	2.2	2.3	2.4	2.5	2.6	2.7	2.8
y	7	7.3	6.5	6.7	6.3	5.8	6	5.6	5.7	5.5	5	4.8	5.3	5	4.3	4.6	4	4.4

c)

x	2.2	2.3	2.4	2.5	2.6	2.7	2.8	2.9	3	3.1	3.2	3.3	3.4	3.5	3.6	3.8	3.9	4
y	4	3.2	3.7	3	4.3	3.2	3.5	2.4	4	1.7	3.4	2.7	3	2	2.5	3	2.6	3

d)

x	3.2	3.3	3.4	3.5	3.6	3.7	3.8	3.9	4	4.1	4.2	4.3	4.4	4.5	4.6	4.7	4.8	4.9
y	4.6	3.7	5.7	5	5.9	4.5	5.6	6.1	5.5	4.8	6.7	7	7	6.4	6	7	6.9	6.3

3 In order to determine whether there is a correlation between a woman's age and ear length, the adjacent data was collected. Is there a linear correlation between the two variables studied? Explain your answer.

Study on women

Age	Ear length (cm)	Age	Ear length (cm)	Age	Ear length (cm)
5	4.9	32	6.4	61	6.3
8	5.3	37	6.1	67	6.4
12	5.2	41	6.5	73	6.7
16	6	44	6.6	75	6.5
20	5.8	47	6.3	78	6.7
21	6.3	51	6.2	82	6.8
25	6	55	6.5	84	7

4 A statistician examined the annual family household income in a city and the percentage of its families who owned at least one vehicle. The data is presented in the graph below.

a) What is the linear correlation coefficient:
   1) if you consider all the results?
   2) if you do not consider the outliers?

b) Determine the equation of the regression line.

c) What should be the percentage of families who own at least one vehicle and whose annual household income is $85,000?

d) What is the percentage of families with an annual income of $72,000 who own at least one vehicle?

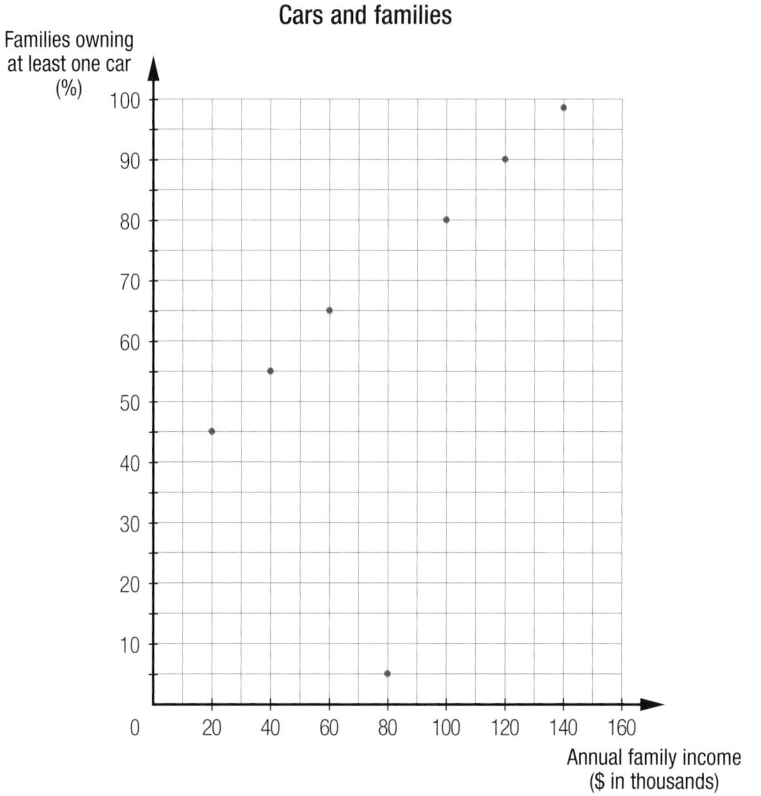

5 The foot length and body height of 17-year-old adolescents were measured.

a) What is the linear correlation coefficient between foot length and height:
   1) among girls?
   2) among boys?

b) Determine the equation of the regression line associated with:
   1) girls
   2) boys

c) What is the difference between the height of girls and boys whose feet measure 29 cm?

d) For what foot length is the height of girls and that of boys the same?

Adolescent population

Girls		Boys	
Length of feet (cm)	Height (cm)	Length of feet (cm)	Height (cm)
20	140	21	149
20	144	22	160
21	145	22	155
21	147	22	158
22	151	23	164
23	153	23	167
23	155	24	169
24	157	25	172
24	162	25	175
25	160	26	179
26	164	27	182
26	166	28	188
26	168	28	191

6 In a clinic, patients with back pain were studied to establish the percentage of pain reduction in relation to the number of days since beginning chiropractic treatment.

Chiropractic

Number of days since beginning of treatment	Pain reduction (%)	Number of days since beginning of treatment	Pain reduction (%)
5	5	41	27
7	6	60	31
12	12	65	35
16	8	70	36
21	13	70	43
30	17	84	42
30	21	90	54

a) After how many days can you expect to see pain reduced by 50%?

b) How many days must treatment be followed in order to be pain free?

c) How long was the treatment for a patient whose pain was reduced by 75%?

d) By what percentage would pain be reduced for a patient who followed 240 days of treatment?

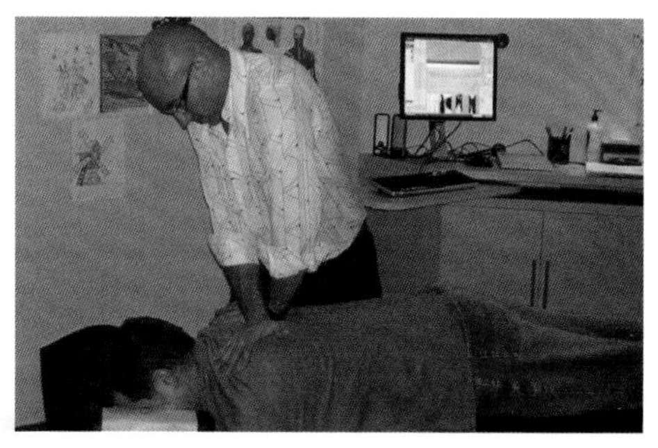

Chiropractors are healthcare professionals who help people recover or maintain their health by diagnosing, treating and preventing functional deficiencies in the body. They focus on the nervous system and particularly the spinal column.

7 The table below shows the relationship between the age of a tree and the circumference of its trunk.

Tree growth

Age	1	2	3	4	5	6	7	8	9	10	11	12	13	14	15
Circumference (cm)	5	8	13	16	22	25	34	40	45	53	60	62	65	75	83

a) What is the correlation coefficient for this situation?

b) Determine the equation of the regression line associated with this situation.

c) What should the circumference of a 25-year-old tree trunk be?

d) If a tree trunk's circumference is 225.9 cm, what should its age be?

The light section of a tree trunk represents spring growth when sap flows heavily. The dark section represents summer growth. The thicker a growth ring, the better conditions were for tree growth.

8 The linear correlation coefficient between two variables is 0.5. Describe a situation in which this correlation may be interpreted as:

a) weak

b) strong

9 In 10 regions where trees are of the same size, data was collected on local air pollution.

Environment

Region	Trees/km^2	Air pollution (%)	Region	Trees/km^2	Air pollution (%)
1	500	30	6	850	15
2	580	27	7	920	15
3	640	24	8	1010	15
4	720	21	9	1000	15
5	780	18	10	1200	15

a) What is the minimum percentage of pollution?

b) What should the percentage of air pollution be in a region:

1) with no trees?

2) having 680 trees/km^2?

c) How many trees per square kilometre should a region with 40% pollution have?

d) How many trees should be planted in a region that has 30% air pollution in order to reduce the pollution to 15%?

CO_2 molecule

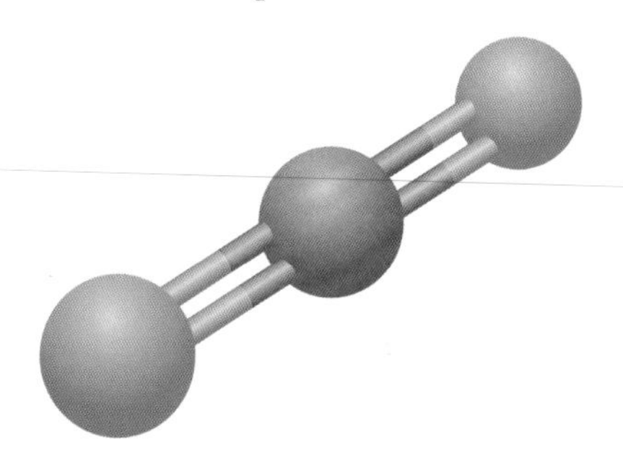

Carbon dioxide (CO_2) is one of the most abundant gases in the atmosphere. It plays a major role in the vital processes of plants and animals, and the human body requires it to breathe. It is also used to remove caffeine from coffee and to make soft drinks, fire extinguishers and dry ice. The amount of CO_2 in the air has increased considerably over the past 150 years due to human activity primarily in the use of fossil fuels and deforestation.

The total amount of CO_2 eliminated from the atmosphere is equal to the quantity of carbon stored in trees and in the ground. The average Canadian tree is capable of storing 2.5 kg of carbon annually in an urban centre and 2.8 kg in a rural setting. Carbon thus stored is freed when trees are cut down.

10 **LIFE EXPECTANCY** Statisticians have studied the life expectancy and the number of years lived in good health of people living in various countries. Below are the results of this research:

Quality of life

Country	Men		Women	
	Life expectancy (years)	Number of healthy years	Life expectancy (years)	Number of healthy years
Afghanistan	42	35	42	36
Algeria	70	60	72	62
Argentina	72	62	78	68
Australia	79	70	84	75
Belgium	76	69	82	73
Bolivia	63	54	67	55
Brazil	68	57	75	62
Cambodia	51	46	57	49
Canada	78	70	83	74
Chile	74	65	81	70
China	71	63	74	65
Egypt	66	58	70	60
Ethiopia	50	41	53	42
France	77	69	84	75

Country	Men		Women	
	Life expectancy (years)	Number of healthy years	Life expectancy (years)	Number of healthy years
Guatemala	65	55	71	60
Haiti	53	44	56	46
India	62	53	64	54
Japan	79	72	86	78
Madagascar	56	47	60	50
Pakistan	61	54	62	52
San Marino	80	71	84	76
South Africa	50	43	52	45
Switzerland	79	71	84	75
Turkey	69	61	74	63
Uganda	48	42	51	44
United States	75	67	80	71
Zambia	40	35	40	35
Zimbabwe	43	34	42	33

Using the table, explain why the following statements are true or false.

a) By 80 years of age, the number of years lived in good health is approximately the same for men as for women.

b) Women live in good health longer than men do.

c) A man whose life expectancy is 55 should have 47 years of healthy living.

d) A woman who lives 59 years in good health should have a life expectancy of approximately 65.

e) The correlation between the life expectancy of a woman and that of a man is negative.

Life expectancy in Canada is one of the highest in the world. However, it varies greatly from region to region. The life expectancy of people in the North and outlying regions, many of whom are Aboriginals, more closely resembles that of people in developing countries than that in industrialized nations. Life expectancy is highest in British Columbia at 83.4 years. On the opposite end of the scale, Nunavik has the lowest life expectancy at 66.7 years.

11 **HEALTH** The table below lists the annual income of healthcare professionals and the mean number of hours worked in a week during 2000 and 2001.

Healthcare professionals

Healthcare professional	2000 Annual income ($)	2000 Mean work week (hours)	2001 Annual income ($)	2001 Mean work week (hours)
Doctor specialist	110,100	46.6	125,700	54.5
General practitioner	97,000	46.4	104,100	53.5
Dentist	80,000	37	95,900	42.3
Veterinarian	50,000	42.8	55,800	49.7
Optometrist	62,000	37.1	70,000	43
Chiropractor	42,000	37.2	50,000	42.9
Diagnostics technician	27,000	33.1	35,000	43
Pharmacist	52,000	35.3	59,600	42.5
Dietician and nutritionist	33,000	30.1	42,500	39.8
Audiologist and Speech therapist	45,000	31.8	50,900	40.2
Physiotherapist	40,600	32.2	48,700	40.4
Occupational therapist	40,000	30.6	46,000	39.2
Therapist	28,000	30.3	35,000	40.8
Head nurse	48,000	33.6	51,400	41
Registered nurse	40,000	31.5	46,000	40.3
Nurse's aide	28,000	30.7	31,200	39.8

Calculate the values requested below, if possible; if not possible, explain why.

a) The annual income of a pharmacist in 2009.

b) The mean number of hours worked weekly in 2001 by a person whose annual income is $160,000.

c) The mean number of hours worked weekly in 2001 by a person who worked 50 h/week in 2000.

d) The annual income in 2001 for a person who earned $80,000 in 2000.

e) The number of hours worked weekly by a doctor specialist in 2009.

Hippocrates (ca. 460 - ca. 370 BCE) is considered the greatest doctor of antiquity. During his many voyages he treated both slaves and citizens, making no distinction between the two. Hippocrates lent scientific support to medicine which had previously been viewed as a religion or something sacred. He defined the goal of medicine: to support nature, and above all not to harm. Today the Hippocratic Oath is the foundation of medical ethics.

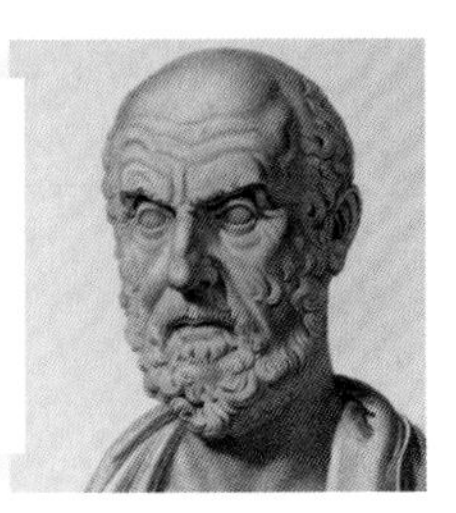

12 **DENTAL CAVITIES** From 1983 to 1997, the percentage of children aged 12 who had at least one cavity in their permanent teeth was recorded. Below are the results:

Dental hygiene

Region	Year	With a cavity (%)	Without a cavity (%)
Europe	1983	3.1	19
	1985	3.1	22
	1987	4.2	—
	1990	2	40
	1993	1.4	47
	1995	1.4	50
	1997	1.1	56
North America	1984	4.4	11
	1987	1.8	42
	1990	3.1	23
	1991	1.4	50
	1997	2.1	36

a) 1) For each of the two regions, construct a scatter plot that represents the relationship between the year and the percentage of 12-year-old children with at least one cavity in their permanent teeth.

2) Determine the equation of the regression line associated with each of the scatter plots.

3) Which region should be last to record 0% of 12-year-old children with cavities?

4) In 1940, which region had the highest percentage of 12-year-old children with at least one cavity in their permanent teeth? Explain your answer.

b) 1) Construct a graph representing the percentage of 12-year-old children with no cavities for each of the two regions, by year.

2) In 1970, which region had the highest percentage of 12-year-old children with no cavities?

3) Will this region still be in first place in 2020? Explain your answer.

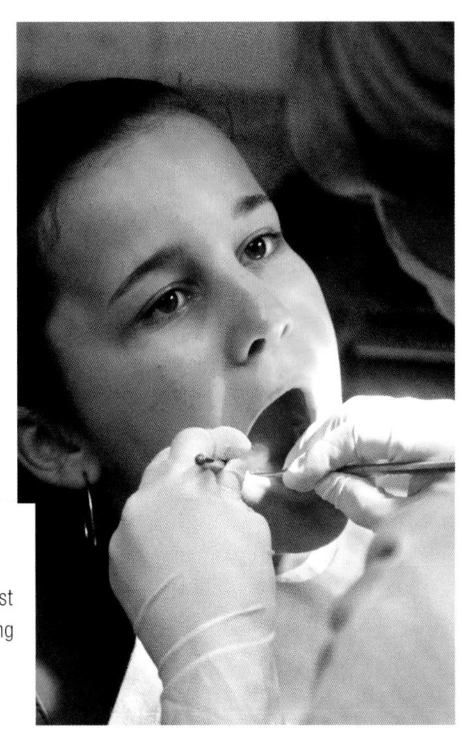

In Québec, cavity reduction among children began in the 80s, and continues today. But social inequalities persist, and cavities are concentrated among a group of high-risk children coming, for the most part, from disadvantaged regions. There is also a large disparity among Québec regions. Studies at the international level have shown that Québec is late in making up for lost ground and that the number of cavities is higher than in most industrialized nations.

Chronicle of the

past

Francis Galton

His life

Francis Galton was born on February 16, 1822, in Sparkbrook, England. He is considered to have been a multidisciplinary genius. Anthropologist, explorer, geographer, meteorologist and statistician, he invented the concepts of correlation and regression, among other things. Galton invented the weather chart and was the first to suggest the existence of anticyclones. He is also the founder of psychometrics and worked on the classification of fingerprints. It is even believed that he invented the sleeping bag! Galton died on January 17, 1911, in Haslemere, England.

Galton published more than 340 works in his lifetime. His cousin is Charles Darwin, creator of the theory of evolution.

The heritability of characteristics

Galton was very interested in heritability which is the probability that a person's descendants will inherit some of his or her characteristics. In 1886, he published a study in which he compared the height of 928 adult chidren to the height of their parents, for a total of 205 couples. In this study, Galton observed the following:

- Parents who are taller than the mean of the population give birth to children who are also taller than the mean of the population but not as tall as their parents.
- Parents who are shorter than the mean of the population give birth to children that are shorter than the mean of the population but are taller than their parents.

In this study, Galton also noticed that the mean height of men is 8% greater than that of women.

Contingency table showing the results of this study.

TABLE 1.

Number of Adult Children of Various Statures Born of 205 Mid-parents of Various Statures.
(All Female heights have been multiplied by 1·08).

Heights of the Mid-parents in inches.	Heights of the Adult Children.														Total Number of		Medians.
	Below	62·2	63·2	64·2	65·2	66·2	67·2	68·2	69·2	70·2	71·2	72·2	73·2	Above	Adult Children.	Mid-parents.	
Above ..	..	..	..	..	..	..	..	..	..	..	..	1	3	..	4	5	..
72·5	..	..	..	..	..	..	..	1	2	1	2	7	2	4	19	6	72·2
71·5	..	..	..	..	1	3	4	3	5	10	4	9	2	2	43	11	69·9
70·5	1	..	1	..	1	1	3	12	18	14	7	4	3	3	68	22	69·5
69·5	..	..	1	16	4	17	27	20	33	25	20	11	4	5	183	41	68·9
68·5	1	..	7	11	16	25	31	34	48	21	18	4	3	..	219	49	68·2
67·5	..	3	5	14	15	36	38	28	38	19	11	4	..	..	211	33	67·6
66·5	..	3	3	5	2	17	17	14	13	4	..	..	..	..	78	20	67·2
65·5	1	..	9	5	7	11	11	7	7	5	2	1	..	..	66	12	66·7
64·5	1	1	4	4	1	5	5	..	2	..	..	..	..	..	23	5	65·8
Below ..	1	..	2	4	1	2	2	1	1	..	..	..	..	..	14	1	..
Totals ..	5	7	32	59	48	117	138	120	167	99	64	41	17	14	928	205	..
Medians ..	..	..	66·3	67·8	67·9	67·7	67·9	68·3	68·5	69·0	69·0	70·0	..	..	..	..	..

Source : GALTON, F. *Regression Towards Mediocrity in Hereditary Stature*, 1886.

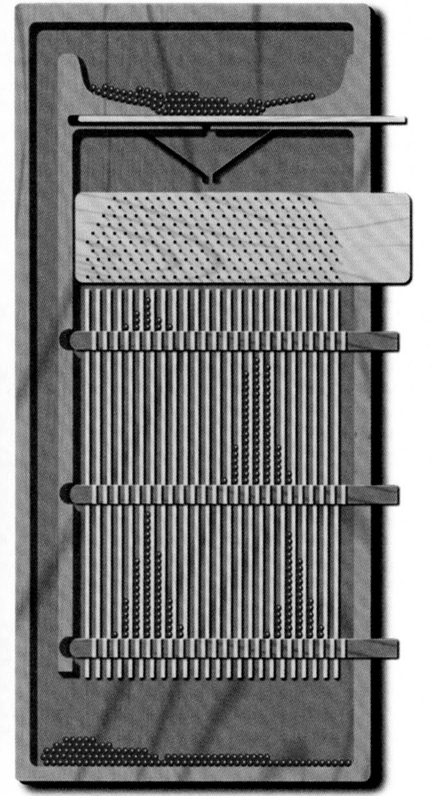

Galton's board

In order to simulate random events, Galton constructed the device shown in the adjacent illustration. A number of nails are arranged in staggered rows on an inclined board. Marbles fall one by one from a funnel-shaped reservoir and roll down the board. Each time a marble hits a nail, it has an equal probability of going to the left or right. Containers, placed below the last row of nails, collect the marbles so that the results can be compiled.

Psychometry psychology

The objective of the science of psychometric psychology is to evaluate the capacity of the human mind. Galton created this subject because he wanted to measure and compare levels of intelligence. The Intelligence Quotient, or IQ test, is probably the best-known tool of psychometrics. Today, tests are used to evaluate different aspects of intelligence, such as learning ability, knowledge, personality and emotional intelligence. However, before using these tests it is important to confirm their validity, that is, their ability to actually evaluate what they seek to measure.

1. In 8 successive trials, 50 marbles are dropped on a Galton board. The centre container located at the bottom of the board successively collects 13, 9, 12, 15, 11, 10, and 15 marbles.

a) What is the mean number of marbles collected in the centre container?

b) What is the mean deviation of the marbles collected in the central container?

2. Galton established the fact that a child's height f upon reaching adulthood is calculated according to the rule $f = 0.6p + 67.2$, where p represents the mean height (in cm) of his or her parents.

a) What is the height of an adult child whose parents' mean height is 175 cm?

b) What is the mean height of the parents whose adult child is 162.5 cm tall?

3. Consider the graphical representation of the relationship between a mother's height and the height of her adult daughter. Graphically estimate the correlation coefficient between the two variables.

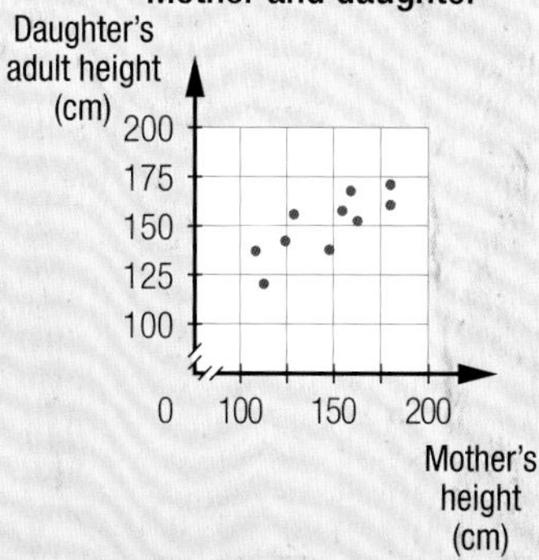

4. A company posts an executive position. During the psychometric test measuring aptitude for making good decisions quickly, scores obtained by applicants were 20, 35, 41, 48, 50, 55, 68, 69, 71, 76, 78, 85 and 86. What is the percentile of the person who scored 71?

In the

workplace

Environmental analysts

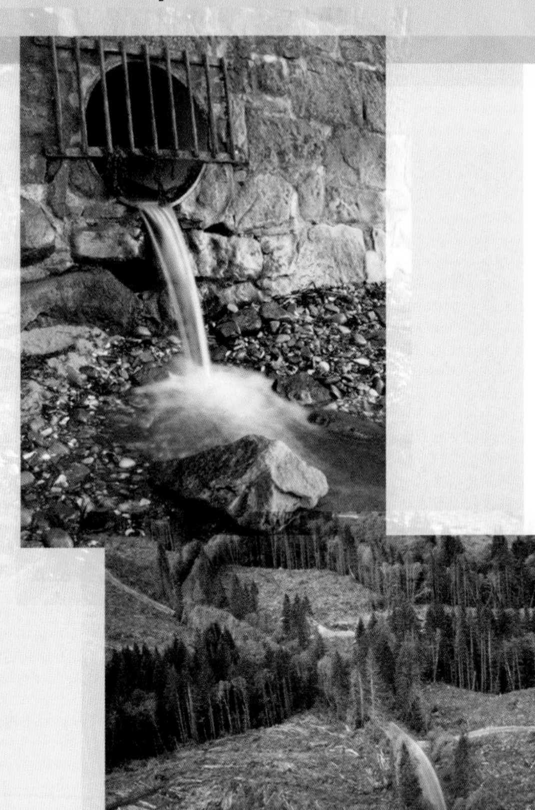

The profession

An environmental analyst is responsible for a variety of environmental projects such as environmental impact studies, compliance procedures for set standards, or the introduction of management systems for dangerous materials that respect the environment.

After completing studies in geography, anthropology or urban planning, an environmental analyst generally can work in the private or public sector.

The analyst's work is to gather data by carrying out field studies, analyses, surveys or documentary research, and to analyze these data to compile reports or authorization requests, as the case may be.

Autonomous and meticulous, the environmental analyst also enjoys working on a team. His or her ability to effectively communicate verbally and in writing helps in the production of reports on the results of studies and analyses.

Data acquisition

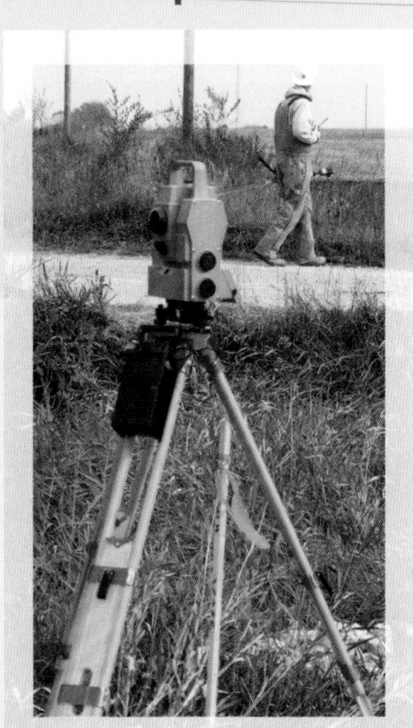

Data used by an environmental analyst is collected in a variety of ways. Depending on whether the study area is in the wild, or in a rural or an urban setting, data may be collected by field technicians or by surveys and on-site censuses. Those mandated to collect this data work under the supervision of the environmental analyst.

An environmental analyst collected the following data on the air quality and on the number of people with respiratory problems admitted to the emergency unit for the same region.

Regional hospital

Day	Air quality index	Number of patients admitted to emergency
Sunday	2	21
Monday	0	32
Tuesday	4	34
Wednesday	6	45
Thursday	11	57
Friday	13	52
Saturday	4	22
Sunday	0	15
Monday	0	25
Tuesday	1	26
Wednesday	2	47
Thursday	4	21
Friday	9	54
Saturday	4	45

Data analysis

Once the data has been collected, the environmental analyst conducts an in-depth study. It is at this stage that his or her professional judgement comes into play as the results must be properly interpreted.

The correlation between two statistical variables may be determined by mathematical tools, but the causal relationship is often more difficult to establish. For example, according to the graphs below, there seems to be a correlation between an increase in cellular telephone sales and a decline in bee populations in North America. While some might associate the cause of this decline with the use of cellphones, others would associate it with factors like global warming.

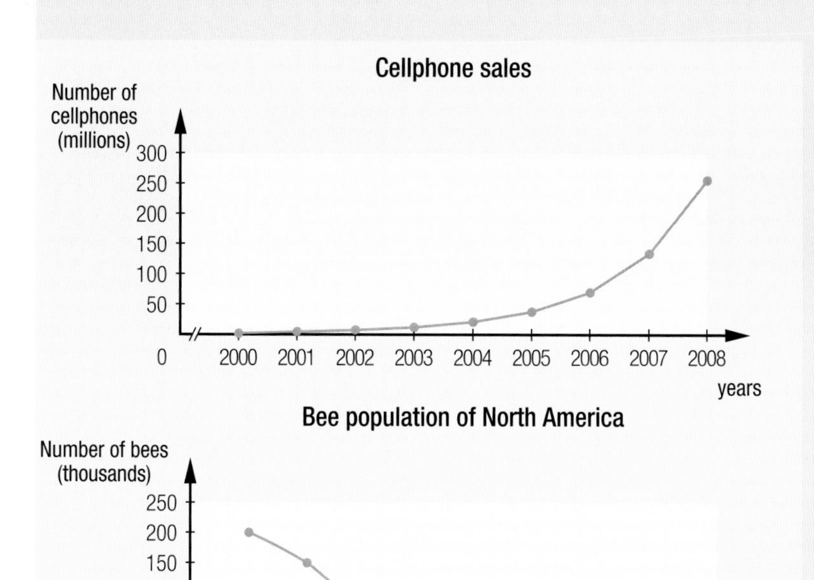

Analyst's report

An environmental analyst must present an analysis in the form of report or sometimes as authorization requests to different governmental organizations. The analyst then presents the reports to committees, commissions of inquiry or study groups.

1. According to the data on this region's air quality and the number of people admitted to emergency with respiratory problems, do the following:

a) Determine the equation of the regression line that corresponds to this situation.

b) Determine how many people with respiratory problems should be admitted to emergency if the air quality indicator is at 60.

2. Refer to the data about the number of cellphones sold and the population of bees in North America since 2000, and the following:

a) Describe the correlation between the number of cellphones sold and the population of bees in North America.

b) Determine if you can conclude that use of cellphones is causing the disappearance of bees or that the disappearance of bees is encouraging people to buy cellphones. Justify your answer.

overview

1 Match each linear correlation coefficient in the left-hand column with a description in the right-hand column.

Linear correlation coefficient		Description of scatter plot	
A	-0.9	1	It is as wide as long.
B	-0.5	2	It is almost square.
C	0	3	It is twice as long as it is wide.
D	0.25	4	It is long and narrow.

2 Place the scatter plots in ascending order according to the linear correlation coefficient that corresponds to each.

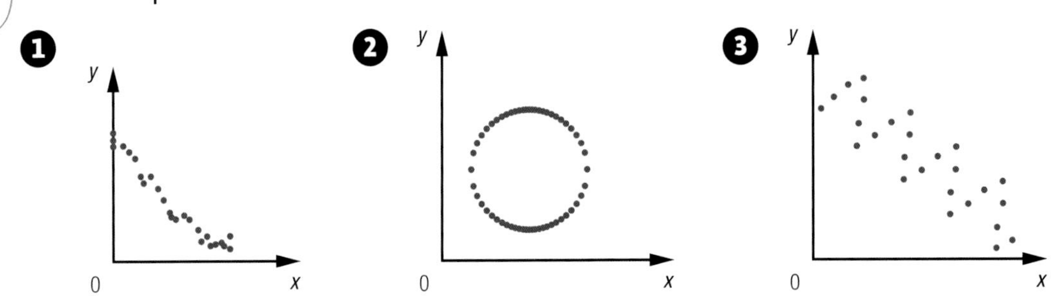

3 a) Which of the scatter plots below has the highest linear correlation coefficient?

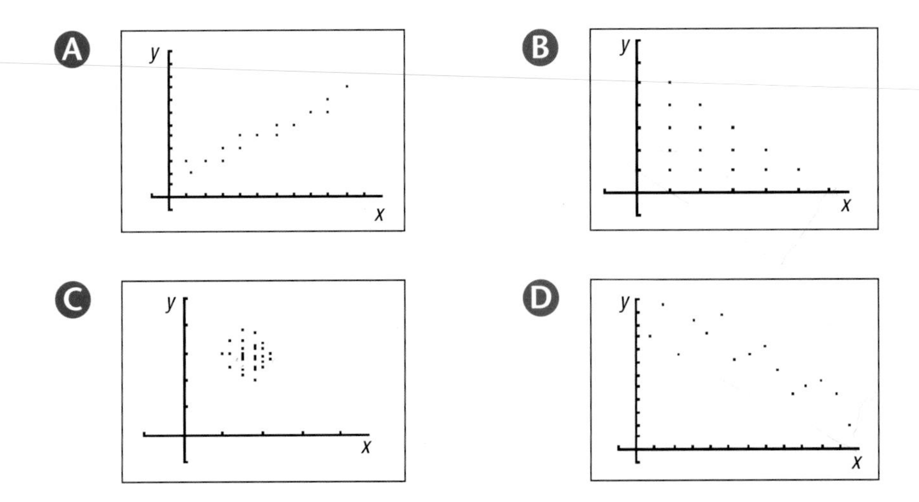

b) Graphically estimate the linear correlation coefficient for each of the scatter plots.

4 Which of the linear correlation coefficients below corresponds to a scatter plot with points that are the most closely aligned?

A -0.85　　B 0　　C 0.8

5 Of the two adjacent distributions, which has the lowest mean deviation?

Trees in a park

Height of wild cherry trees (cm)					Height of birch trees (cm)				
	9	8	3	**21**	0	0	1		
	4	2	2	**22**	0	7	7	8	9
			0	**23**	2	9			
8	6	5	4	**24**	9				
9	8	8	7	**25**					
			0	**26**	6	8			

6 The number of errors made on the same test was recorded for two groups of students. Indicate which group had the lowest mean deviation.

Group A

0, 0, 0, 1, 2, 2, 2, 2, 2, 3, 3, 3, 3, 4, 4, 4, 4, 4, 4, 4, 4, 5, 6, 7, 8, 8, 9, 10, 10, 11

Group B

Number of errors	Frequency	Number of errors	Frequency
2	5	7	0
3	8	8	2
4	6	9	2
5	2	10	1
6	4	11	1

7 Below is the number of absences for each student in a group for a semester:

0, 0, 0, 0, 0, 0, 0, 0, 0, 0, 1, 1, 2, 4, 5, 6, 7, 12, 13, 15, 16, 20, 25, 25, 26, 27, 28, 30, 34, 37, 37, 38, 40, 41

Represent this distribution with a stem-and-leaf plot.

8 The adjacent scatter plot shows the time spent by factory workers at work and on break in one day.

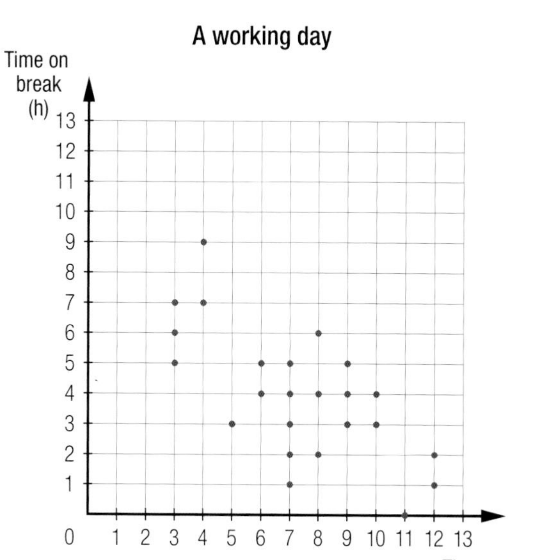

Which of the equations below best represents the scatter plot?

A $y = -0.5x + 8$ **B** $y = -0.3x + 11$

C $y = 0.5x + 8$ **D** $y = 0.3x + 11$

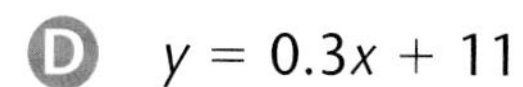

9 **HEIGHT** Below is the height (in metres) of 66 well-known tall people:

2.72	2.30	2.235	2.235	2.21	2.185	2.185	2.185	2.185	2.185	2.16
2.13	2.13	2.09	2.08	2.06	2.06	2.06	2.03	2.03	2.006	2.006
2.006	2.006	2.006	2.006	2	2	1.981	1.981	1.981	1.981	1.981
1.981	1.981	1.981	1.981	1.981	1.981	1.981	1.98	1.98	1.97	1.96
1.96	1.955	1.955	1.955	1.955	1.955	1.955	1.955	1.955	1.955	1.955
1.955	1.955	1.955	1.95	1.95	1.95	1.95	1.943	1.943	1.94	1.94

a) Magic Johnson, a former basketball player, is 2.06 m tall. What is his percentile?

b) Given that his percentile is 67, how tall is the singer Garou?

c) The wrestler Hulk Hogan is 2.03 m tall. The percentile of the basketball star Michael Jordan is 52. Who is taller?

His real name is Pierre Garand, but this Québec singer is known as Garou. His theatrical performance as Quasimodo in Notre-Dame-de-Paris was his springboard to stardom.

10 The life cycle (in h) of hibiscus flowers was recorded. Below is how long each lasted:

15 15 16 17 17 18 19 19 20 20 21 22 24 24 25 26 26 26 27

In this distribution:

a) what is the range?

b) what is the median?

c) what is the mean lifespan of the flowers?

d) which value has the percentile rank of 79?

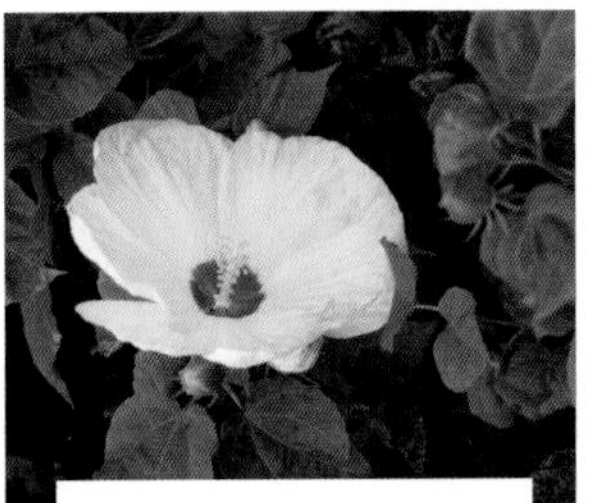

The hibiscus is a tropical plant, but some varieties can be cultivated in our region.

The fruit of the species *Hibiscus esculentus*, called gumbo or okra, is used in many Creole and African dishes.

11 This contingency table displays the number of brothers and sisters of 50 people. Is there a relationship between the two variables studied? Explain your answer.

Familles

Number of sisters \ Number of brothers	0	1	2	3	Total
0	4	8	5	4	21
1	3	8	1	2	14
2	1	4	0	0	5
3	4	2	4	0	10
Total	12	22	10	6	50

12 In a statistical distribution, what is the percentile of a value if 60% of the data are less than this value and 5% are equal to this value?

13 Wendy is absent for the term exam. Her teacher determines her grade using a regression line. What is this grade?

First term

Student	Term grade (on 80)	Exam mark (on 20)	Student	Term grade (on 80)	Exam mark (on 20)	Student	Term grade (on 80)	Exam mark (on 20)
Julie	67	16	George	75	17	Vincent	46	14
Katie	78	20	Bryan	73	16	Sandy	36	10
Eva	56	16	Greg	53	15	Melanie	36	9
Julianne	34	9	Jules	65	15	Wendy	70	
Elvira	54	12	William	73	18	Sebastien	71	14
Alondra	67	16	Fred	63	14	Ahmed	69	13
Charlotte	48	13	Marco	73	16	Samir	68	13
Susie	74	17	Steven	76	16	Amy	74	16
Chloe	63	14	Robert	47	14	Michelle	45	10

14 MARATHON On August 26, 2007, Pierre-Luc Goulet finished the Deux-Rives Marathon in 8th place among a total of 958 participants. On September 9, 2007, he finished 21st among 1190 participants during the Montréal Oasis Marathon. In which marathon did he obtain the higher percentile?

15 The distribution table below shows the price fluctuations for an item over several years.

Year	1995	1996	1997	1998	1999	2000	2001	2002	2003	2004	2005	2006	2007
Price ($)	2.50	3.00	5.00	4.50	5.00	3.00	3.50	5.00	4.00	6.00	5.50	6.50	5.50

In which year should the item cost $8.50?

16 In order to form "development" and "elite" swimming teams in a region, swimmers' times are clocked for the 400 m race. Below are the results (in minutes and seconds):

4:34	4:37	4:45	4:51	4:53	4:55	5:01	5:05	5:05	5:10	5:12	5:13	5:18	5:23	5:24
5:29	5:32	5:34	5:38	5:43	5:45	5:50	5:58	6:02	6:02	6:10	6:12	6:19	6:22	6:25

Swimmers whose results correspond to the 80th to 90th percentile are selected for the "development" team. Those whose results correspond to the 91st and higher percentile ranks are selected for the "elite" team. Identify the times for swimmers selected for:

a) the "development" team

b) the "elite" team

17 What is the percentile of a person who ranks:

a) 8th out of 25? b) 1st out of 20? c) 6th out of 9? d) 30th out of 30?

18 Using these three scatter plots, complete the table below.

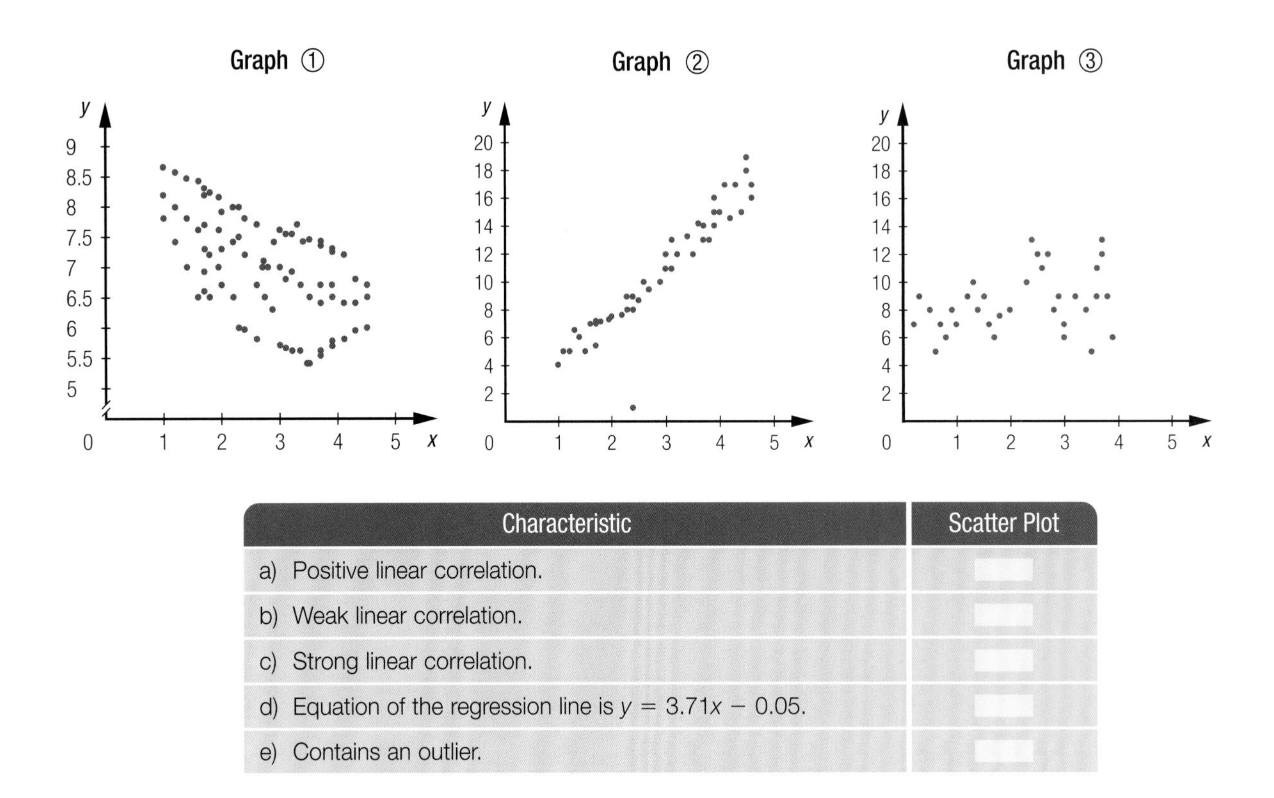

Characteristic	Scatter Plot
a) Positive linear correlation.	
b) Weak linear correlation.	
c) Strong linear correlation.	
d) Equation of the regression line is $y = 3.71x - 0.05$.	
e) Contains an outlier.	

19 **BASKETBALL** The following table indicates the number of minutes played by 10 Los Angeles Lakers players and the number of points they each scored in 2007.

Is there a relationship between the amount of time played and the number of points scored? Explain your answer using appropriate statistical calculations.

Los Angeles Lakers

Player	Time played (min)	Points
Kobe Bryant	1768	855
Derek Fisher	858	397
Andrew Bynum	894	396
Lamar Odom	981	380
Jordan Farmar	656	291
Vladimir Radmanovic	560	214
Luke Walton	688	212
Ronny Turiaf	471	175
Sasha Vujacic	336	164
Trevor Ariza	422	154

Kobe Bryant is considered by most experts as the most talented player the NBA has seen since Michael Jordan.

20 At a family reunion, Emily compares her age and height to that of her nine cousins. Below are the results:

The Smith family

Name	Age	Height (cm)	Name	Age	Height (cm)
Amy	17	167	Emily	6	124
Anna	15	160	Margarita	6	106
Bianca	14	171	Simona	9	129
Cynthia	12	163	Siu Ting	18	175
Diana	2	85	Valerie	16	164

Growth curve for girls aged 2 to 20

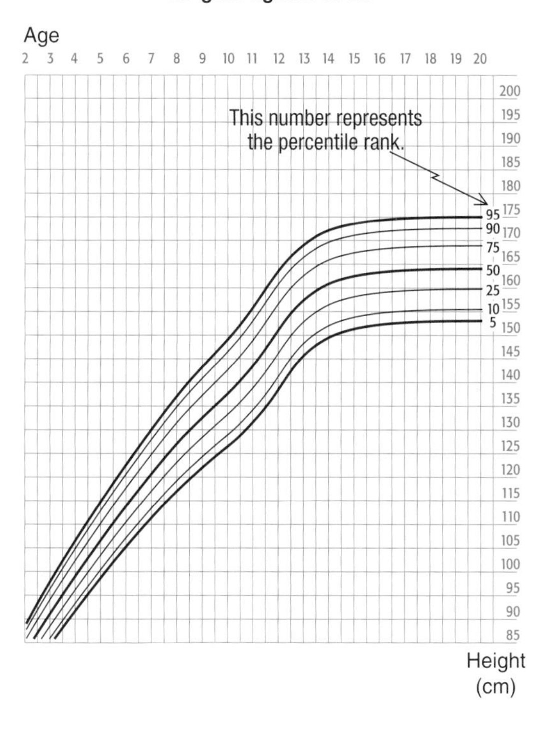

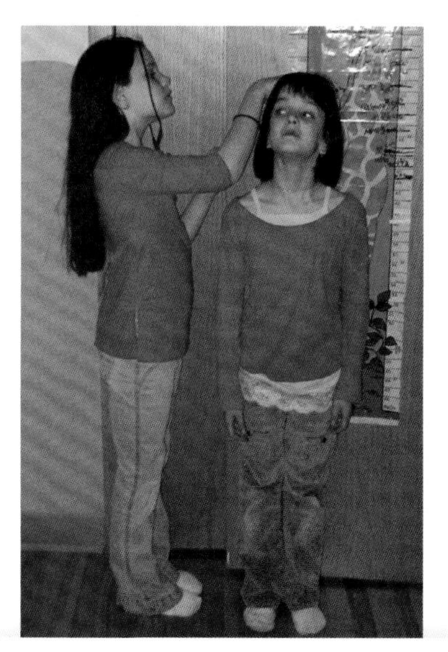

It used to be that a girl who began puberty before the age of 8 was considered precocious. The threshold has now been reduced to 6-7 years of age and appears to be quite common. It seems that better overall diet is one of the factors that triggers puberty.

a) What is the percentile of Emily's height in comparison to her cousins?

b) Is there a linear correlation between the cousins' age and height? Explain your answer.

c) According to the growth curve graph for girls aged 2 to 20, answer the following:
   1) What is Emily's percentile?
   2) How many cousins have the same percentile as Emily?
   3) How many cousins have a percentile lower than 50?

d) How tall should Emily be at 20?

21 Below is a graphical representation of the relationship between the annual income of families of four who live in the same city and the percentage of income set aside for their basic needs.

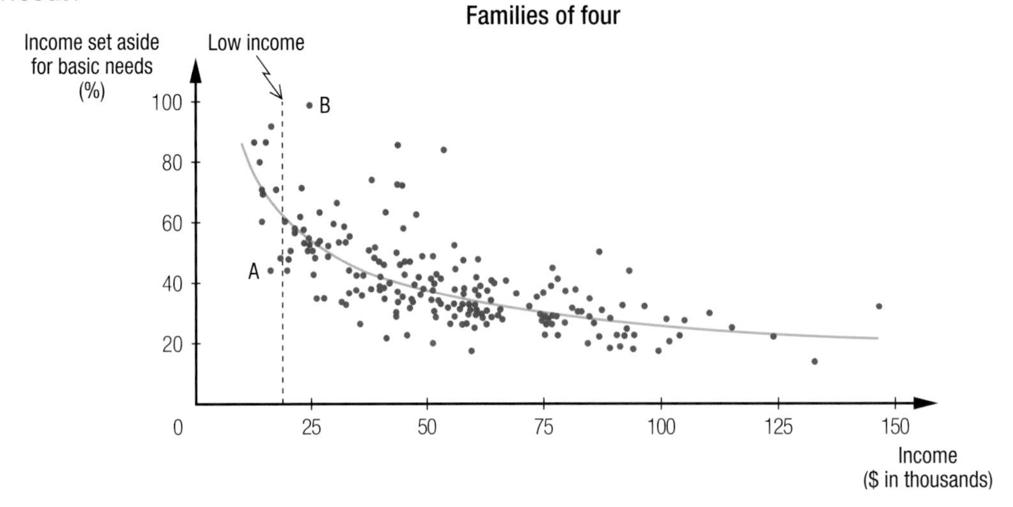

a) Would a regression line be more appropriate than the orange curve to study the relationship between the income set aside for basic needs and the annual income of these families? Explain your answer.

b) Can you state that the higher the annual income the lower the percentage of income set aside for basic needs? Explain your answer.

c) What do the following coordinates represent:

1) point A?

2) point B?

d) How many families are considered to have low income in this city?

Statistics Canada describes low income cutoff (LICO) as the level of income where a family tends to set aside a much larger portion for food, lodging and clothing than the average family. Thus, a family that sets aside 70% of its income for basic needs can be considered low income.

22 **ALOUETTES** A sports reporter predicts that to increase the number of yards run by all players on the Montréal Alouettes football team, lighter players should be recruited. What do you think about this statement?

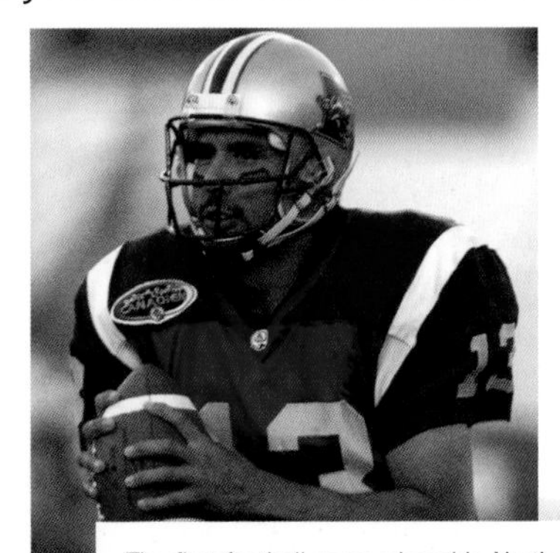

The first football game played in North America took place on a cricket field in downtown Montréal in 1868. Four years later, the Montréal Foot Ball Club was created, and in 1946 the Montréal Alouettes football team came into existence.

Montréal Alouettes

Player	Mass (kg)	Number of yards run in 2007
Brian Bratton	84.1	1184
Ben Cahoon	85.5	1127
Ashlan Davis	81.4	1401
Jarrett Payton	102.3	970
Elijah Thurmon	95.5	926
Kerry Watkins	85.9	1114

23 HEIGHT Below are statistics on the height of 20-year-old men in several countries:

Height of men aged 20 years

Denmark	
Year	Height (cm)
1880	167.7
1895	168.4
1913	169.5
1920	169.2
1930	169.9
1938	171
1950	173.9
1960	175.4
1970	177.1
1980	179.8
1980	179.3
1990	179.4

France	
Year	Height (cm)
1880	169
1890	169.6
1900	170.4
1910	171.1
1920	171.4
1930	172.8
1938	173.8
1956	176.9
1960	177.1

Norway	
Year	Height (cm)
1880	168.6
1890	169.4
1900	170.3
1910	171.7
1920	172.7
1930	173.6
1940	174.4
1950	175
1960	177
1970	178.3
1980	179.3
1990	179.4

Sweden	
Year	Height (cm)
1880	165.4
1890	165.4
1900	165.8
1910	166.4
1920	165.7
1930	167.4
1940	168.5
1950	168.3
1960	170
1974	172.6
1979	173.9

Note these statistics, and answer the following:

a) In 2030, which country will have the tallest 20-year-old men?

b) In 1800, which country had the shortest 20-year-old men?

c) In 1945, how tall were 20-year-old men in Sweden?

d) How tall will men in France be in 2040?

e) In what year did 20-year-old men in Denmark measure 165 cm?

f) What is the equation of the regression line for the scatter plot representing 20-year-old Norwegians?

g) What is the equation of the regression line for the scatter plot representing all 20-year-old men?

h) In what year will all 20-year-old men measure at least 2 m?

Scandinavia includes Norway, Sweden and Denmark. The people in these three countries are very similar and speak languages that have similar roots. These countries have a long, common history that is not limited to the Viking Era. They have often been united under one crown.

bank of problems

24 LIGHTBULBS In a factory there are two types of lightbulbs manufactured: incandescent and fluorescent. Below is the life expectancy (in h) of 20 lightbulbs of each type chosen at random:

Incandescent bulb: 800 810 810 830 840 860 865 870 900 900 910 920 920 925 930 940 950 1000 1000 1050

Fluorescent bulb: 2300 2300 2350 2355 2355 2360 2380 2390 2395 2400 2410 2410 2450 2450 2450 2460 2460 2460 2460 2460

Below is additional data on these two types of bulbs:

Incandescent bulb	Fluorescent bulb
• Price: $1.38/bulb.	• Price: $4.50/bulb.
• Can be used with a dimmer.	• Cannot be used with a dimmer.
• Is not recyclable.	• Is recyclable.
• Can be handled with bare hands.	• Cannot be handled with bare hands.
• Low import cost.	• High import cost.
• Produces a lot of heat.	• Produces little heat.
• Excellent colour rendering.	• Adequate colour rendering.

Using the information available and a statistical approach, indicate which type of bulb is the best choice.

25 The number of words that boys and girls of different ages could read was compared. Below are the results:

Number of words read (boys)

(6, 50) (6, 62) (6, 75) (7, 200) (7, 210) (7, 225) (7, 240) (7, 245) (8, 370) (8, 390) (8, 410) (8, 440) (9, 587) (9, 590) (9, 621) (9, 635) (10, 670) (10, 734) (10, 742) (10, 770) (10, 789) (11, 842) (11, 870) (11, 919) (11, 956) (11, 989) (12, 1050) (12, 1095) (12, 1105) (13, 1187) (13, 1230) (13, 1300) (14, 1450) (14, 1342) (14, 1500)

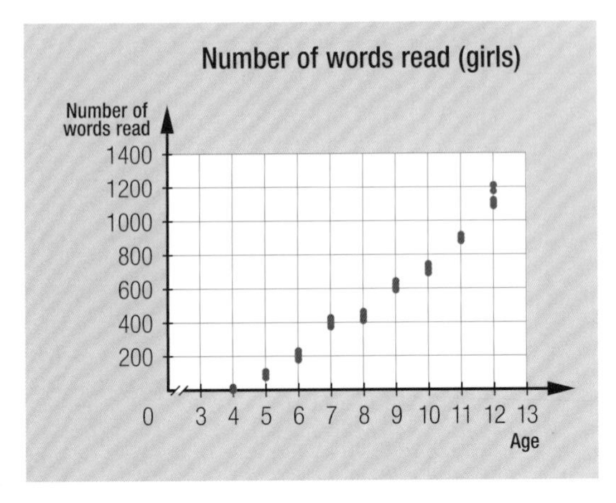

Each of these ordered pairs represents a boy's age and the number of words he can read.

According to these results, at what age will the boys and girls be able to read the same number of words?

26 The table below contains data on sandwiches available in a restaurant. Determine the energy value of a 100 g portion of a Regal sandwich, given that it contains 22 g of fat and 9 g of protein.

Fast food

Sandwich (100 g portion)	Fat content (g)	Protein content (g)	Energy value (kJ)
Chef's special	22	11	258
Fisherman's classic	22	11	293
Hunter's classic	22	10	272
Chicken wrap	22	12	283
Regular	12	12	250
Bacon and cheese	18	12	245
Grilled chicken	12	14	220
Hungry man's special	23	17	210
Grilled cheese	16	13	261

Kilocalories (kcal) and kilojoules (kJ) are units that represent the quantity of energy provided by carbohydrates, lipids and proteins found in food. Although the term kilocalorie is frequently used, the official international unit is the kilojoule: 1 kcal - 4.18 kJ.

27 The person in charge of a mountain excursion runs 30 candidates through a series of tests to determine which candidates will be part of the excursion. Only those who score a percentile of 75 or greater in all the tests will get the opportunity to participate. William is part of this group; his results are indicated below. Using a statistical approach, indicate whether he will be part of the excursion.

Number of chin-ups in 1 min: 1, 1, 1, 2, 3, 4, 5, 6, 7, 12, 12, 14, 15, 16, 17, 17, 17, 17, 18, 19, 20, 22, 22, 22, 23, 24, 25, 26*
* 2 runners dropped out of the race

Race: 4th out of 26*
* 4 candidates dropped out of the race

Number of push-ups in 1 min

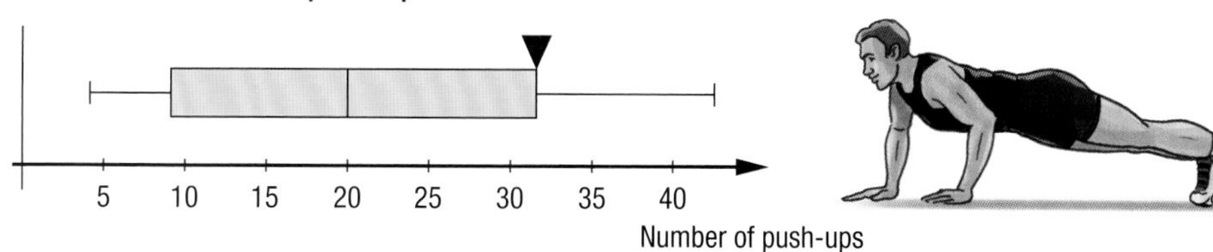

Number of sit-ups in 1 min*

1	5	6								
2	6	6	7							
3	0	2	4	4	8	9				
4	1	1	1	2	2	2	3	3	8	9
5	0	0	2	6	7					
6	1	4	7							

* 1 candidate dropped out

Vision 3

From congruent to similar figures

How do you know if two triangles are congruent? What method can be used to verify if the slope of a roof corresponds to its construction plan? How can you calculate the height of the Eiffel Tower? How can you determine the width of a ravine without crossing it? In "Vision 3" you will discover the geometric statements that can be used to determine if triangles are congruent or similar. You will use statements to find various missing measurements in polygons. You will also study the metric relations in right triangles and explore Thale's theorem.

Arithmetic and algebra	Geometry	Statistics	Probability
	• Congruent triangles • Similar triangles • Metric relations in right triangles • Finding missing measurenrts		

REVISION 3

PRIOR LEARNING 1 A mathematical mosaic

A mosaic is a decorative art form which uses fragments of rock, tile, glass and ceramic that are assembled to create designs or figures.

In a fine arts class, students constructed a mosaic:

- ACF is an isosceles triangle.
- Quadrilateral BDFG is a parallelogram.

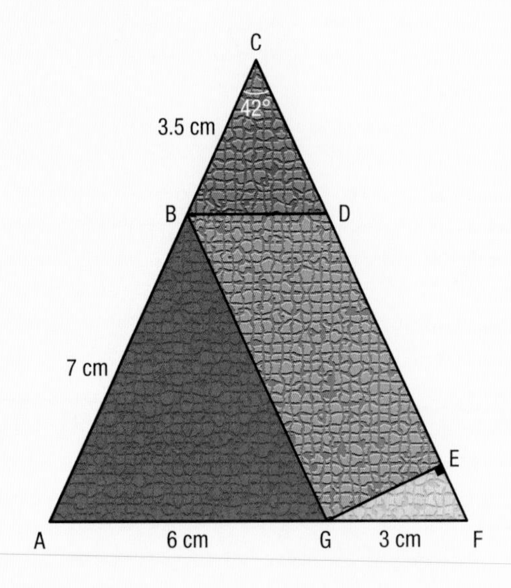

a. What type of triangle is ABG?

b. What is the measure of angle F?

c. What is the area:

1) of triangle ACF?
2) of triangle BCD?
3) of triangle ABG?
4) of parallelogram BDFG?
5) of triangle EFG?

This mosaic decorated the floor of a Roman-style villa in Corinth (approximately 200 BCE). The god Dionysus is depicted at its centre.

knowledge summary

CLASSIFICATION OF TRIANGLES

	Sides	
Illustration	Characteristics	Name
	No congruent sides	Scalene triangle
	Two congruent sides	Isosceles triangle
	Three congruent sides	Equilateral triangle

	Angles	
Illustration	Characteristics	Name
	One obtuse angle	Obtuse triangle
	Three acute angles	Acute triangle
	One right angle	Right triangle
	Three congruent angles	Equiangular triangle

PROPERTIES OF QUADRILATERALS

		Properties			
Illustration	Name	Sides	Angles	Diagonals	Axis of symmetry
	Irregular trapezoid	One pair of parallel sides			
	Isosceles trapezoid	One pair of parallel sides and one pair of congruent sides	Two pairs of congruent angles		
	Right trapezoid	One pair of parallel sides	Two right angles		
	Parallelogram	Two pairs of opposite sides parallel and congruent	Congruent opposite angles Supplementary adjacent angles		
	Rectangle	Two pairs of opposite sides parallel and congruent	Four right angles Supplementary adjacent angles		

(*continued on next page*)

Illustration	Name	Properties			
		Sides	Angles	Diagonals	Axis of symmetry
	Rhombus	Two pairs of parallel opposite sides Four congruent sides	Congruent opposite angles Supplementary adjacent angles		
	Square	Two pairs of parallel opposite sides Four congruent sides	Four right angles Supplementary adjacent angles		

REGULAR POLYGONS

A polygon is **regular** if all sides have the same length and all interior angles are congruent.

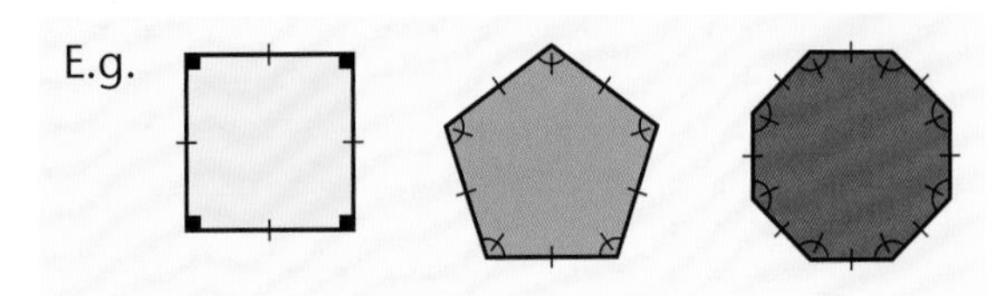

AREA: TRIANGLE, QUADRILATERAL, REGULAR POLYGON AND CIRCLE

Shape	Area
Triangle (b, h)	$A_{triangle} = \frac{b \times h}{2}$
Trapezoid (b, B, h)	$A_{trapezoid} = \frac{(B + b) \times h}{2}$
Parallelogram (b, h)	$A_{parallelogram} = b \times h$
Rhombus (D, d)	$A_{rhombus} = \frac{D \times d}{2}$

Shape	Area
Rectangle (b, h)	$A_{rectangle} = b \times h$
Square (s)	$A_{square} = s^2$
Apothem	$A_{regular\ polygon} = \frac{\text{perimeter} \times \text{apothem}}{2}$
Circle (r)	$A_{circle} = \pi r^2$

PYTHAGOREAN THEOREM

In a right triangle, note the following:

- The **hypotenuse** is the side opposite the right angle. It is the longest of the three sides.
- The **legs** define the right angle.
- The square of the length of the hypotenuse is equal to the sum of the squares of the length of the two legs.

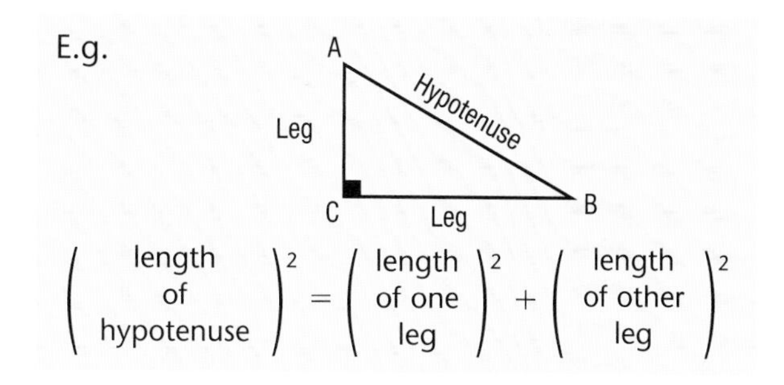

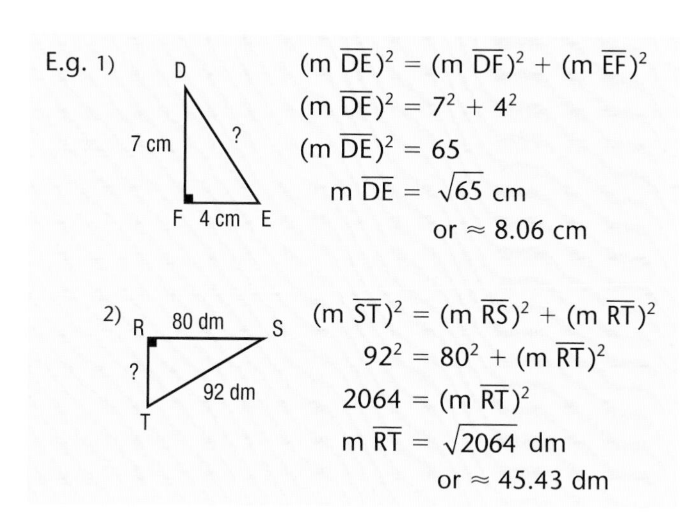

ANGLES FORMED BY A TRANSVERSAL INTERSECTING TWO PARALLEL LINES

When two parallel lines are intersected by a transversal, note the following:

- The alternate interior angles are congruent:
  $\angle 4 \cong \angle 6$ and $\angle 3 \cong \angle 5$
- The alternate exterior angles are congruent:
  $\angle 1 \cong \angle 7$ and $\angle 2 \cong \angle 8$
- The corresponding angles are congruent:
  $\angle 1 \cong \angle 5$ and $\angle 2 \cong \angle 6$
  $\angle 4 \cong 8$ and $\angle 3 \cong \angle 7$

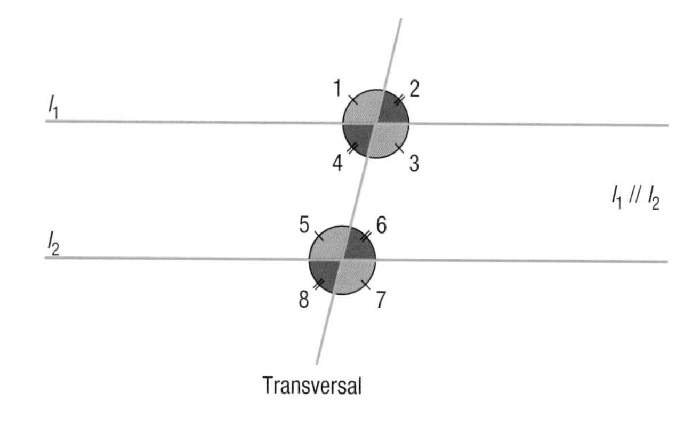

In summary $\angle 1 \cong \angle 3 \cong \angle 5 \cong \angle 7$ and $\angle 2 \cong \angle 4 \cong \angle 6 \cong \angle 8$.

knowledge in action

1 Name the following triangles:

1) according to their angles
2) according to their sides

a) 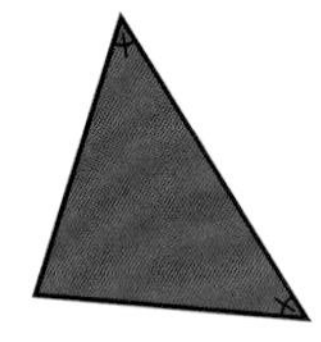

b) 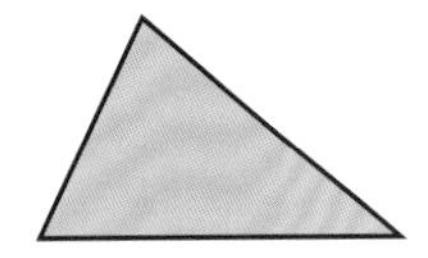

c) 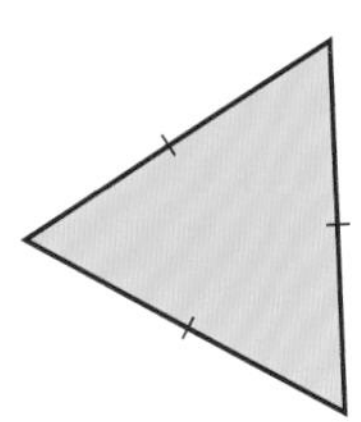

d) 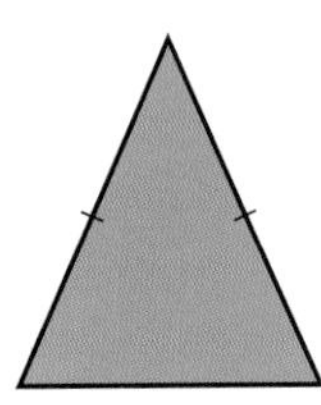

e) 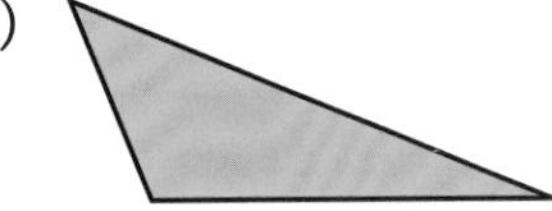

f) 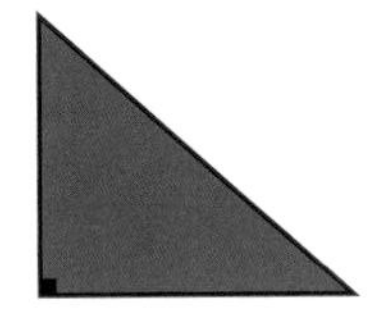

2 The lengths of three sides of a triangle are provided below. Which ones allow you to construct a right triangle?

A 3, 4, 5 **B** 13, 14, 15 **C** 7, 24, 25 **D** 4.5, 6, 7.5

E 11, 24, 26 **F** 8, 15, 17 **G** 50, 50, 75 **H** 64, 225, 289

3 Calculate the area of the following shapes.

a) Triangle

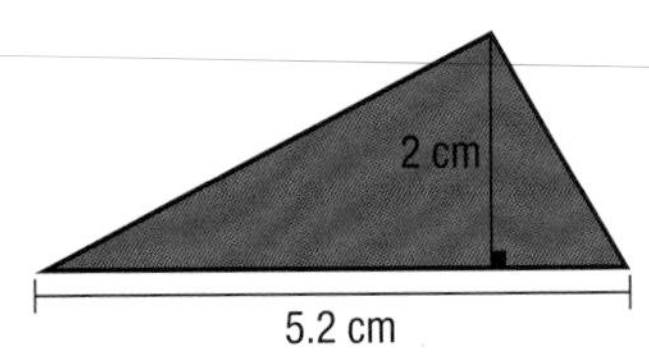

b) Rhombus

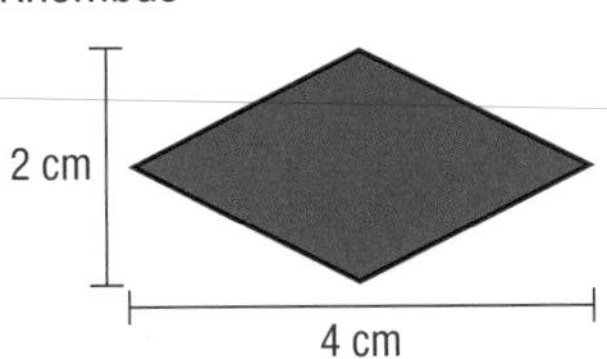

c) Parallelogram

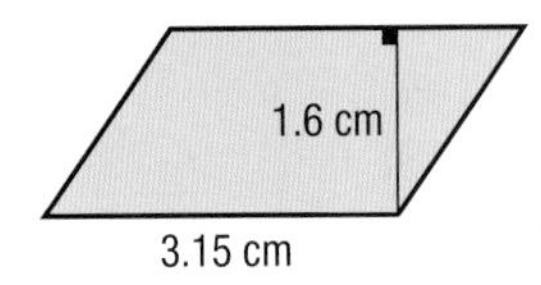

d) Trapezoid

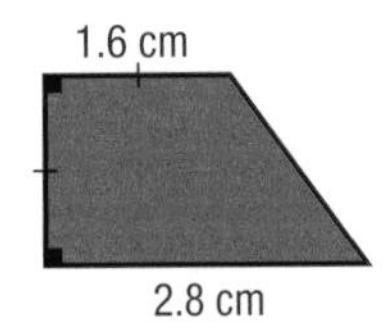

e) Circle

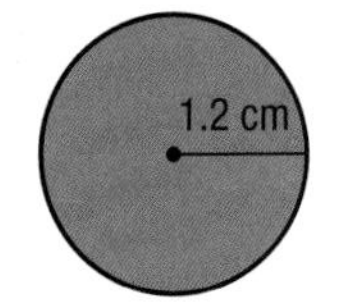

f) Regular hexagon

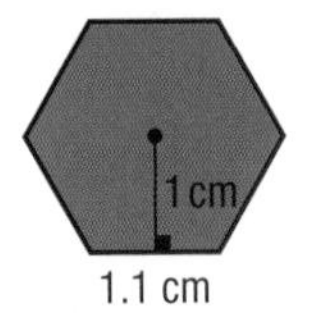

4 For each of the shapes below determine the area of the coloured sections.

a)

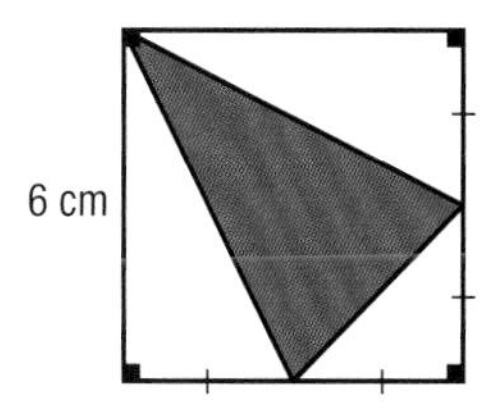

b)

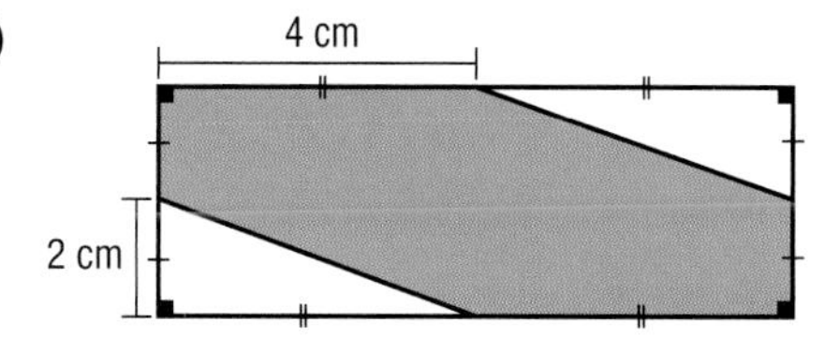

c)

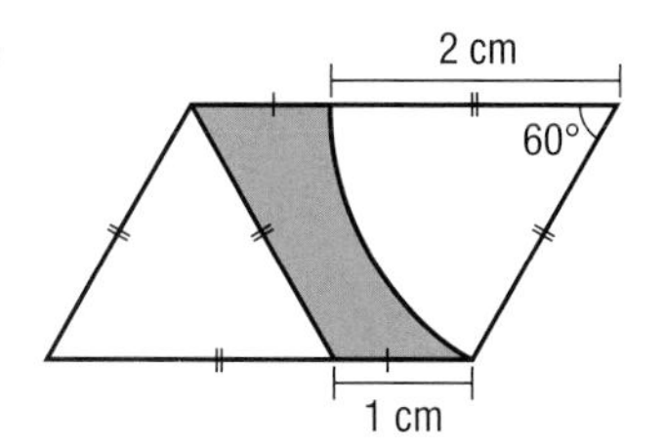

d)

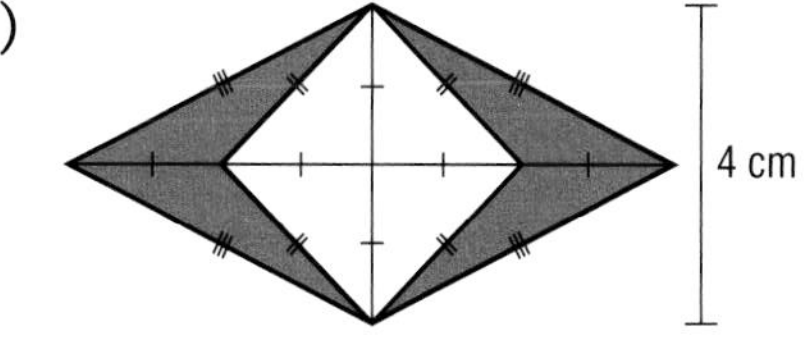

5 For each of the following shapes:

1) determine the value of x
2) provide the geometric statement that supports your calculations

a) **Square**

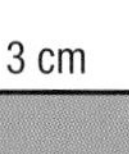

b) **Parallelogram**

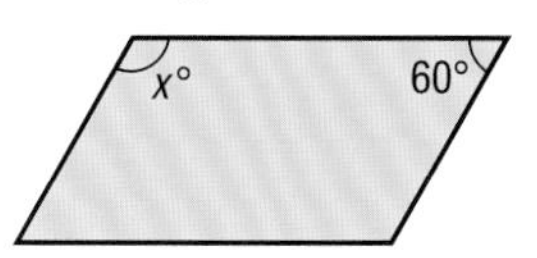

c) **Rhombus**

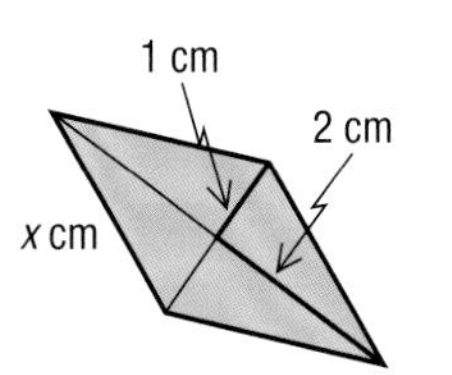

d) **Trapezoid**

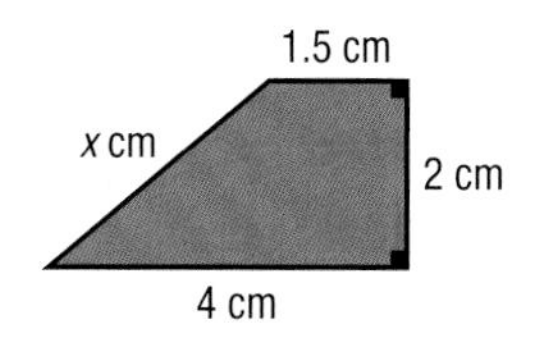

6 A stuntman throws a 45 m long cable from the roof of a 20 m tall building to another stuntman on the ground. The cable is held tightly in place. How far is the stuntman on the ground from the building?

7 A right triangle ABC is inscribed in a circle with centre O and a radius of 12 cm. The hypotenuse of the triangle is equal to the diameter of the circle. If the length of side AB is equal to the radius of the circle, what is the area of triangle AOC?

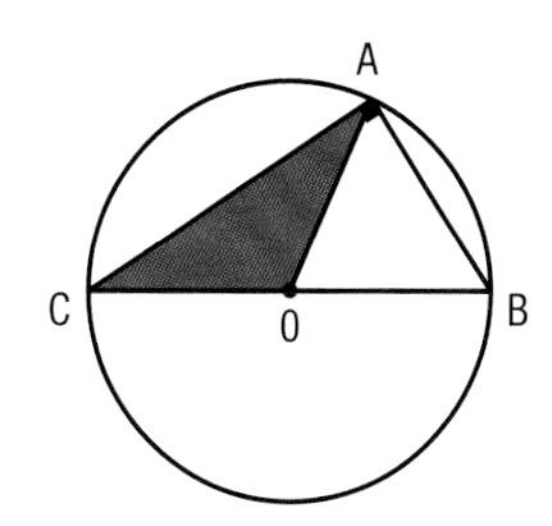

8 a) Determine the area of parallelogram KMOQ inscribed in the rectangle below.

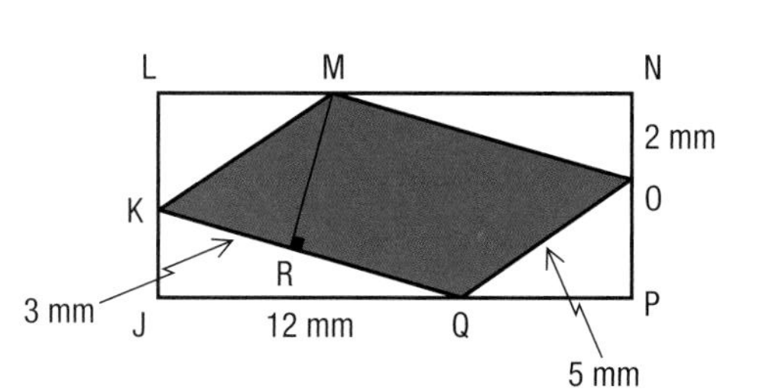

b) Determine the perimeter of parallelogram BDFH inscribed in the rectangle below.

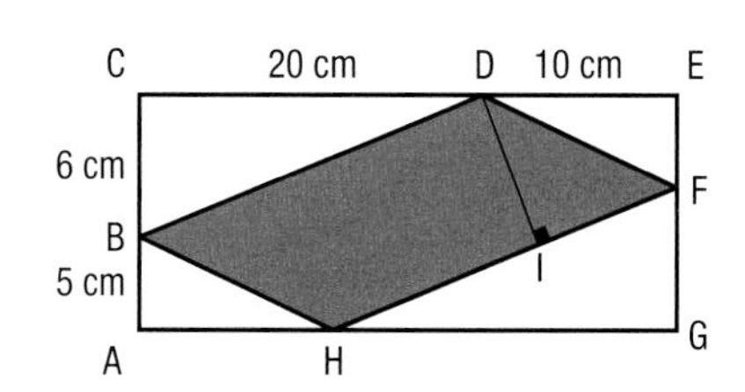

9 GYMNASTICS The adjacent figure represents the high bar in men's gymnastics. If each of the 400 cm long cables holding the bar are set at 290.47 cm from the base of the post, at what height from the floor is the high bar?

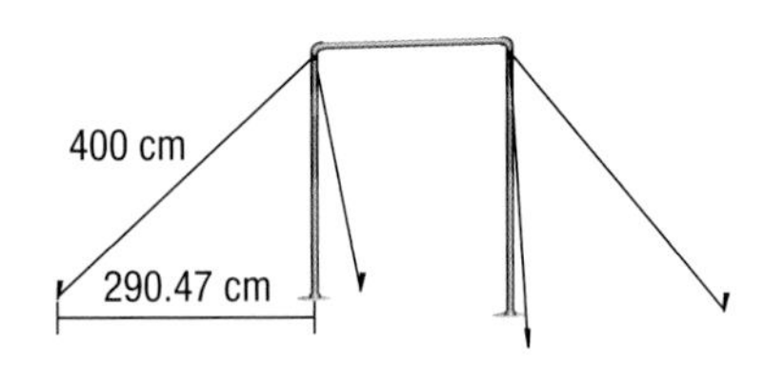

The men's gymnastic competition program consists of six events: floor exercise, pommel horse, still rings, vault, parallel bars and high bar.

10 For each of the pairs of polygons below, identify a property that describes Polygon **2** but not Polygon **1**.

	Polygon 1	Polygon 2
a)	Trapezoid	Parallelogram
b)	Parallelogram	Isosceles trapezoid
c)	Rectangle	Rhombus
d)	Rhombus	Square

11 When Khurram flew his kite, its string formed a 60° angle with the horizon. If the string was 40 m long and was held 1.5 m from the ground, how high was the kite?

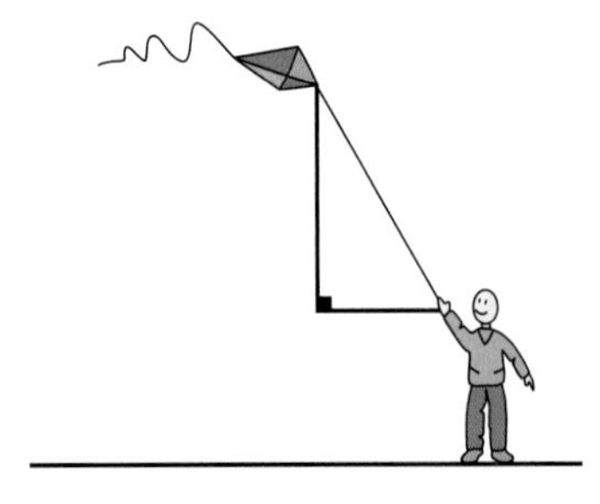

12 Calculate the area of each of the regular polygons below.

a)

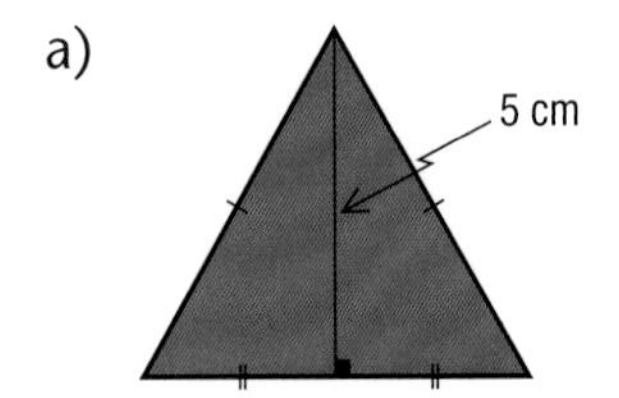

b)

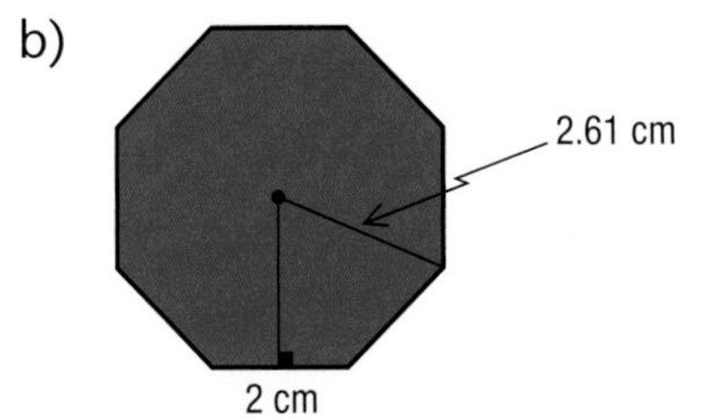

c)

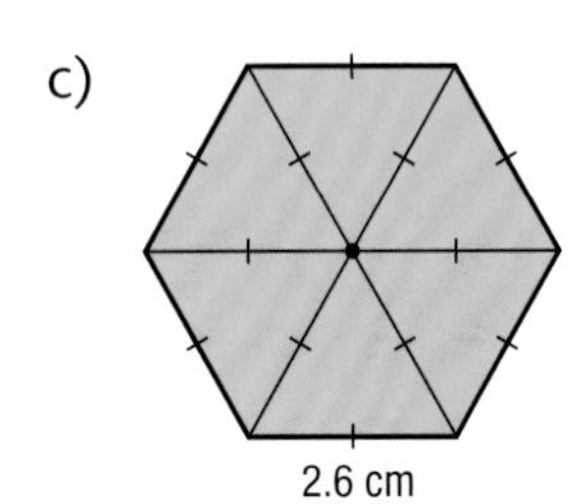

13 In the figure below, lines AE and FI are parallel.

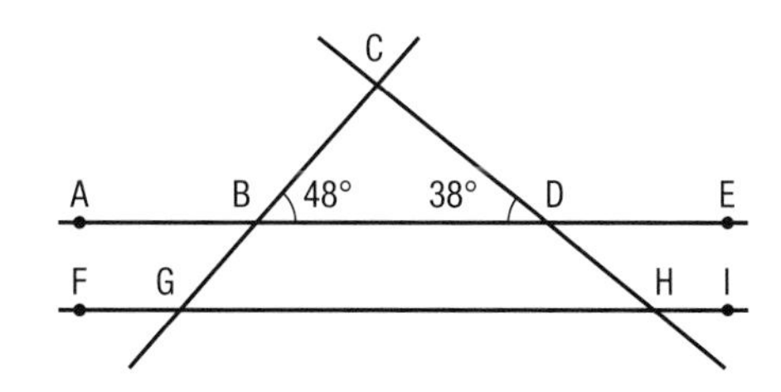

What geometric statement is needed to justify each of the following?

a) m∠BCD = 94°

b) m∠GHD = 38°

c) m∠ABG = 48°

d) m∠BGF = 132°

14 In each case:

1) determine the measure of angles 1 to 6
2) identify the geometric statement that supports your calculation

a) **Figure 1**

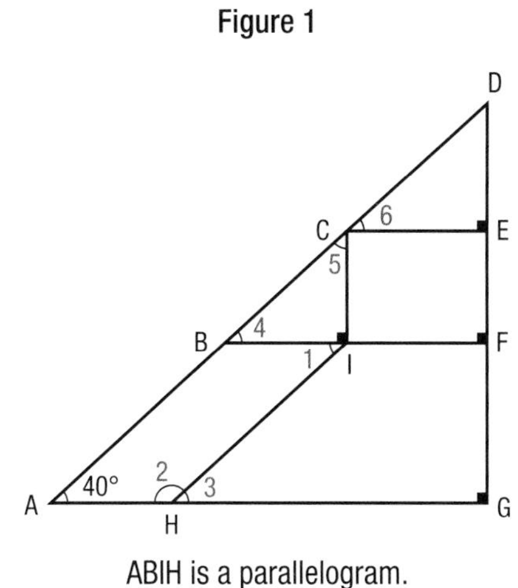

ABIH is a parallelogram.

b) **Figure 2**

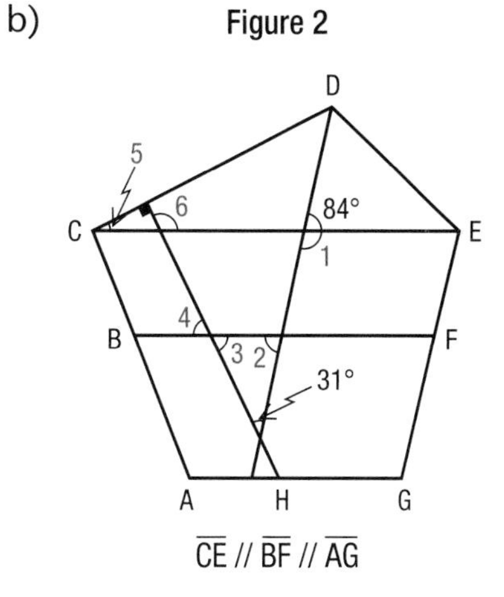

$\overline{CE} \parallel \overline{BF} \parallel \overline{AG}$

15 A factory producing highway signs needs to manufacture as many directional arrows as possible with the dimensions given below. What is the maximum number of arrows that can be cut from one rectangular piece of metal measuring 5 m by 850 cm?

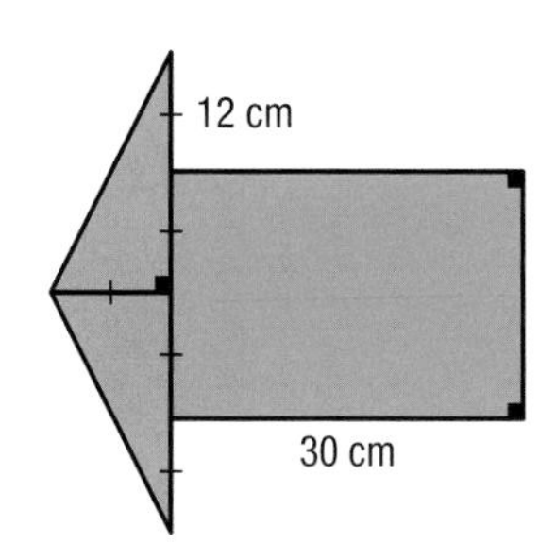

SECTION 3.1 Congruent triangles

This section is related to LES 5.

PROBLEM Billiards

Pool or billiards is a game played on a table surrounded by bands. The players sometimes use the bands to execute their shots. The success of these manoeuvres requires the knowledge and application of certain mathematical notions.

During a practice game, a white ball and a red ball were arranged on a billiard table in the following way.

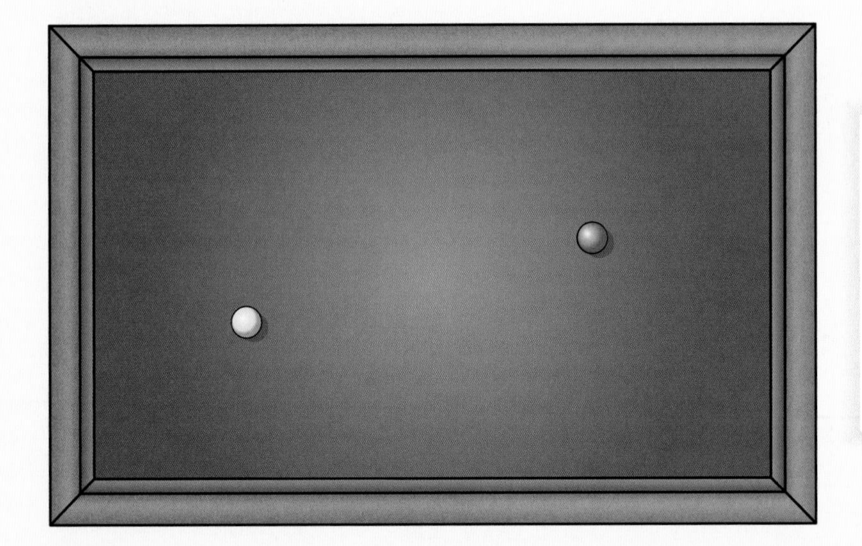

There are many variations of the game of pool or billiards: American, English, French, Italian, Russian, snooker, etc. Depending on the type of billiards, the table may or may not have pockets, and it may be played with a different number of coloured balls. French billiards, also called *billard carambole*, is played on a table without holes with one red ball and two white balls.

Describe a trajectory that the white ball must travel in order to touch two bands before touching the red ball.

Albert Einstein said, "Billiards represents the supreme art of anticipation. It is not merely a game but a complete artistic sport that requires, in addition to good physical condition, the logical reasoning of a chess player and the touch of a concert pianist."

ACTIVITY 1 It's all about triangles!

Whether it is to provide structural support or to create a special effect, artists and architects sometimes choose to reproduce the same shapes.

To create the work of art below, the artist reproduced triangle ABC many times by moving it in a systematic way.

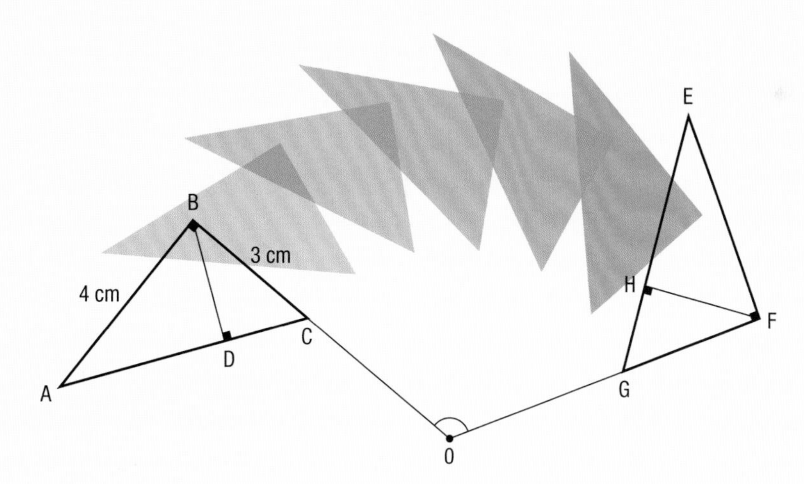

a. What geometric transformation is used in the above artwork?

b. In triangle EFG, which side is congruent to:

1) AB? 2) BC? 3) AC?

c. Calculate the length of side:

1) EF 2) FG 3) EG

d. What can you say about:

1) the altitudes BD and FH? 2) the perimeter of each triangle? 3) the area of each triangle?

ACTIVITY 2 The chimney

A chimney creates a draft that helps the combustion of a fire while allowing the safe release of smoke.

Two people are standing near a chimney. When the first person stands at point A, he or she can see the top of the chimney at an angle of elevation of 40°. When the person stands at point B, he or she can see the top of the chimney at an angle of elevation of 52°.

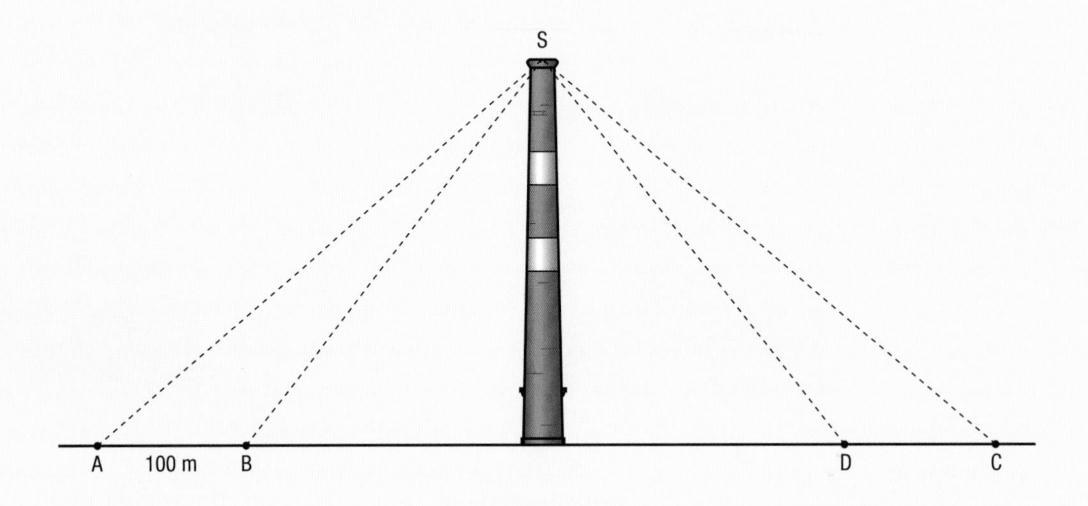

a. What are the measures of the angles in triangle ABS?

b. The second person standing at point C sees the top of the chimney with an angle of elevation of 40°. When the person stands at point D, he or she can see the top of the chimney at an angle of elevation of 52°.

1) What is the length of segment CD?

2) What are the measures of the angles in triangle CDS?

c. What conclusions can be made about triangles ABS and CDS?

d. How many different triangles can be constructed if you know the length of one side of a triangle and the measures of two angles situated at each end of this side?

A chimney can also be found in homes with a heat-radiating fireplace. Traditionally a hearth was the social centre of a home as a fire was used to heat and cook. During the depths of winter people would gather in front of the fire to exchange news, tell stories...

Techno math

Dynamic geometry software allows you to compare geometric figures. By using the tools, TRIANGLE, DISTANCE and ANGLE MEASUREMENT, you can construct triangles and verify if they are congruent.

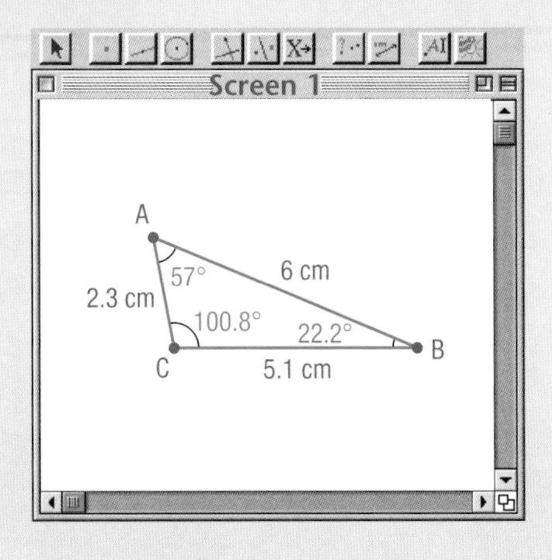

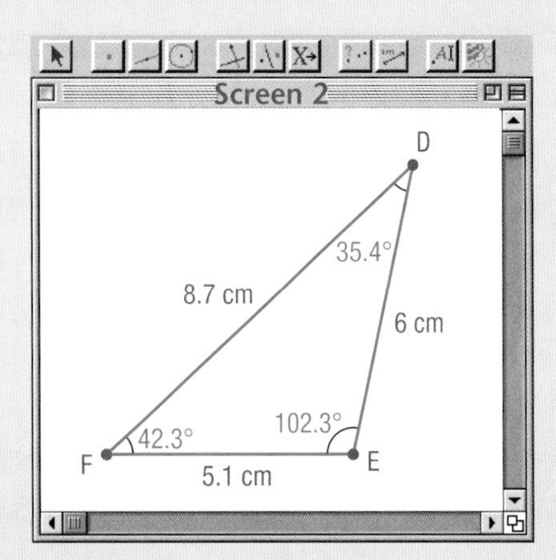

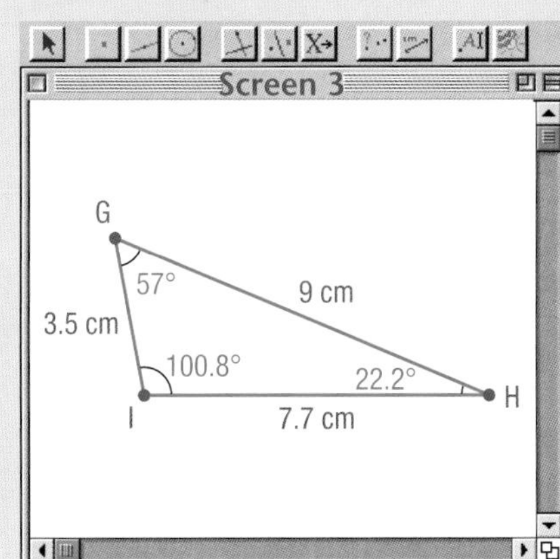

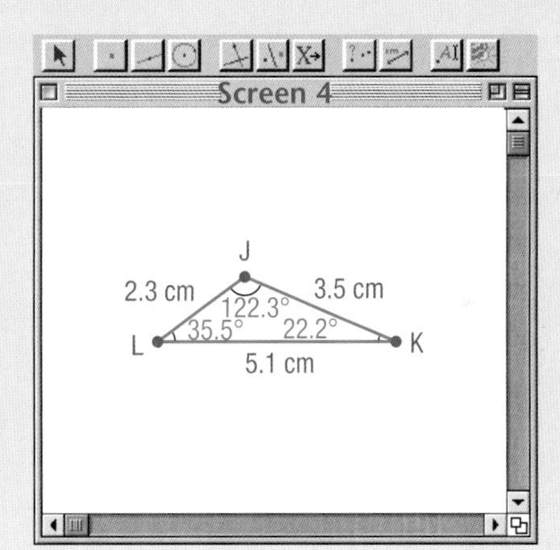

a. By comparing the measurements of angles and sides, what do the triangles in the following screens have in common:

1) Screens **1** and **2**? 2) Screens **1** and **3**? 3) Screens **1** and **4**?

b. Among the triangles on Screens **1, 2, 3** and **4,** is there a pair of congruent triangles?

c. Can you conclude that two triangles are congruent if:

1) two sides of one triangle are congruent to two sides of the other triangle?

2) three angles of one triangle are congruent to three angles of the other triangle?

d. Two sides and an angle of triangle MNO are congruent to two sides and an angle of triangle PQR. With the help of dynamic geometry software, do the following:

1) Construct triangles MNO and PQR.

2) Explain under what conditions triangles MNO and PQR are congruent.

MINIMUM CONDITIONS FOR CONGRUENT TRIANGLES

Congruent triangles are triangles whose **corresponding angles** and **corresponding sides** are congruent.

The geometric statements below describe the minimum conditions necessary to state that two triangles are congruent.

1. If the corresponding sides of two triangles are congruent, then the triangles are congruent (SSS).

The abbreviation SSS (Side-Side-Side) is used to simplify the written form of this statement.

E.g. $\overline{AC} \cong \overline{DE}$
$\overline{AB} \cong \overline{DF}$
$\overline{BC} \cong \overline{EF}$

Thus, $\Delta ABC \cong \Delta DEF$

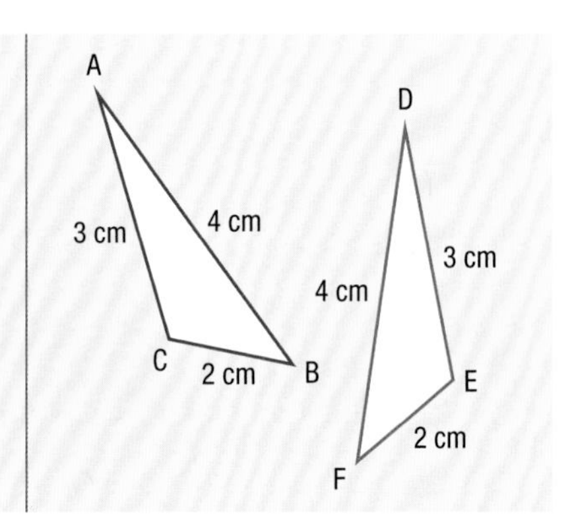

2. If two angles and the contained side of one triangle are congruent to the corresponding two angles and contained side of another triangle, then the triangles are congruent (ASA).

The abbreviation ASA (Angle-Side-Angle) is used to simplify the written form of this statement.

E.g. $\angle B \cong \angle E$
$\overline{BC} \cong \overline{EF}$
$\angle C \cong \angle F$

Thus, $\Delta ABC \cong \Delta DEF$.

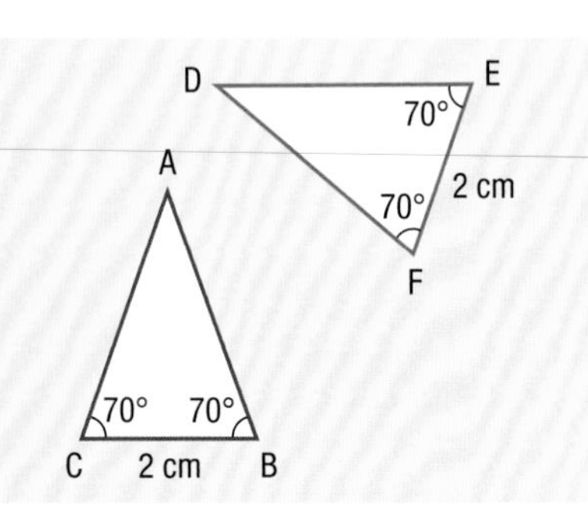

3. If two sides and the contained angle of one triangle are congruent to the corresponding two sides and contained angle of another triangle, then the triangles are congruent (SAS).

The abbreviation SAS (Side-Angle-Side) is used to simplify the written form of this statement.

E.g. $\overline{AC} \cong \overline{DF}$
$\angle C \cong \angle F$
$\overline{BC} \cong \overline{EF}$

Thus, $\Delta ABC \cong \Delta DEF$.

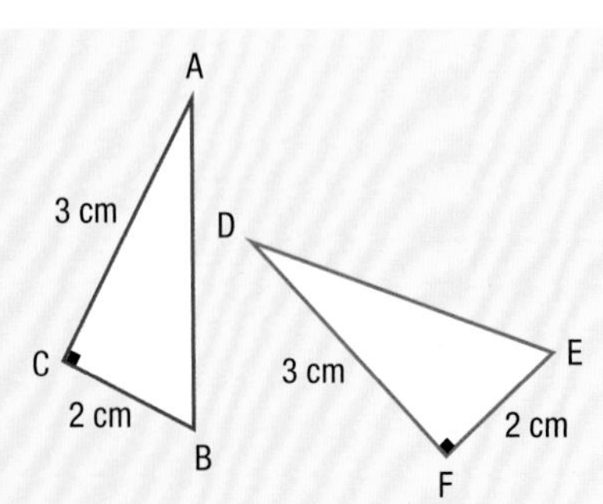

DEDUCTIVE REASONING

In geometry, there are various types of statements that allow you to organize a line of deductive reasoning.

Conjecture

A **conjecture** is a statement that has either been proved or refuted.

Theorem

A **theorem** is a conjecture that has been proved.

Counter-example

A **counter-example** is an example that disproves a conjecture.

Proof

A **proof** is a logical line of reasoning that allows you to make statements based on previously established or accepted properties.

E.g. The following is a method used to prove that $\angle BCD \cong \angle CBE$ as shown in the adjacent figure.

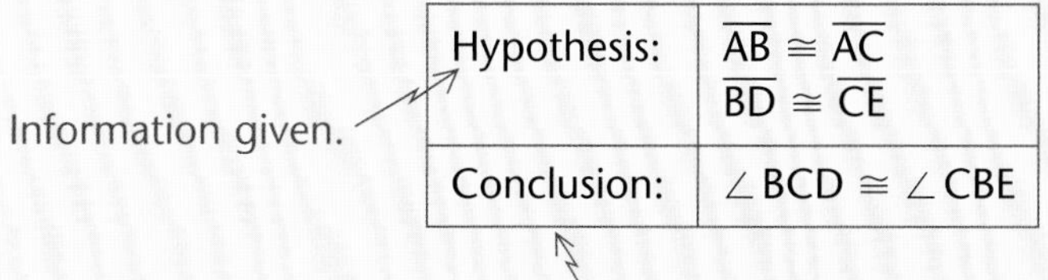

Hypothesis:	$\overline{AB} \cong \overline{AC}$ $\overline{BD} \cong \overline{CE}$
Conclusion:	$\angle BCD \cong \angle CBE$

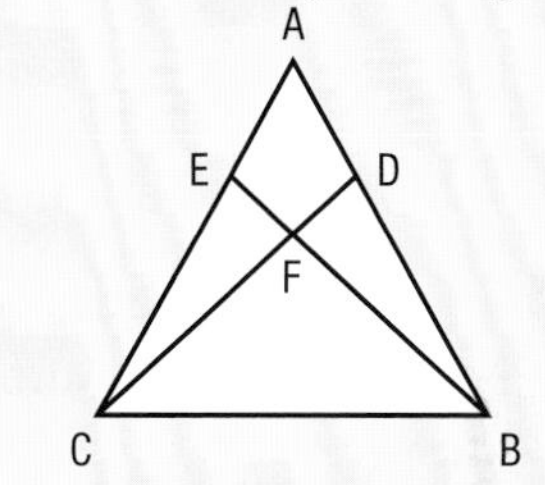

The basis of the proof.

The explanation that supports the statement provided.

STATEMENT	JUSTIFICATION
Triangle ABC is isosceles.	By hypothesis ($\overline{AB} \cong \overline{AC}$).
$\angle ACB \cong \angle ABC$	Angles opposite the congruent sides of an isosceles triangle are congruent.
$\overline{BD} \cong \overline{CE}$	By hypothesis ($\overline{BD} \cong \overline{CE}$).
$\overline{BC} \cong \overline{BC}$	Common side.
$\Delta BCE \cong \Delta BCD$	Two triangles that have one congruent angle contained by corresponding congruent sides are congruent (SAS).
$\angle BCD \cong \angle CBE$	In congruent triangles, corresponding angles are congruent.

practice 3.1

1 In each case, identify the geometric statement that states that the blue and yellow triangles are congruent.

a)

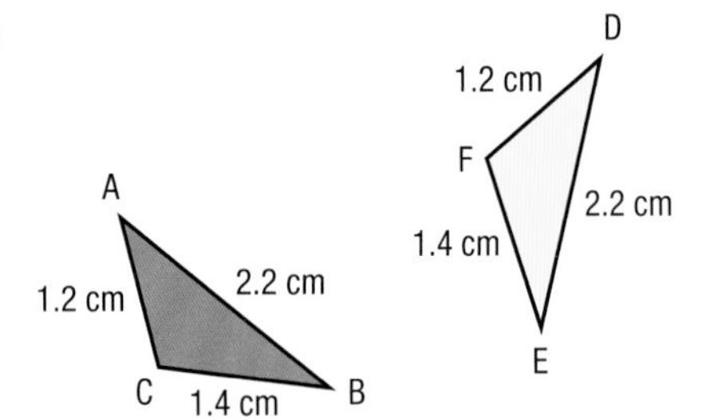

b)

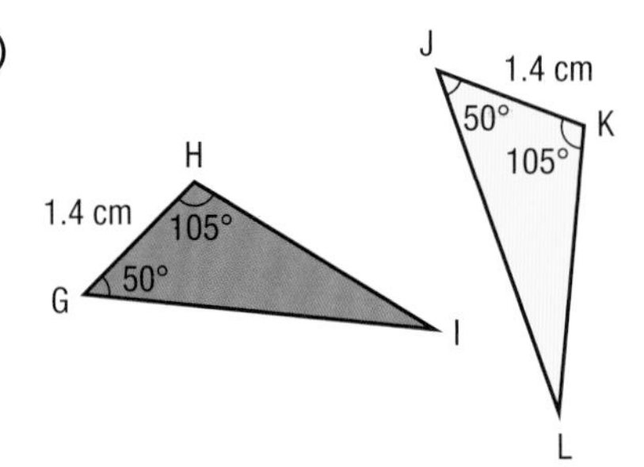

c)

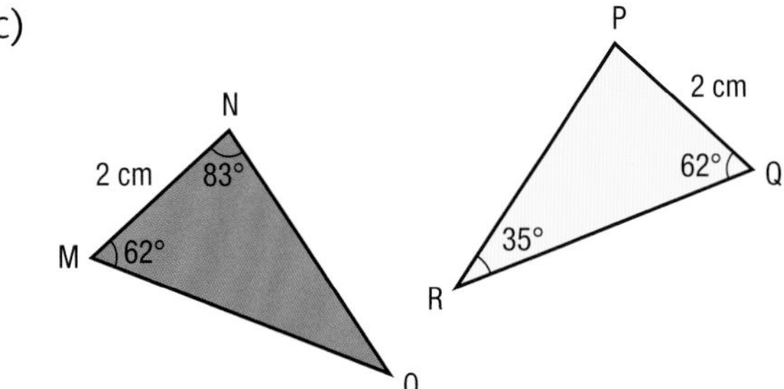

d)

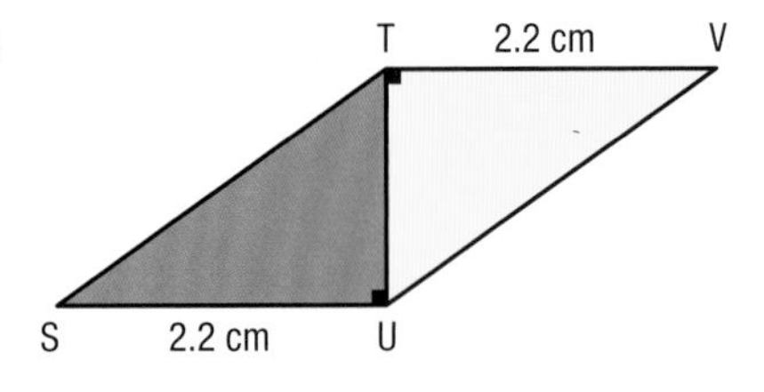

2 In each case below, what concludes that the blue and green triangles are congruent?

a)

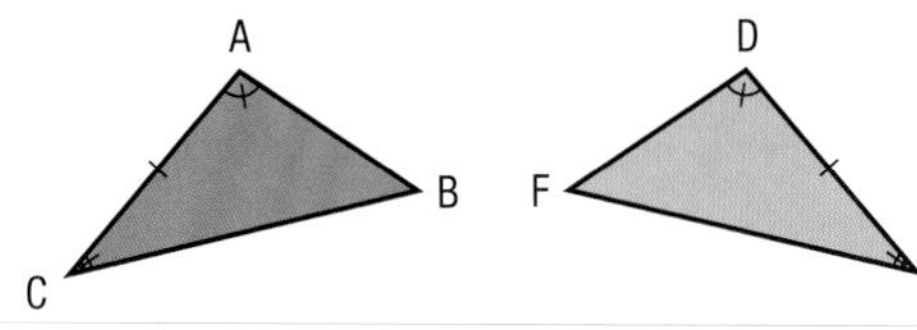

b)

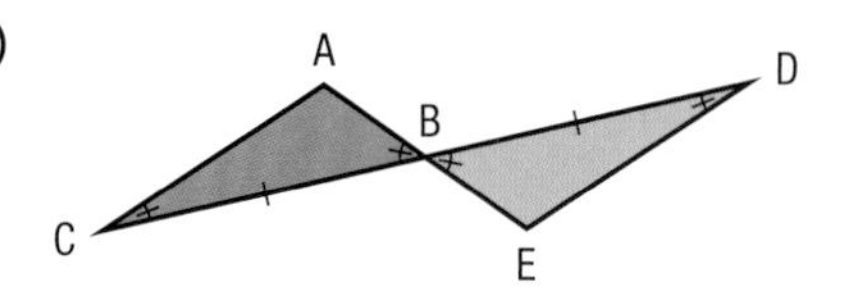

c)

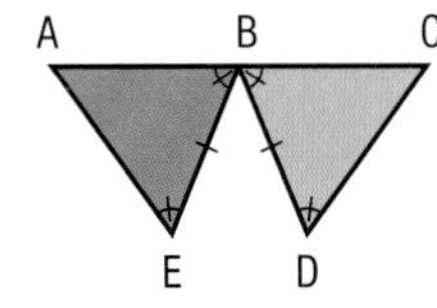

d)

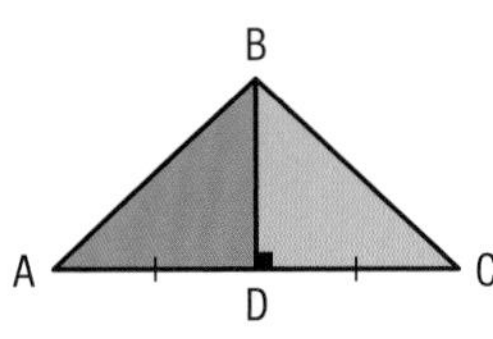

e)

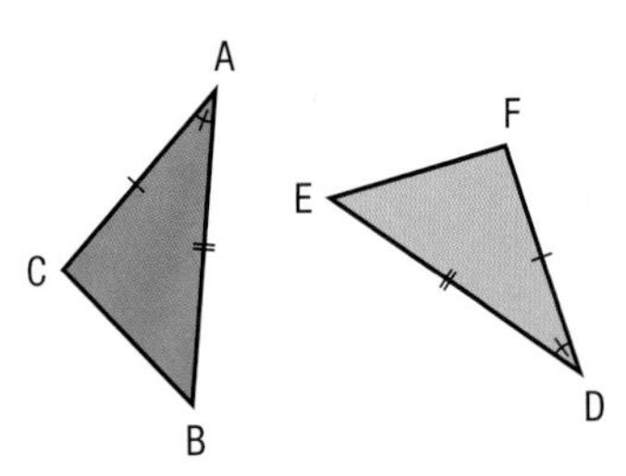

f)

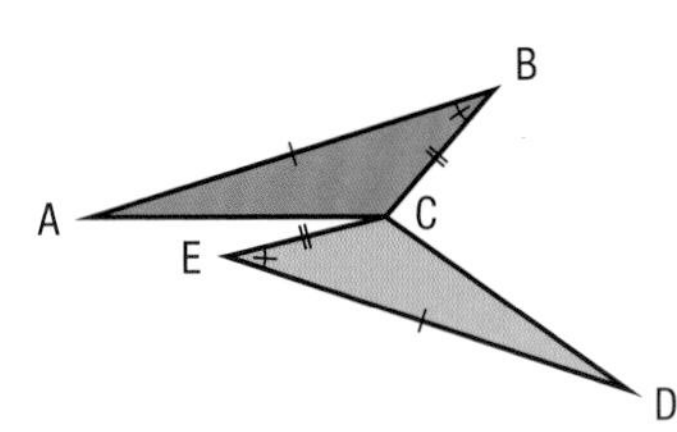

g)

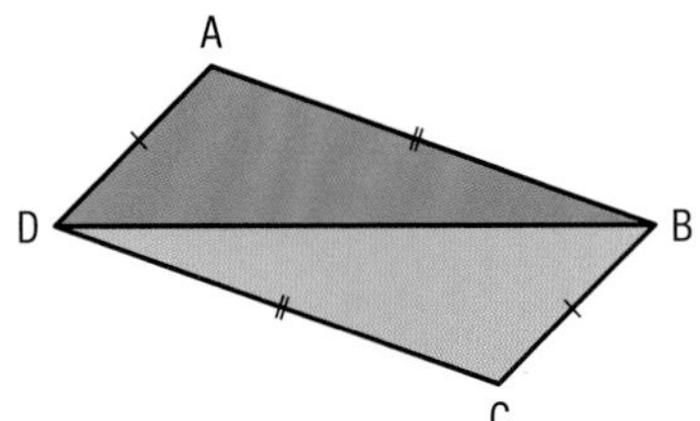

h)

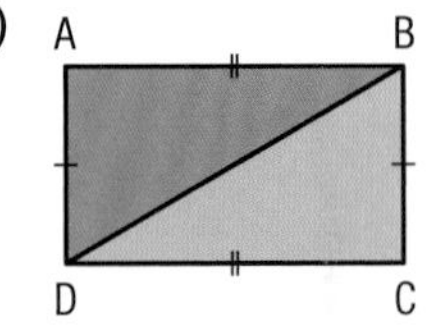

3 For each of the following conjectures identify the hypothesis and the conclusion.

a. If the three sides of one triangle are congruent then the triangle is equilateral.

b. If the diagonals of a quadrilateral are congruent and intersect at their midpoints, then the quadrilateral is a square.

c. If two angles share a vertex and have a common side, then they are adjacent angles.

d. If a polygon is a square, then all its interior angles are right angles.

e. If a polygon is a parallelogram, then its diagonals intersect at their midpoint.

4 Indicate if the following conjectures are true or false. If false, provide a counter-example.

a. If the diagonals of a quadrilateral are congruent, then it is a rectangle.

b. If an isosceles triangle has an angle measuring 40°, then it must be an obtuse triangle.

c. Opposite angles in a quadrilateral are supplementary.

d. The diagonals and sides of a rhombus form four congruent triangles.

e. Two parallel lines intersected by a transversal result in congruent corresponding angles.

f. In a circle, an inscribed triangle is equilateral.

g. Every scalene triangle has an obtuse angle.

5 For each of the following conjectures, do the following:

1) Determine the hypothesis and the conclusion.
2) Verify if the conjecture is true or false. If false, provide a counter-example.

a) If you double the lengths of the two legs of a right triangle, then you also double the length of the hypotenuse.

b) If the altitude of a right triangle is double its base, then one of the acute angles is twice the size of the other acute angle.

c) In a right triangle, the altitude drawn from the vertex of the right angle separates the triangle into two congruent triangles.

6 From the triangles below, identify pairs of congruent triangles. Justify your answers.

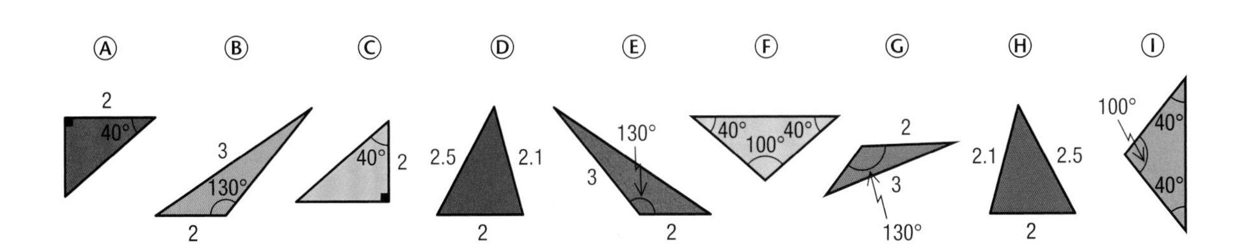

7 Certain triangles found on a roof structure are congruent. How can you prove that they are congruent without measuring the angles of these triangles?

8 In each case, indicate the number of triangles ABC that can be constructed using the given measurements:

a) m $\overline{AB}$ = 10 km, m $\overline{BC}$ = 13 km, m $\overline{AC}$ = 16 km

b) m $\overline{AB}$ = 8 km, m∠A = 65°, m∠B = 25°

c) m∠A = 60°, m∠B = 45°, m∠C = 75°

9 In each case, indicate if the given triangles are congruent. If so, justify your answer with a geometric statement.

	Triangle	Characteristic
a)	ΔABC and ΔDEF	∠A ≅ ∠F, $\overline{AC}$ ≅ $\overline{FD}$, $\overline{BC}$ ≅ $\overline{DE}$
b)	ΔABC and ΔBDE	
c)	ΔGFH and ΔFHI	

10 The adjacent quadrilateral ABCD is a parallelogram.

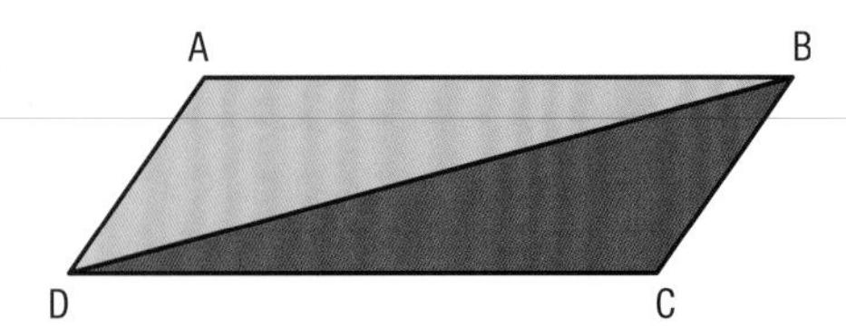

a) What geometric statement allows you to justify that:

1) $\overline{AD}$ ≅ $\overline{BC}$?

2) ∠ADB ≅ ∠DBC?

3) ΔABD ≅ ΔBCD?

b) What geometric transformation associates triangle ABD to triangle BCD?

11 Among the pairs of triangles below, identify which pairs are congruent.

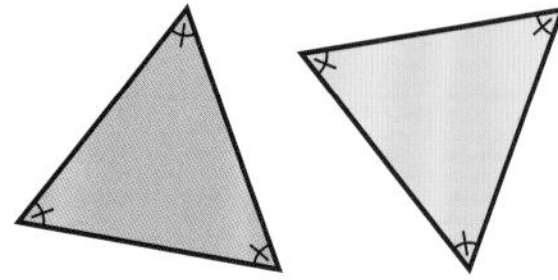

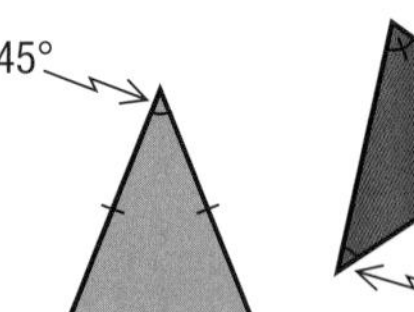

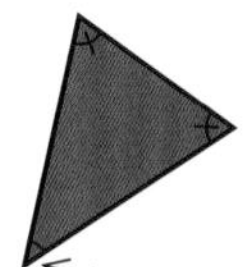

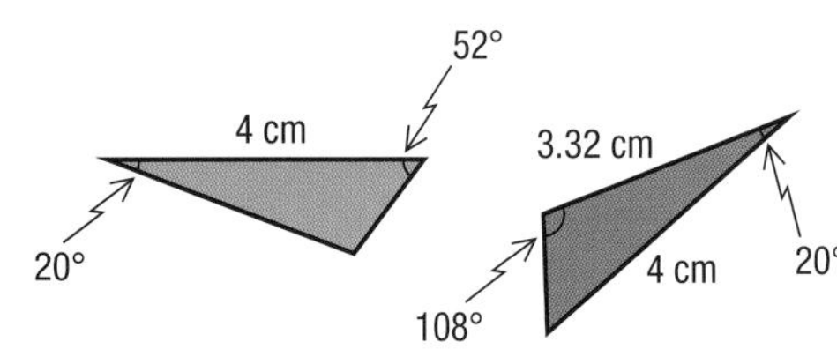

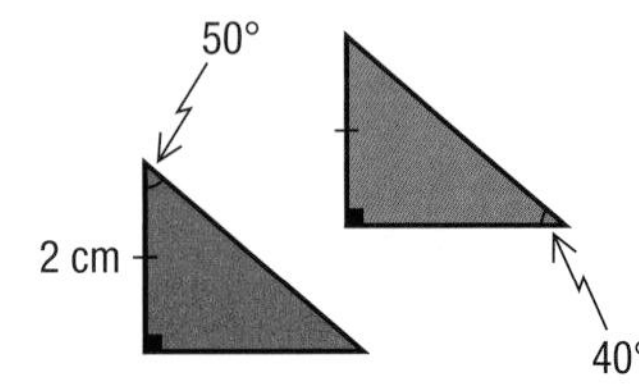

 **12** The adjacent diagram represents the position of a security guard who is viewing a door with the help of a flashlight. Where should a second security guard stand so that the triangle formed by his light beam is congruent to that of the first guard?

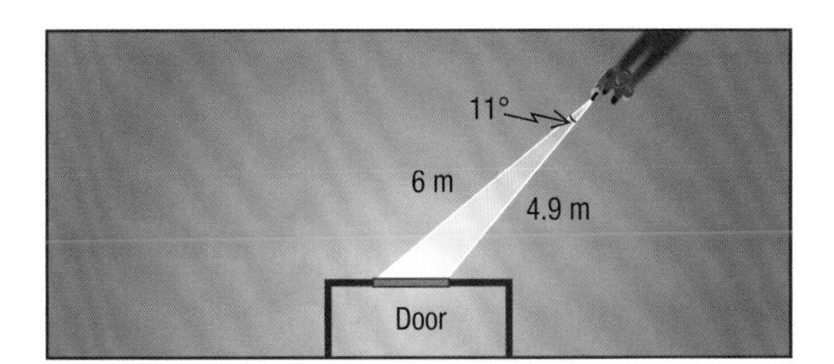

 **13** In the adjacent diagram, the lines $\overline{MP}$ and $\overline{NQ}$ are parallel. The midpoint of $\overline{MN}$ is O.

Complete the following proof to show that point O is also the midpoint of $\overline{PQ}$.

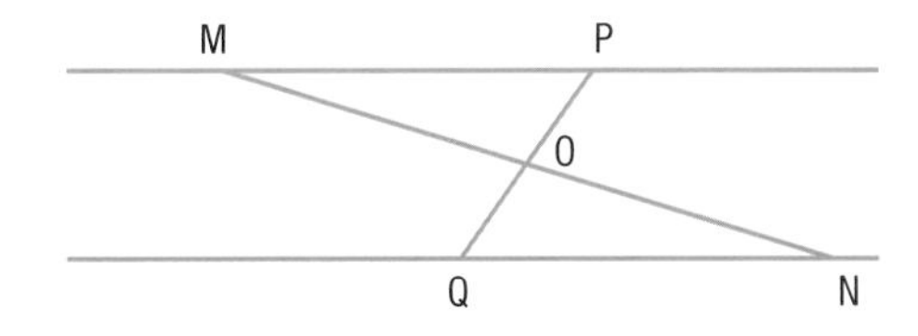

Hypotheses:	1) ▒ 2) ▒
Conclusion:	O is the midpoint of $\overline{PQ}$.

STATEMENT	JUSTIFICATION
$\angle POM \cong \angle QON$	3) ▒
$\overline{MO} \cong \overline{NO}$	4) ▒
$\angle PMO \cong \angle QNO$	5) ▒
$\Delta MOP \cong \Delta NOQ$	6) ▒
$\overline{OP} \cong \overline{OQ}$	In congruent triangles, corresponding sides are congruent.
O is the midpoint of PQ.	7) ▒

 **14** In the diagram below, $\angle ABC \cong \angle ACB$ and $\overline{DA} \cong \overline{AE}$. Complete the following proof to conclude that $\overline{CD} \cong \overline{BE}$.

Hypotheses:	1) ▒ 2) ▒
Conclusion:	$\overline{CD} \cong \overline{BE}$

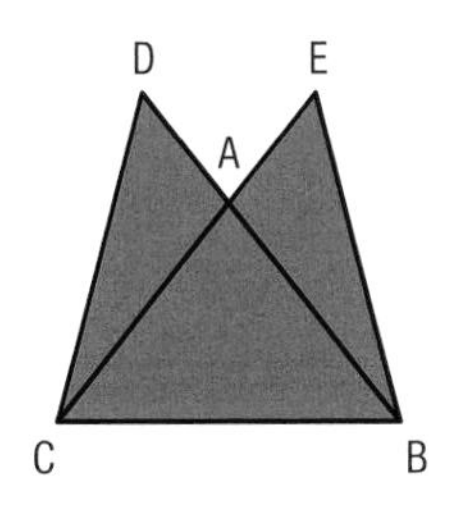

STATEMENT	JUSTIFICATION
ΔABC is an isosceles triangle.	3) ▒
$\overline{AC} \cong \overline{AB}$	4) ▒
$\angle DAC \cong \angle EAB$	5) ▒
$\overline{AD} \cong \overline{AE}$	6) ▒
$\Delta DAC \cong \Delta EAB$	7) ▒
$\overline{CD} \cong \overline{BE}$	8) ▒

15 In a forest there are two cross-country ski trails. Considering that point C is the midpoint of both segments AE and BD, show that segments AB and DE are parallel.

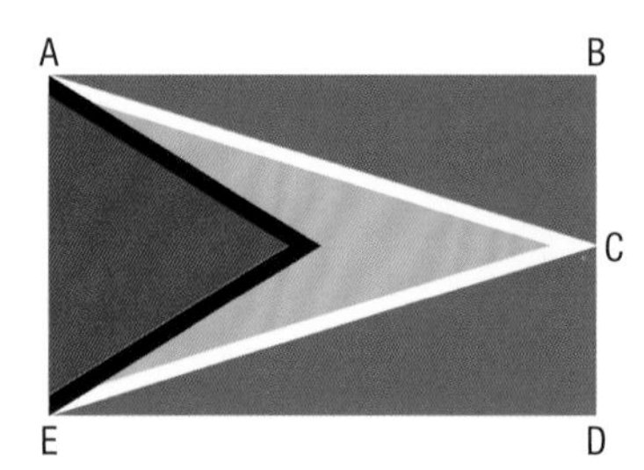

16 GUYANA The flag below is that of Guyana, the only member of the Commonwealth in South America. Considering that point C is the midpoint of BD, and that the flag is rectangular, show that $\Delta ABC \cong \Delta EDC$.

A B C E D

17 In the adjacent diagram, segment DB is the bisector of angle ADC. If $\overline{AD} \cong \overline{CD}$, show that $\overline{AB} \cong \overline{BC}$.

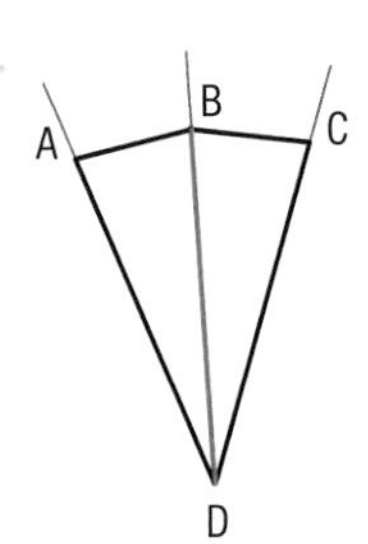

18 The following are some characteristics of a geometric figure:

- P, Q, and R are vertices of a triangle.
- Point S is the midpoint of $\overline{QR}$.
- $\overline{PS}$ is a median.
- Points P, S, and T are colinear.
- $\overline{PT}$ is outside the triangle.
- $\overline{PS} \cong \overline{ST}$.

a) Draw this figure. b) Show that $\overline{PR} \cong \overline{QT}$. c) Show that $\overline{PQ} \cong \overline{RT}$.

19 In triangle ABC, point E is the midpoint of $\overline{AC}$ and point D is the midpoint of $\overline{BC}$. Segment DE is extended to point F such that $\overline{ED} \cong \overline{EF}$.

a) Draw this figure.

b) Show that $\overline{AF} \cong \overline{CD}$.

20 Show that triangles EFG and LMN are congruent. Justify your answer.

Regular pyramid 1

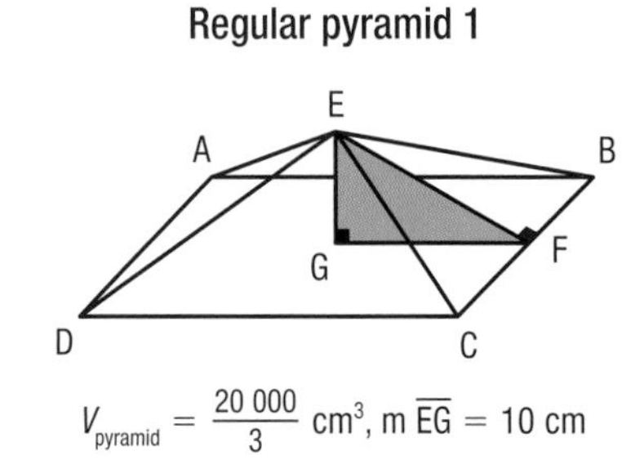

$V_{pyramid} = \frac{20\ 000}{3}$ cm^3, m $\overline{EG}$ = 10 cm

Regular pyramid 2

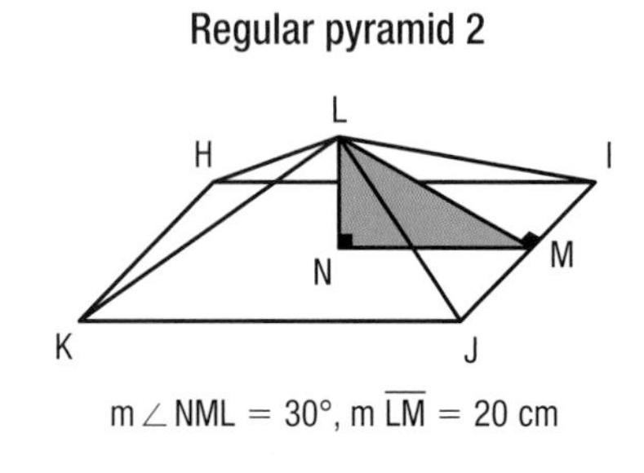

m ∠ NML = 30°, m $\overline{LM}$ = 20 cm

SECTION 3.2 Similar triangles

This section is related to LES 5 and 6.

PROBLEM Glider airplanes

For a contest, participating teams must make a glider airplane out of a cardboard. The winning team will be the one whose glider flies the longest time after being launched from a height of 2 m.

One team chooses to make an airplane which:

- is an enlargement of the adjacent triangle
- is cut from a single sheet of cardboard measuring 152 cm by 190 cm

5 cm
5 cm
4 cm

What are the maximum dimensions of the top of the plane?

In aeronautics, the triangular form is widely used in the design of aircraft wings.

The Concorde

Fighter jet

Stealth jet

ACTIVITY 1 A safe method

In order to find the width of a ravine, measurements were taken as indicated in the drawing below.

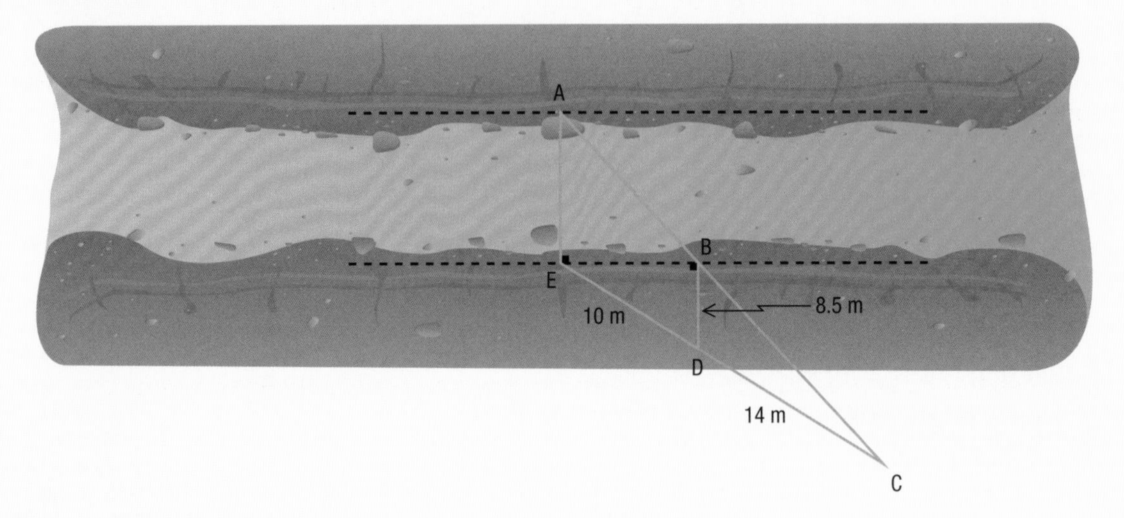

a. With respect to each other, what is the relative position of segments AE and BD?

b. What geometric statement allows you to state that $\angle AEC \cong \angle BDC$?

c. What can you conclude about angles ACE and BCD?

d. Triangle ACE is an enlargement of triangle BCD. What is the ratio of similarity between these two triangles?

e. Use the ratio found in **d.** to determine the width of the ravine.

The Grand Canyon in Arizona, USA, is a spectacular place. The canyon was formed by the erosion of the Colorado River so named by the Spaniards for the reddish color of its waters. The dimensions of the canyon are enormous: about 450 km in length, an average depth of 1300 m, and a width varying from 5 to 30 km. It is formed from some of the oldest rocks on Earth.

ACTIVITY 2 Thales' theorem

Thales of Miletus (ca. 625 - ca. 546 BCE) is regarded by many as the first philosopher, scientist and mathematician in Greece. He was particulary interested in geometric figures and he deduced a number of theorems.

One of the theorems discovered by Thales concerns transversals intersected by parallel lines. This theorem, however, was only later proved by Euclid.

Born in Miletus, Asia Minor, Thales was the founder of Greek philosophy. He is also known for his knowledge of astronomy; he probably introduced geometry to the Greeks.

Using dynamic geometry software, transversals t_1 and t_2, as well as parallel lines l_1, l_2 and l_3 were drawn. By changing the position of one or more lines on Screen **1**, changes in the length of certain segments were observed.

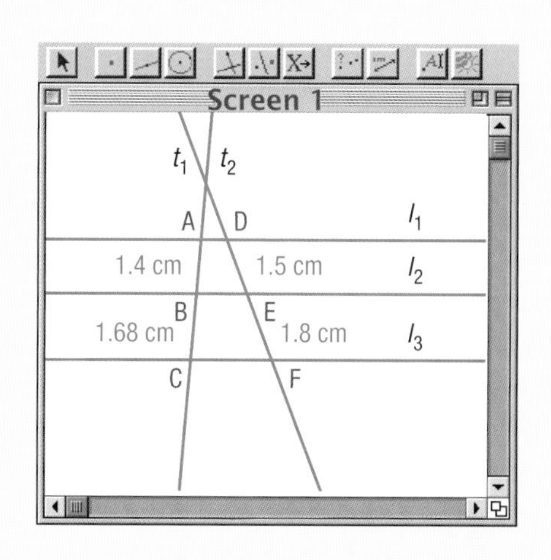

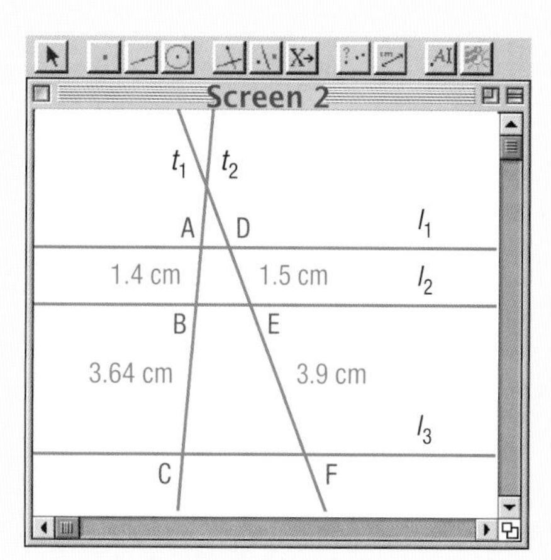

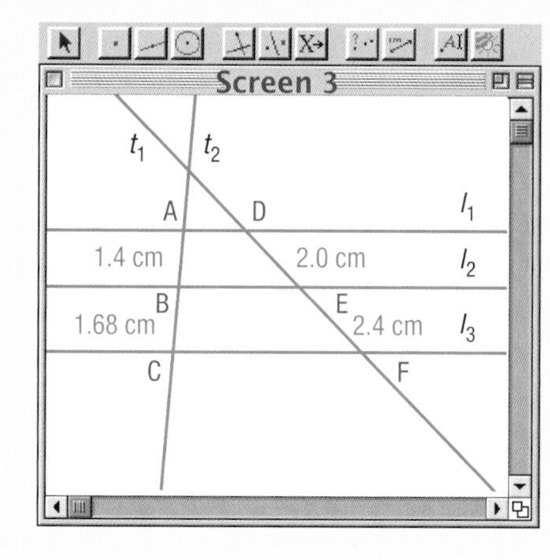

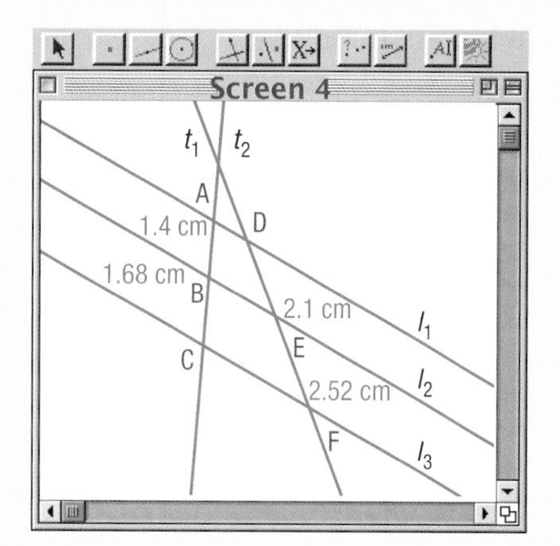

a. Referring to Screen **1**, what change was made in:

1) Screen **2**? 2) Screen **3**? 3) Screen **4**?

b. For Screens **1**, **2**, **3**, and **4** calculate the ratios below:

1) $\frac{m\ \overline{DE}}{m\ \overline{AB}}$ 2) $\frac{m\ \overline{EF}}{m\ \overline{BC}}$

c. Based on the results above, what conjecture would you formulate?

Techno math

Dynamic geometry software allows you to compare geometric figures. By using the tools: TRIANGLE, DISTANCE and ANGLE MEASURE you can draw triangles and check whether they are similar.

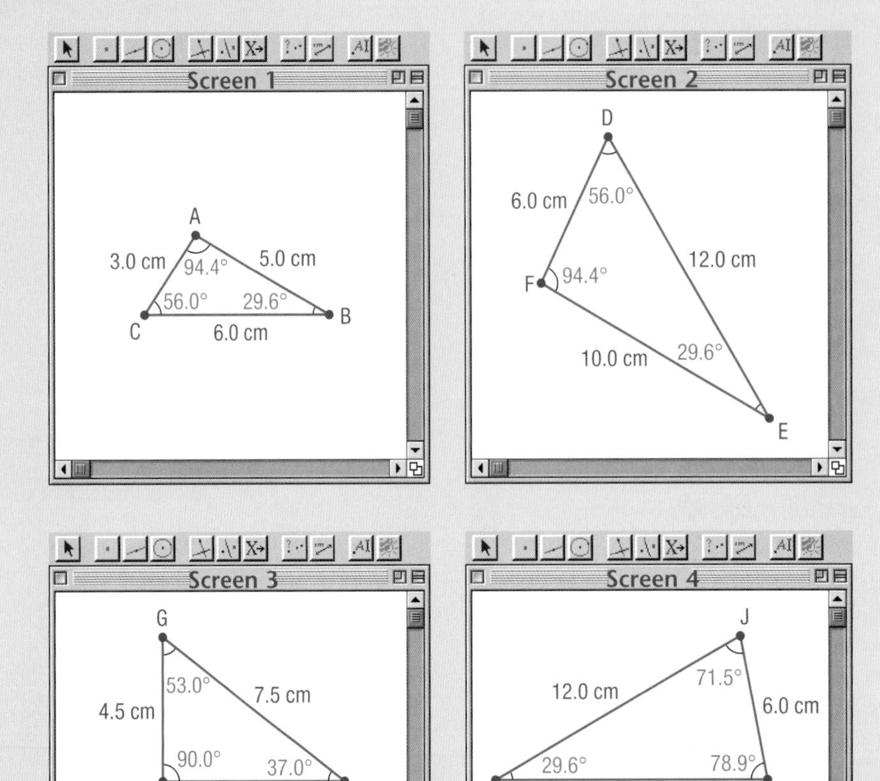

a. How many pairs of congruent angles are there if you compare the triangles in:

1) Screens **1** and **2**? 2) Screens **1** and **3**? 3) Screens **1** and **4**?

b. Verify that:

1) $\frac{m\ \overline{EF}}{m\ \overline{AB}} = \frac{m\ \overline{DE}}{m\ \overline{BC}} = \frac{m\ \overline{DF}}{m\ \overline{AC}}$ 2) $\frac{m\ \overline{GH}}{m\ \overline{AB}} = \frac{m\ \overline{GI}}{m\ \overline{AC}}$ 3) $\frac{m\ \overline{JK}}{m\ \overline{AC}} = \frac{m\ \overline{JL}}{m\ \overline{BC}}$

c. Among the triangles in Screens **1**, **2**, **3** and **4** is there a pair of similar triangles?

d. Can you state that two triangles are similar if:

1) the lengths of two sides of one triangle are proportional to the lengths of two sides of the other triangle?

2) one angle of a triangle is congruent to one angle of the other triangle?

e. The lengths of two sides of triangle MNO are proportional to the lengths of two corresponding sides of triangle PQR. One angle of triangle MNO is congruent to one angle of triangle PQR. Using dynamic geometry software, do the following:

1) Construct triangles MNO and PQR.

2) Examine many possible configurations and explain under what conditions triangles MNO and PQR can be similar.

MINIMUM CONDITIONS FOR SIMILAR TRIANGLES

Similar triangles are triangles whose corresponding angles are congruent and whose corresponding sides are proportional in length.

The geometric statements below describe the minimum conditions necessary to state that two triangles are similar.

1. **Two triangles that have two congruent corresponding angles are similar (AA).**

The abbreviation AA (Angle-Angle) is used to simplify the written form of this statement.

E.g. $m\angle A = m\angle D = 85°$
$m\angle B = m\angle E = 39°$
Thus, $\Delta ABC \sim \Delta DEF$.

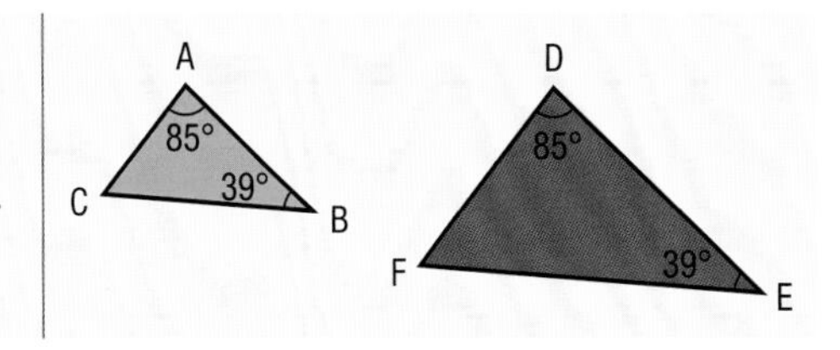

2. **Two triangles that have one congruent angle contained between corresponding sides of proportional length are similar (SAS).**

The abbreviation SAS (Side-Angle-Side) is used to simplify the written form of this statement.

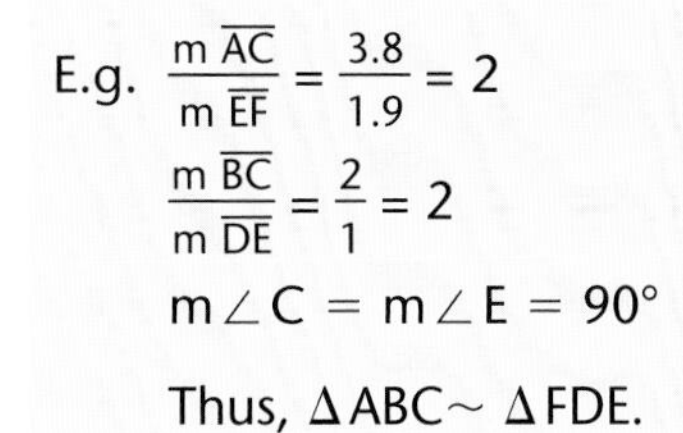

E.g. $\frac{m\,\overline{AC}}{m\,\overline{EF}} = \frac{3.8}{1.9} = 2$

$\frac{m\,\overline{BC}}{m\,\overline{DE}} = \frac{2}{1} = 2$

$m\angle C = m\angle E = 90°$

Thus, $\Delta ABC \sim \Delta FDE$.

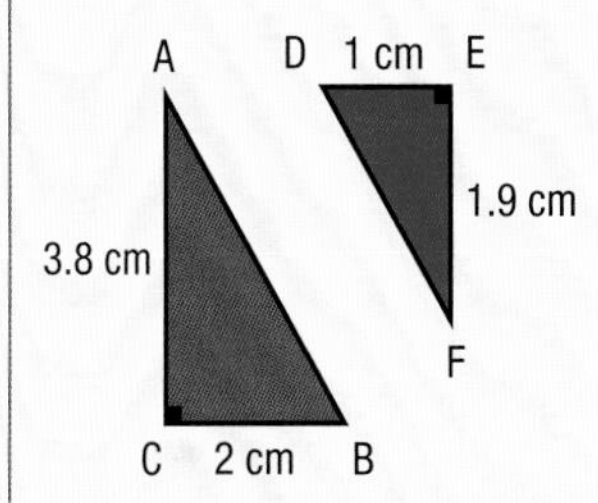

3. **Two triangles whose corresponding sides are proportional are similar (SSS).**

The abbreviation SSS (Side-Side-Side) is used to simplify the written form of this statement.

E.g. $\frac{m\,\overline{AC}}{m\,\overline{DF}} = \frac{8}{3.2} = 2.5$

$\frac{m\,\overline{BC}}{m\,\overline{EF}} = \frac{5}{2} = 2.5$

$\frac{m\,\overline{AB}}{m\,\overline{DE}} = \frac{4.5}{1.8} = 2.5$

Thus, $\Delta ABC \sim \Delta DEF$.

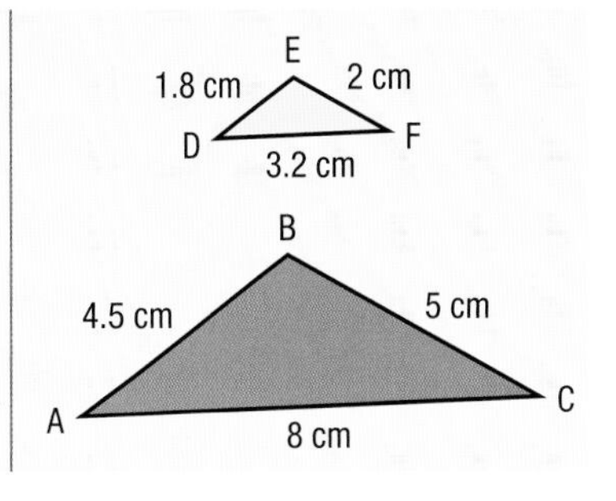

When two triangles are similar, it is possible to find missing measurements.

E.g. Considering that the two triangles below are similar, the length of side AB can be calculated as follows:

$$\frac{m\,\overline{AB}}{m\,\overline{DE}} = \frac{m\,\overline{BC}}{m\,\overline{EF}} = \frac{m\,\overline{AC}}{m\,\overline{DF}}$$

Substitute the known measurements: $\frac{m\,\overline{AB}}{52.5} = \frac{9}{31.5} = \frac{12}{42}$.

Therefore, $m\,\overline{AB} = 15$ cm.

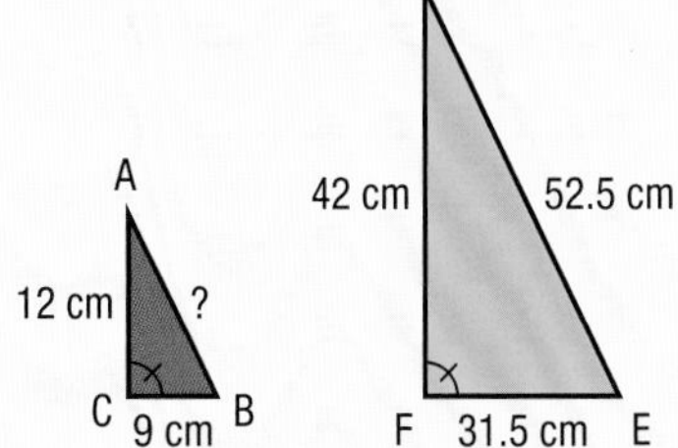

practice 3.2

1 In each case, identify the geometric statement that allows you to conclude that the triangles are similar.

a) 2.4 cm, 1.5 cm, 1.25 cm, 2 cm

b)

c)

d) 62°, 62°

e) 3.2 cm, 122°, 1.6 cm, 122°, 2 cm, 4 cm

f)

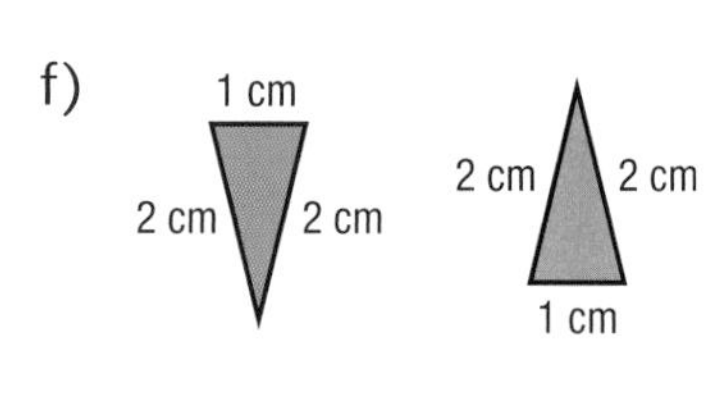

2 In each case, find the missing value.

a)

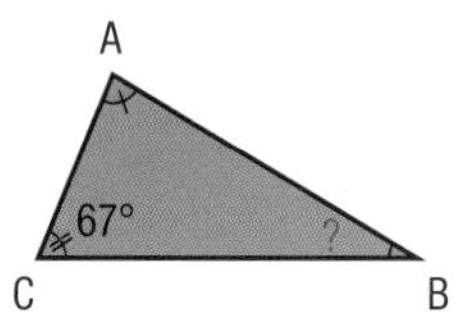

D, 83°, F, E

b)

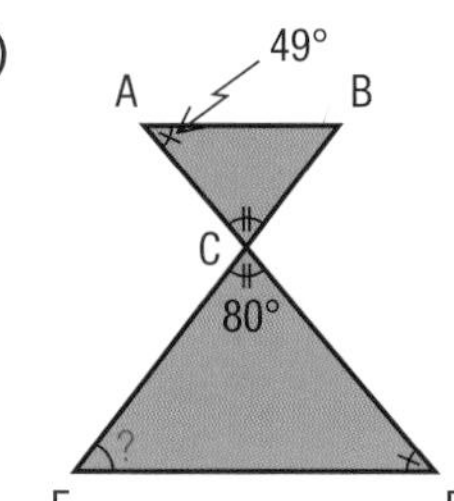

c)

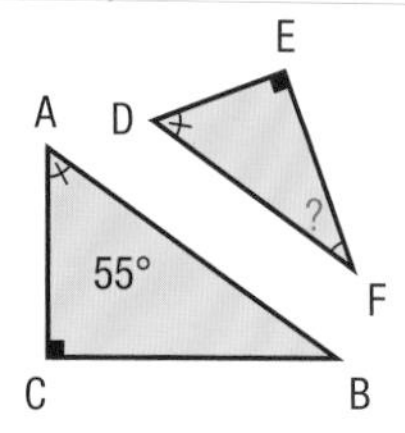

d)

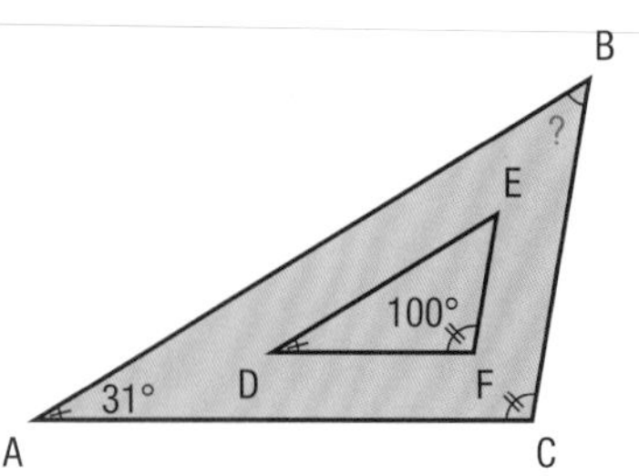

e)

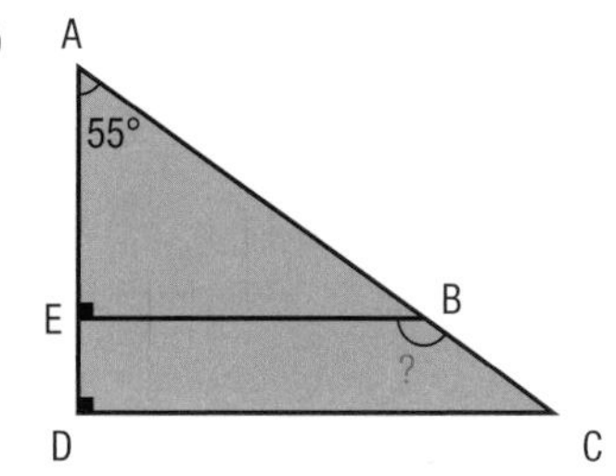

f)

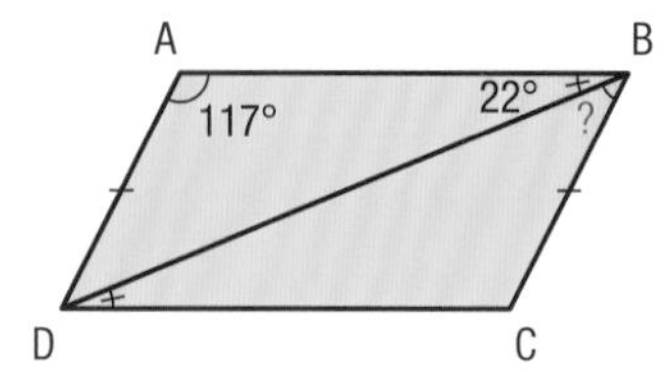

3 In each case determine the lengths that correspond to x and y.

a)

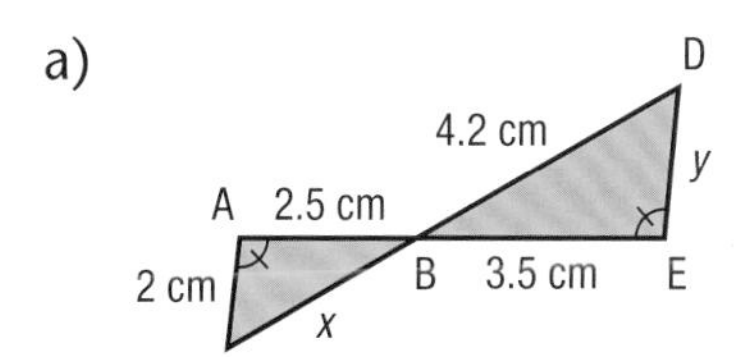

b)

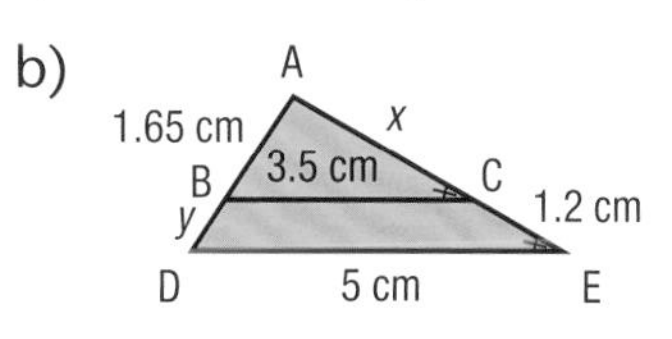

c)

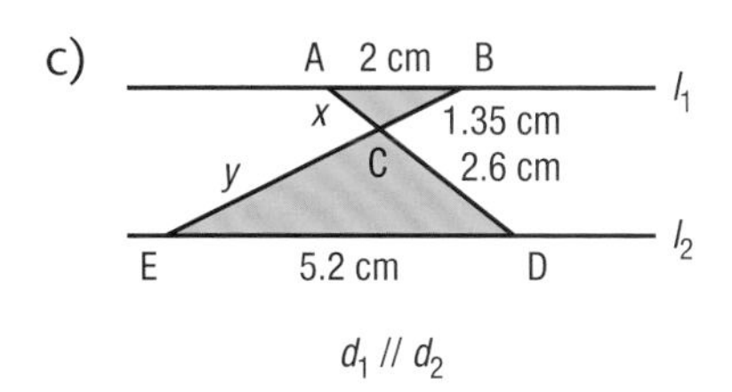

d)

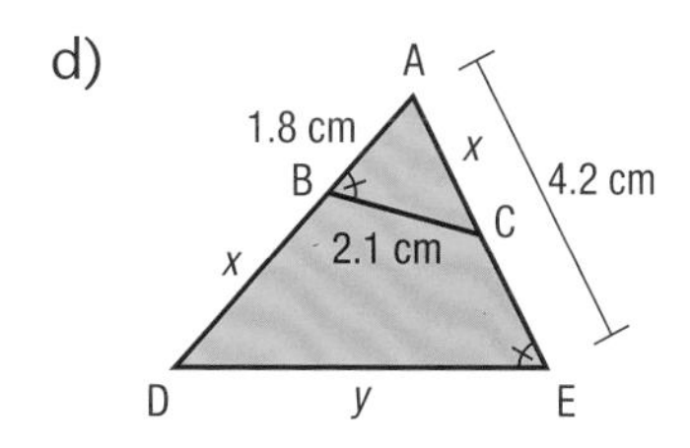

e)

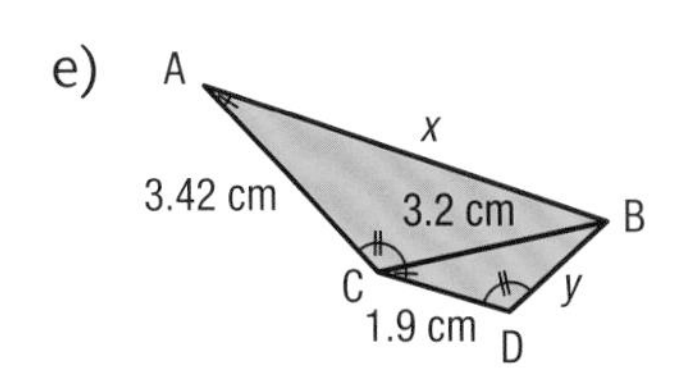

f)

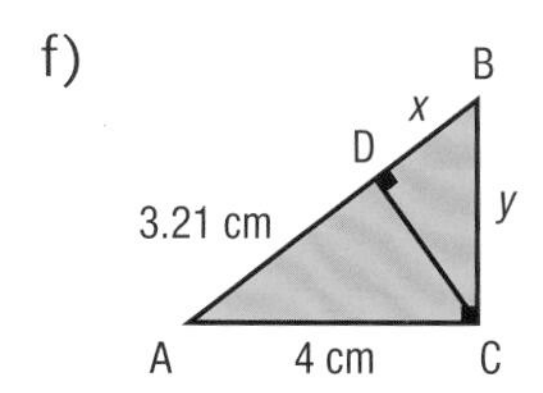

4 Su Minh states that any two isosceles triangles that each have an angle measuring 40° must be similar. Is he right? Justify your answer.

5 In the adjacent figure, $\overline{AD}$ // $\overline{BC}$. Identify the pair of congruent corresponding angles. In each case, justify your answer with a geometric statement.

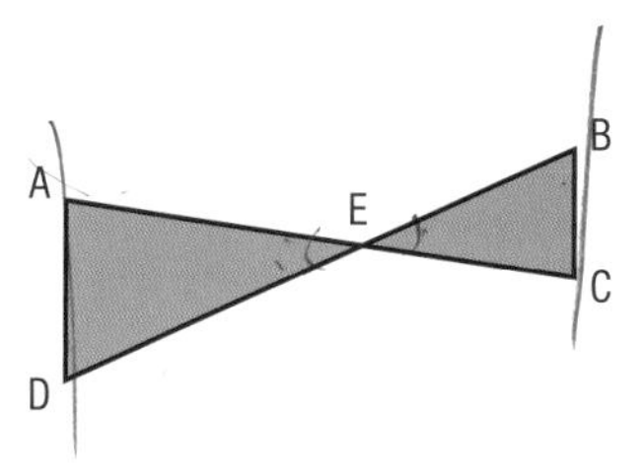

6 In the adjacent figure, name the geometric statement that allows you to conclude that $\Delta ABC \sim \Delta AED$.

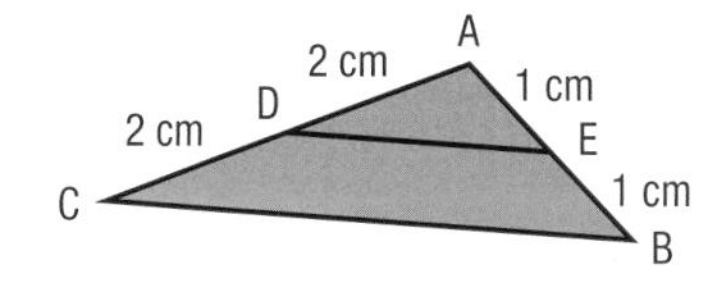

7 In the adjacent figure, $\overline{AB}$ // $\overline{CD}$. What geometric statement allows you to conclude that $\Delta\ ABE \sim \Delta CDE$?

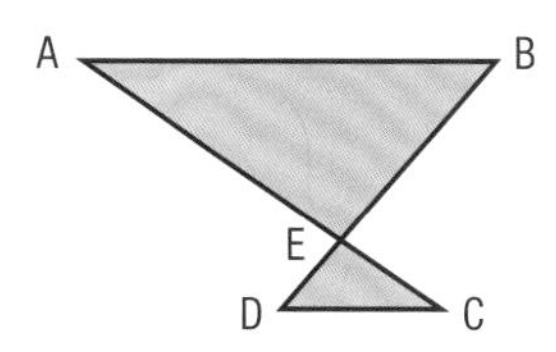

8 Using the information below that refers to the adjacent figure, name the geometric statement that allows you to conclude that $\Delta ABC \sim \Delta AED$.

m $\overline{AD}$ = 9 mm

m $\overline{DB}$ = 3 mm

m $\overline{AE}$ = 6 mm

m $\overline{EC}$ = 12 mm

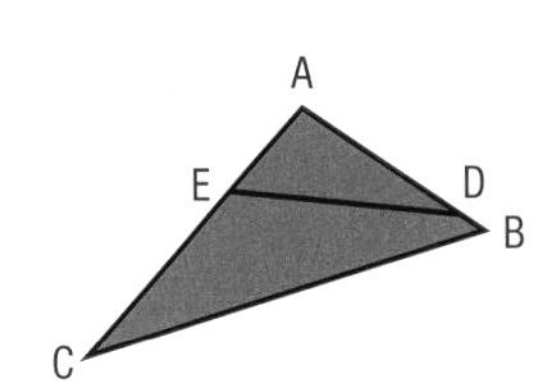

9 The cross-country ski trail shown in the adjacent figure is made of two similar triangles. Calculate the length of this trail.

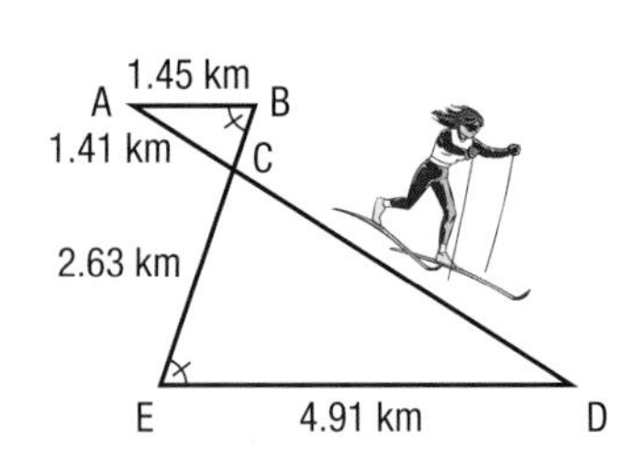

10 In the adjacent triangle ABC, $\overline{DE} \parallel \overline{BC}$.

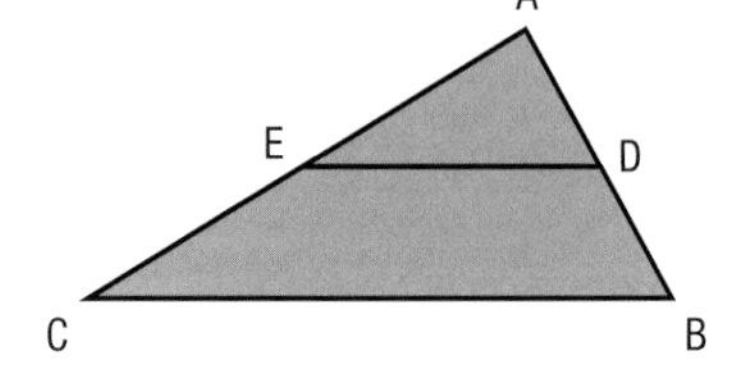

a) If m $\overline{AB}$ = 2 cm, m $\overline{AC}$ = 1.5 cm and m $\overline{AE}$ = 0.8 cm, find m $\overline{BD}$.

b) If m $\overline{AD}$ = 8 cm, m $\overline{BD}$ = 5 cm and m $\overline{AE}$ = 4 cm, find m $\overline{CE}$.

c) If m $\overline{AE}$ = 4.8 cm, m $\overline{AC}$ = 7.2 cm and m $\overline{AD}$ = 6.2 cm, find m $\overline{BD}$.

d) If m $\overline{CE}$ = 3 cm, m $\overline{AE}$ = 5 cm and m $\overline{BD}$ = 2 cm, find m $\overline{AB}$.

e) If m $\overline{AB}$ = 7.2 cm, m $\overline{AC}$ = 8.4 cm and m $\overline{AD}$ = 4.2 cm, find m $\overline{CE}$.

11 Triangle ABC has side lengths of 2 cm, 2.5 cm and 4 cm respectively. If triangle DEF is similar to triangle ABC and the longest side of triangle DEF measures 7 cm, what is the perimeter of triangle DEF?

12 Considering that $\overline{AN} \parallel \overline{BC}$, show that $\frac{m\,\overline{NQ}}{m\,\overline{MQ}} = \frac{m\,\overline{AQ}}{m\,\overline{CQ}}$ by completing the following proof.

Hypothesis:	1)
Conclusion:	$\frac{m\,\overline{NQ}}{m\,\overline{MQ}} = \frac{m\,\overline{AQ}}{m\,\overline{CQ}}$

STATEMENT	JUSTIFICATION
$\angle AQN \cong \angle CQM$	Common angle
$\angle NAQ \cong \angle MCQ$	2)
$\Delta ANQ \sim \Delta CMQ$	3)
$\frac{m\,\overline{NQ}}{m\,\overline{MQ}} = \frac{m\,\overline{AQ}}{m\,\overline{CQ}}$	4)

13 In the adjacent figure, the altitudes of triangle ABC are drawn. Using geometric statements, show that $\frac{m\,\overline{AE}}{m\,\overline{CD}} = \frac{m\,\overline{BE}}{m\,\overline{BD}}$.

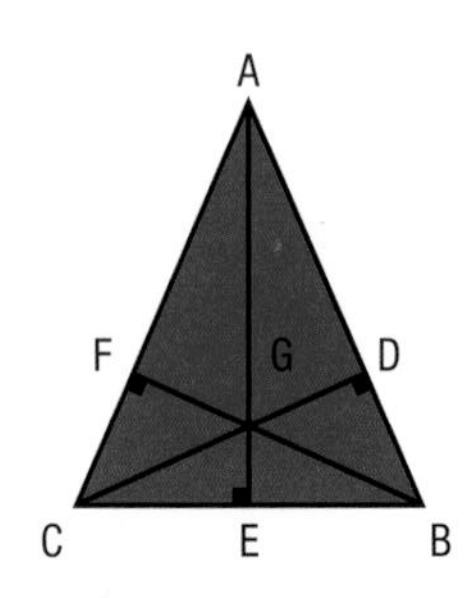

14 In the adjacent trapezoid, show that $\frac{m\,\overline{BE}}{m\,\overline{ED}} = \frac{m\,\overline{AE}}{m\,\overline{CE}}$.

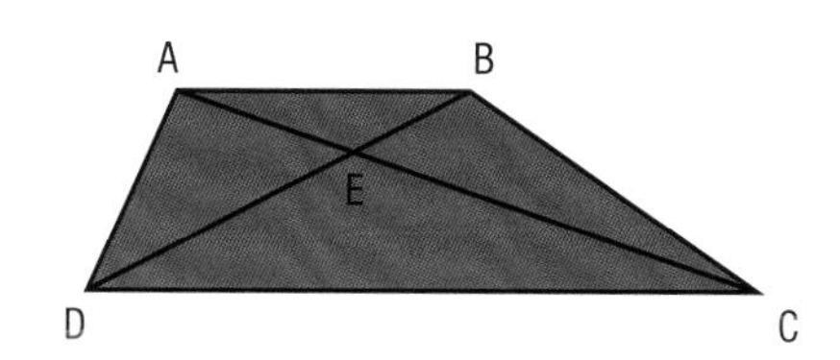

15 Two similar triangular lots are for sale. The sides of Lot **A** measure 5 m, 12 m, and 13 m respectively. The sides of Lot **B** are three times as long as those of Lot **A**. The price of Lot **B** is 5 times that of Lot **A**. Which lot is the better bargain per square m?

16 The parallelogram ADEF is inscribed in triangle ABC.

a) What geometric statement allows you to conclude that:

1) $\Delta ABC \sim \Delta DBE$?
2) $\Delta ABC \sim \Delta FEC$?
3) $\Delta FEC \sim \Delta DEB$?

b) Calculate the length of the following segments:

1) $\overline{DE}$ 2) $\overline{BE}$ 3) $\overline{EF}$ 4) $\overline{AD}$

17 THALES OF MILETUS The story is told of how, on a voyage to Egypt, Thales wished to calculate the height of the pyramid of Cheops. He would have used the height of his own body, the length of his shadow, and the shadow of this regular pyramid. He would have made the following observation:

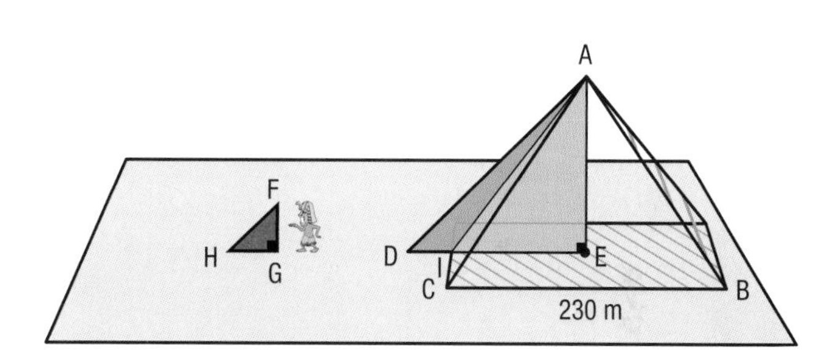

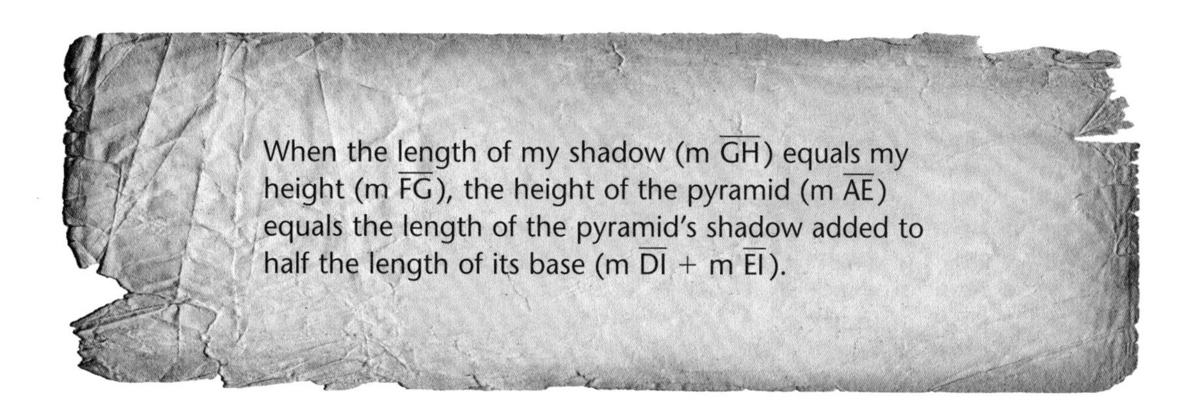

When the length of my shadow ($m\,\overline{GH}$) equals my height ($m\,\overline{FG}$), the height of the pyramid ($m\,\overline{AE}$) equals the length of the pyramid's shadow added to half the length of its base ($m\,\overline{DI} + m\,\overline{EI}$).

a) If $m\,\overline{DI} = 22$ m, what is the height of the pyramid?

b) If, to measure the pyramid, Thales had used a 1 m long stick which cast a shadow of 0.78 m, what would have been the length of the shadow of the pyramid?

18 The operator of the forklift below sits 77 cm above the ground. At their maximum height the forks are 171 cm from the ground. Calculate m $\overline{BE}$, the distance between the tip of the forks and the operator's seat.

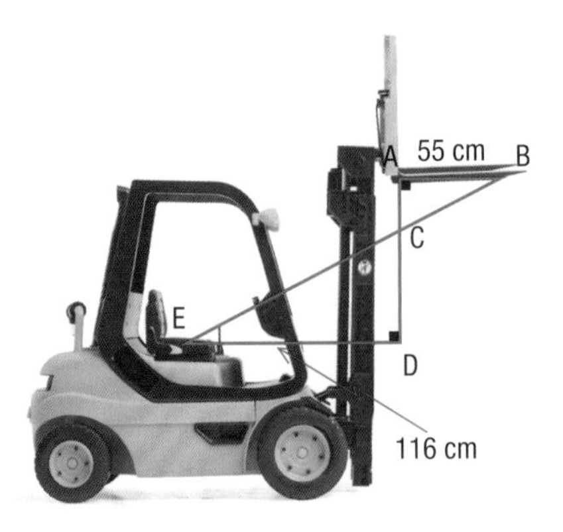

19 A path in a zoo is represented in the illustration below. What is the length of this path?

Scientific research is a goal of zoos and aquariums around the world. This permits certain animals, particularly threatened species, to be better known and protected. In their own way, zoos help with global animal preservation.

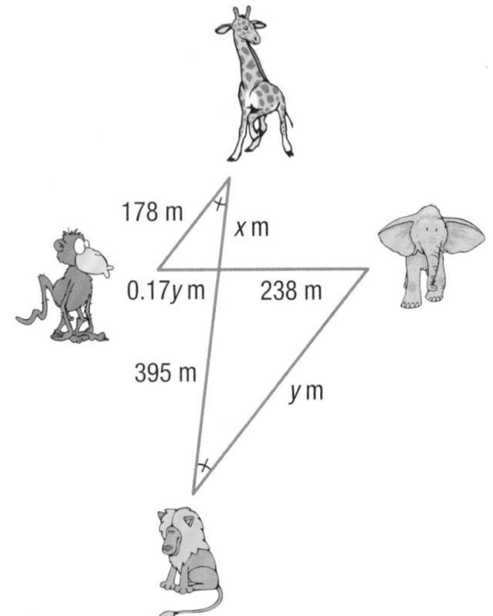

20 It is possible to estimate the height of a building using a mirror. Place the mirror on the ground a certain distance from the building. Stand next to the mirror and then move away from the building until the top of the building is visible in the mirror as shown in the illustration below. What is the height of the building as shown?

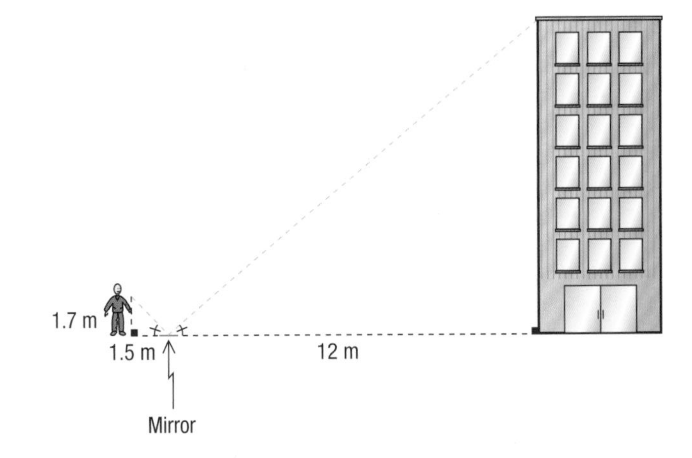

21 This illustration represents the scale model of a stained glass window.

a) Find:

1) m $\overline{BG}$
2) m $\overline{EG}$
3) m $\overline{ED}$
4) m $\overline{DG}$

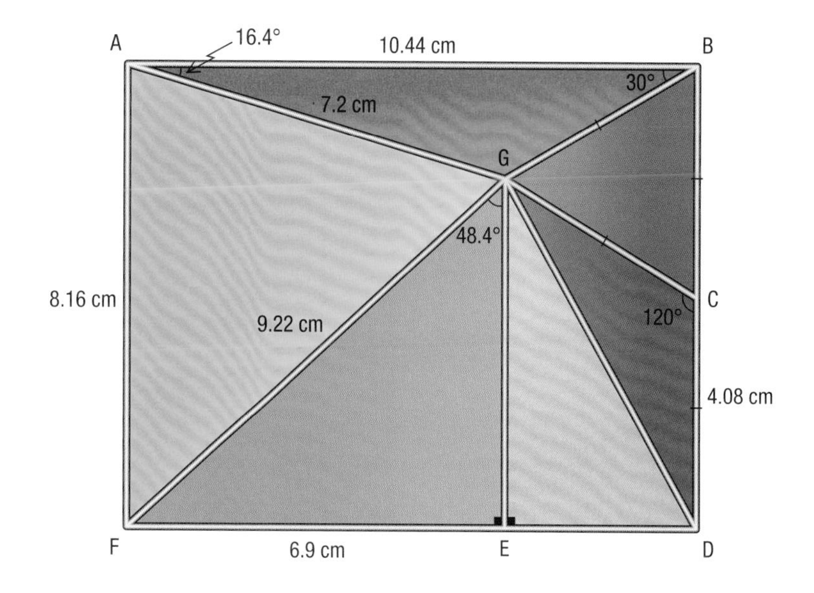

b) Below are the actual dimensions of the pieces of the stained glass window. Find the measures of the angles and the lengths of the sides for each piece.

1) 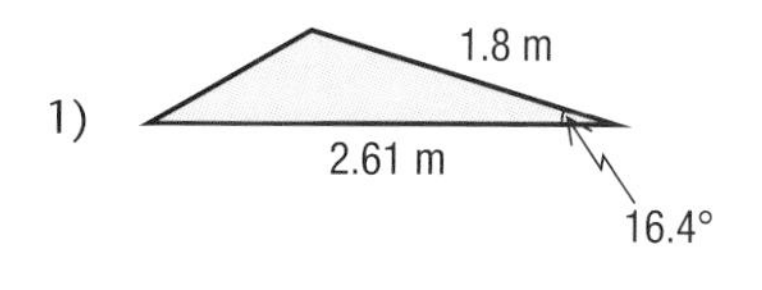

2) 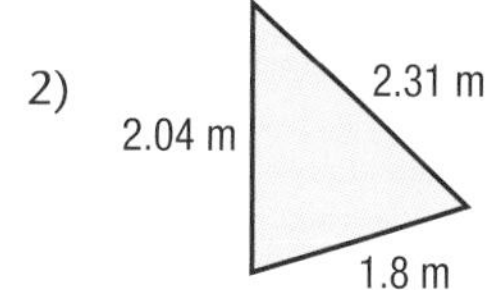

3) 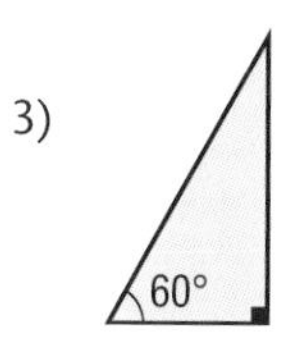

4) 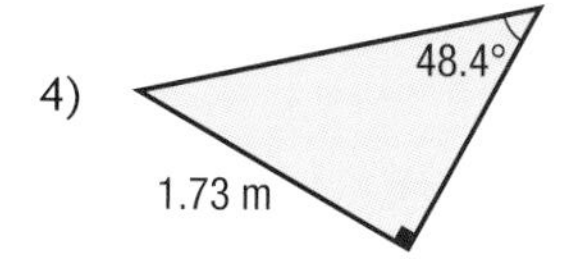

5) 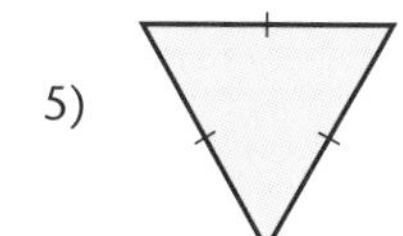

6) 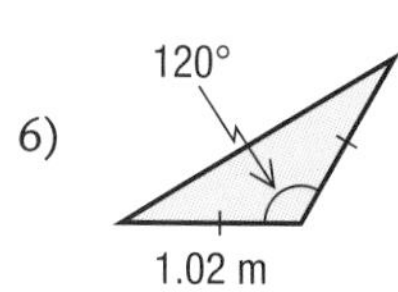

22 At the top of a skyscraper, four tightrope walkers cover the distance separating them from another skyscraper. If upon arrival, two tightrope walkers, F_1 and F_4, find themselves 24 m apart, determine the lengths associated with *x*, *y* and *z*.

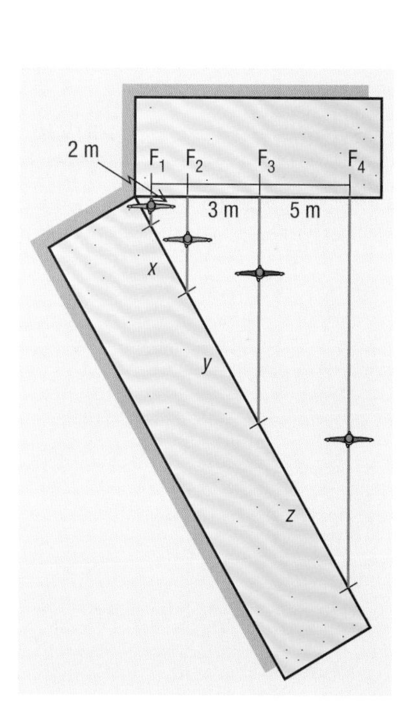

23 The diagram below represents the top view of three kennel enclosures. Determine the length of *c*.

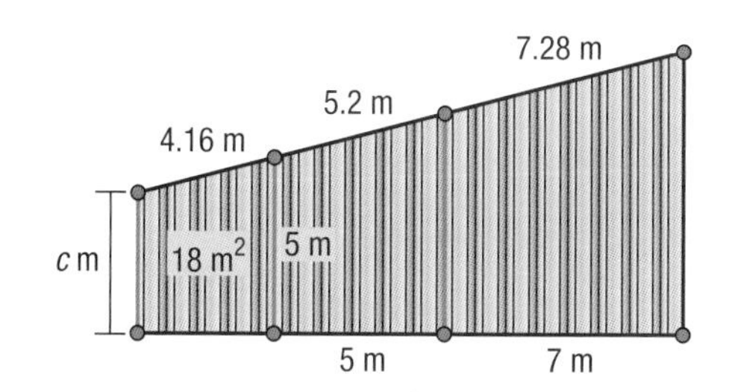

SECTION 3.3 Metric relations

This section is related to LES 6.

PROBLEM Mother Nature enraged!

Meteorologists try to predict the development of potentially violent and dangerous storms as well as the type of danger they represent and the locations that will be affected.

Following a violent storm, a tree fell onto an electrical pole as shown as below.

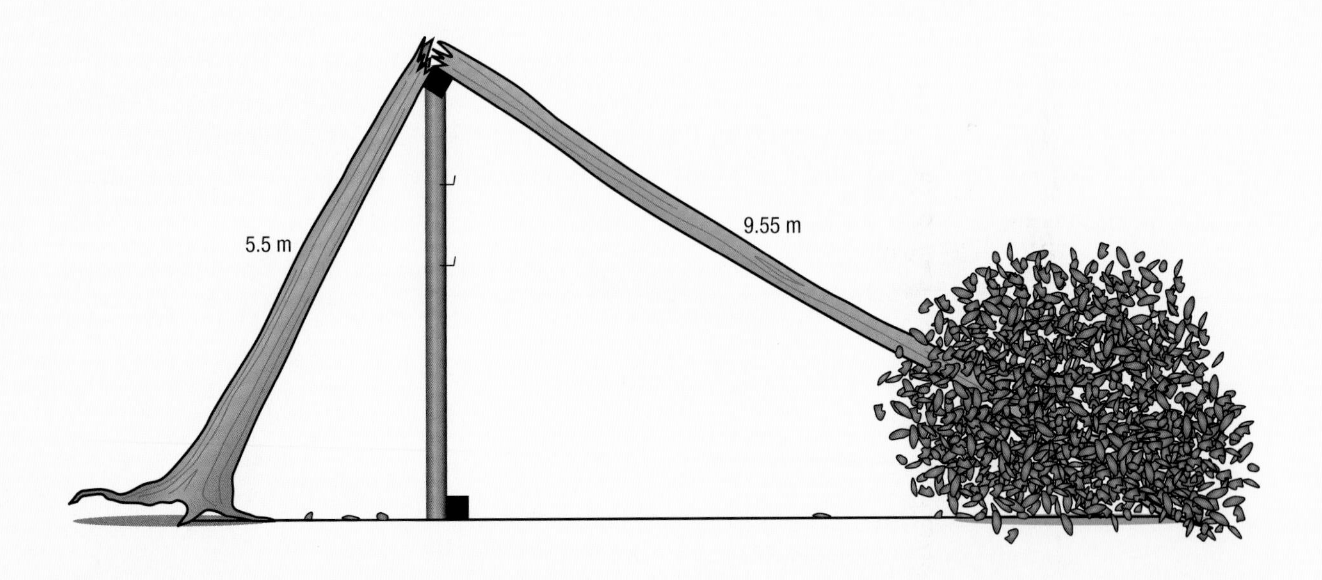

What is the height of the electrical pole?

In general, a storm is considered violent if it produces winds with gusts of 90 km/h or more and is accompanied by hailstones 2 cm or more in diameter since all of this can seriously damage homes. Tornadoes or torrential rains that can cause flooding are other conditions associated with violent storms.

ACTIVITY 1 Birds of a feather stick together!

By drawing the altitude from the right angle of a right triangle, two other triangles are formed. These three triangles are re-oriented and arranged in the same direction as shown below.

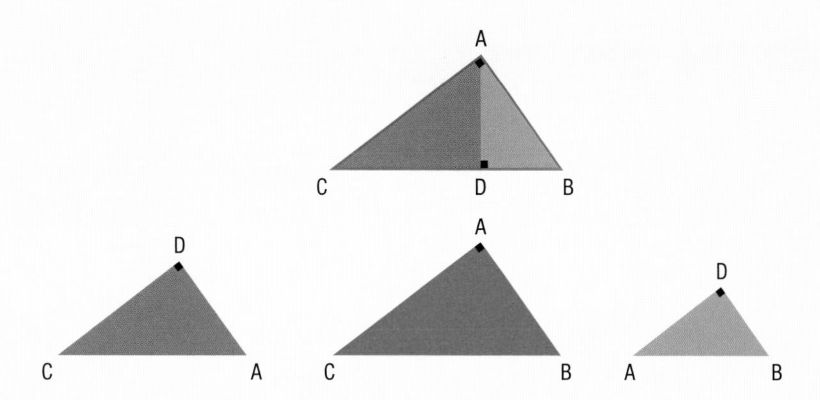

a. What geometric statement allows you to state that:

1) blue $\Delta \sim$ pink Δ?
2) green $\Delta \sim$ pink Δ?
3) blue $\Delta \sim$ green Δ?

b. Complete the following proportions:

1) using the blue and pink triangles

$$\frac{\text{blue } \Delta:}{\text{pink } \Delta:} \quad \frac{m\,\overline{CD}}{m\,\overline{AC}} = \frac{m\,\overline{AC}}{\rule{1cm}{0.3cm}} = \frac{\rule{1cm}{0.3cm}}{m\,\overline{AB}}$$

2) using the pink and green triangles

$$\frac{\text{pink } \Delta:}{\text{green } \Delta:} \quad \frac{m\,\overline{AC}}{m\,\overline{AD}} = \frac{m\,\overline{BC}}{\rule{1cm}{0.3cm}} = \frac{\rule{1cm}{0.3cm}}{m\,\overline{DB}}$$

3) using the blue and green triangles

$$\frac{\text{blue } \Delta:}{\text{green } \Delta:} \quad \frac{\rule{1cm}{0.3cm}}{m\,\overline{AB}} = \frac{m\,\overline{AD}}{\rule{1cm}{0.3cm}} = \frac{\rule{1cm}{0.3cm}}{m\,\overline{AD}}$$

c. With the help of the figure below, determine:

1) $m\,\overline{LN}$
2) $m\,\overline{MN}$
3) $m\,\overline{LM}$

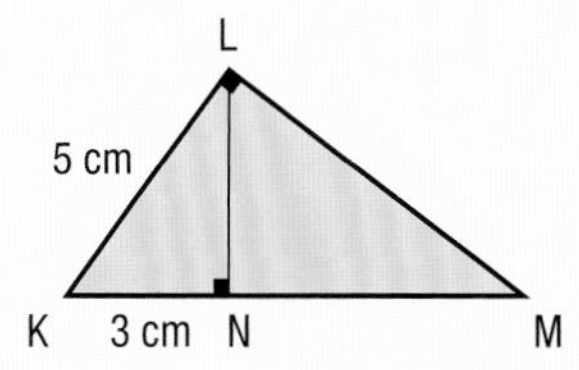

On April 11, 2003, a Sumerian tablet disappeared (1800 BCE) from the Iraq Museum of Archaeology. It contains one of the first versions of Pythagora's theorem.

Techno math

Dynamic geometry software allows you to explore and to verify metric relations in right triangles. By using the tools LINE, PERPENDICULAR LINE, TRIANGLE and DISTANCE, you can construct a right triangle and draw the altitude from the vertex of the right angle.

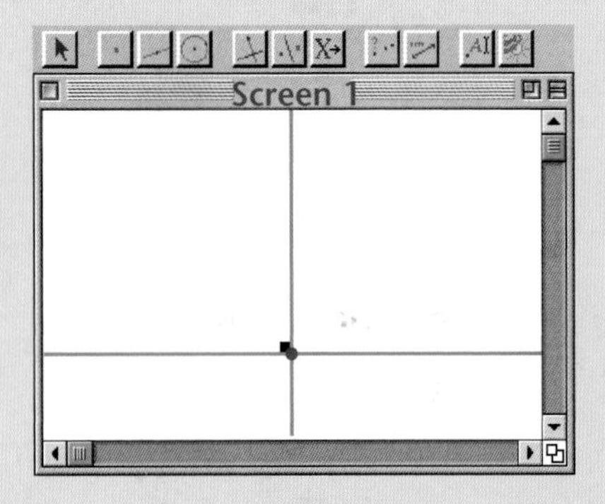

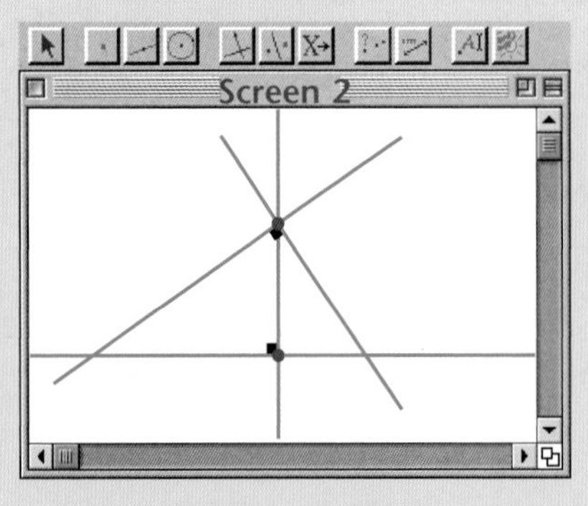

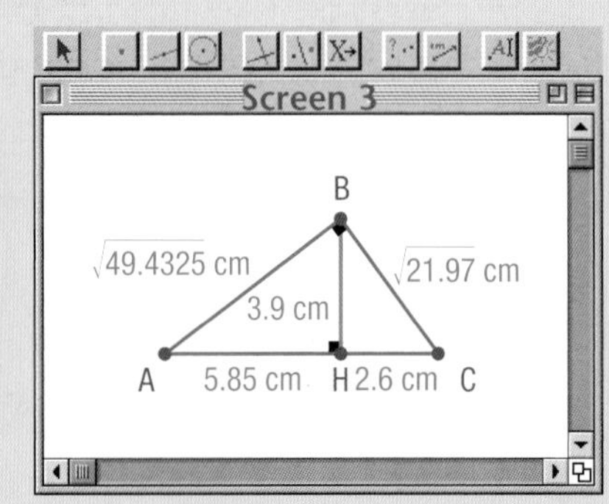

By changing the position of the vertices of the triangle, you can observe certain effects related to the lengths of the sides and altitude of the triangle.

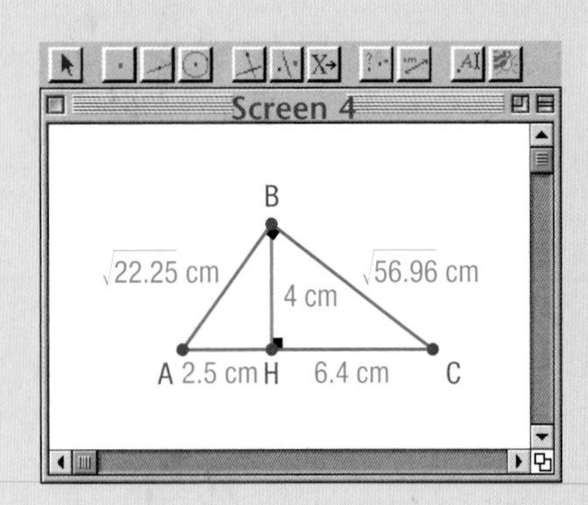

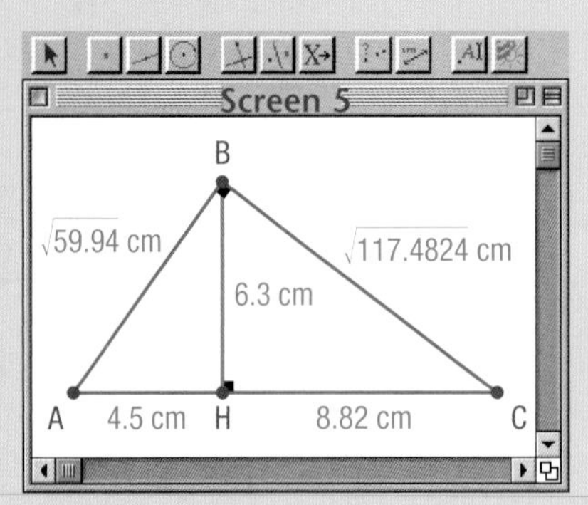

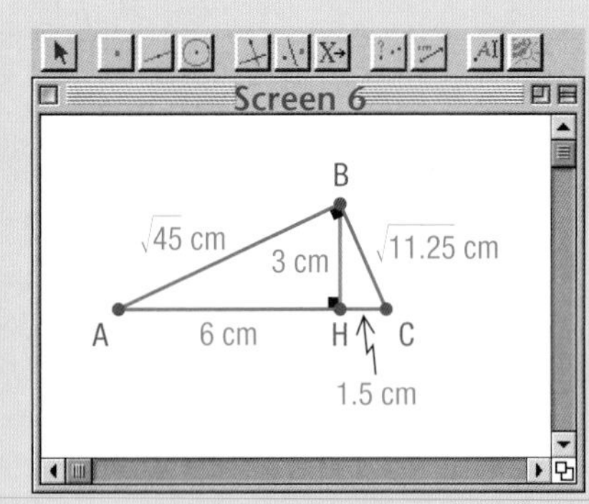

a. Using cacluations and geometric statements, explain why:

1) Δ ABH ~ Δ BCH
2) Δ ABC ~ Δ AHB
3) Δ ABC ~ Δ BHC

b. For each of Screens **3**, **4**, **5** and **6**, calculate:

1) m $\overline{AC}$ × m $\overline{BH}$
2) m $\overline{AB}$ × m $\overline{BC}$

c. Based on your results, what conjecture can be formulated?

d. Using dynamic geometry software, verify whether this conjecture can be applied:

1) to isosceles triangles
2) to equilateral triangles

METRIC RELATIONS IN A RIGHT TRIANGLE

By drawing the altitude from the vertex of a right angle in a right triangle, three similar right triangles are formed. Using the lengths of the corresponding sides of the triangles formed, you can establish proportions and deduce various geometric statements.

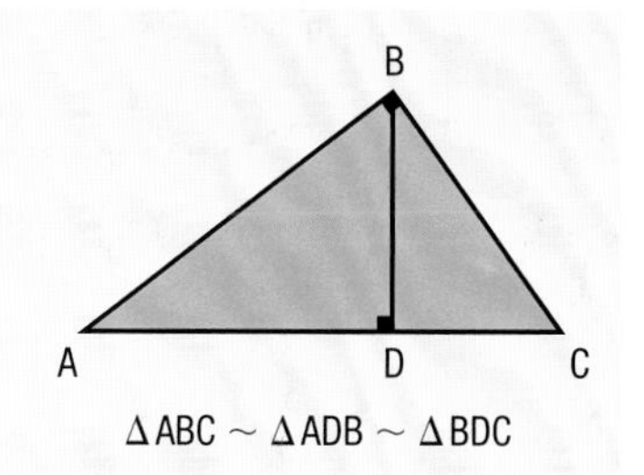

1. In a right triangle, the length of a leg of a right triangle is the geometric mean between the length of its projection on the hypotenuse and the length of the hypotenuse, that is:

$$\frac{m\,\overline{AD}}{m\,\overline{AB}} = \frac{m\,\overline{AB}}{m\,\overline{AC}} \text{ or } (m\,\overline{AB})^2 = m\,\overline{AD} \times m\,\overline{AC}$$

$$\frac{m\,\overline{CD}}{m\,\overline{BC}} = \frac{m\,\overline{BC}}{m\,\overline{AC}} \text{ or } (m\,\overline{BC})^2 = m\,\overline{CD} \times m\,\overline{AC}$$

E.g.

1) B, 8 cm, A, ?, D, C, 10 cm

$$\frac{m\,\overline{AD}}{8} = \frac{8}{10}$$

$$64 = 10 \times m\,\overline{AD}$$

$$m\,\overline{AD} = 6.4 \text{ cm}$$

2)

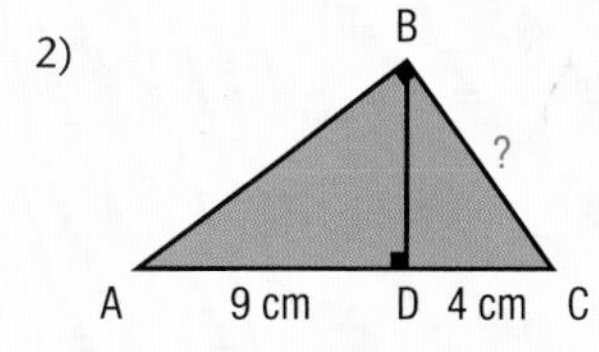

$$\frac{4}{m\,\overline{BC}} = \frac{m\,\overline{BC}}{13}$$

$$(m\,\overline{BC})^2 = 52$$

$$m\,\overline{BC} = \sqrt{52} \text{ cm or } \approx 7.21 \text{ cm}$$

2. In a right triangle, the length of the altitude drawn from the right angle is the geometric mean of the length of the two segments that determine the hypotenuse, that is:

$$\frac{m\,\overline{AD}}{m\,\overline{BD}} = \frac{m\,\overline{BD}}{m\,\overline{CD}} \text{ or}$$

$$(m\,\overline{BD})^2 = m\,\overline{AD} \times m\,\overline{CD}$$

E.g. B, ?, A, 8 cm, D, 3 cm, C

$$\frac{8}{m\,\overline{BD}} = \frac{m\,\overline{BD}}{3}$$

$$(m\,\overline{BD})^2 = 24$$

$$m\,\overline{BD} = \sqrt{24} \text{ cm or } \approx 4.9 \text{ cm}$$

3. In a right triangle, the product of the length of the hypotenuse and its corresponding altitude is equal to the product of the lengths of the legs, that is:

$$m\,\overline{AC} \times m\,\overline{BD} = m\,\overline{AB} \times m\,\overline{BC}$$

E.g.

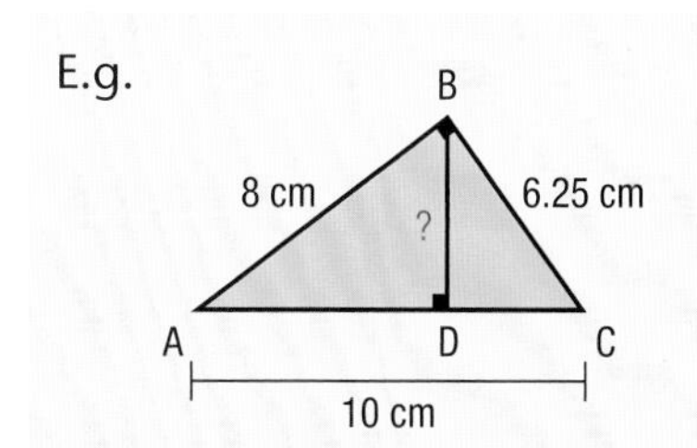

$$10 \times m\,\overline{BD} = 8 \times 6.25$$

$$m\,\overline{BD} = 5 \text{ cm}$$

practice 3.3

1 For each of the right triangles below:

1) determine the value of x
2) identify the geometric statement that supports this calculation

a)

b) 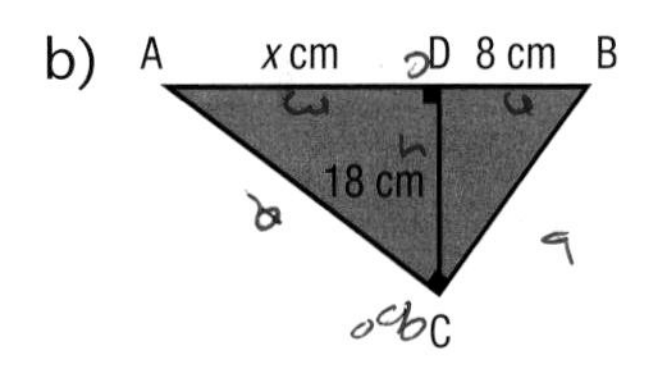

c) 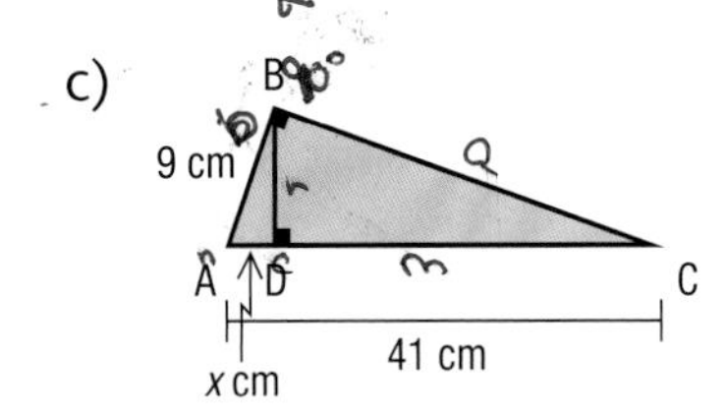

d) 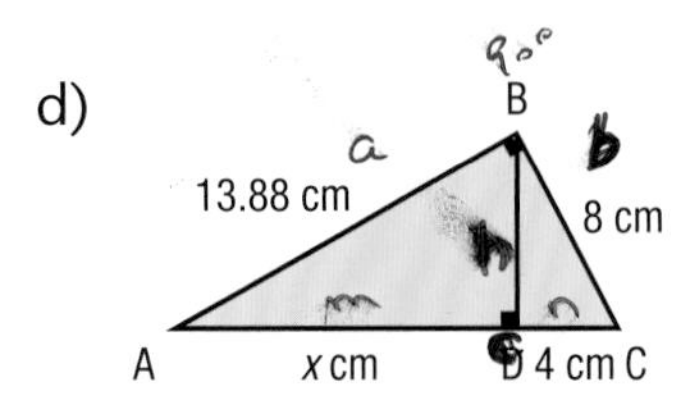

2 In the adjacent triangle ABC,
m $\overline{BC}$ = 10 cm and m $\overline{AB}$ = 26 cm.
Determine:

a) m $\overline{AC}$

b) m $\overline{CD}$

c) m $\overline{BD}$

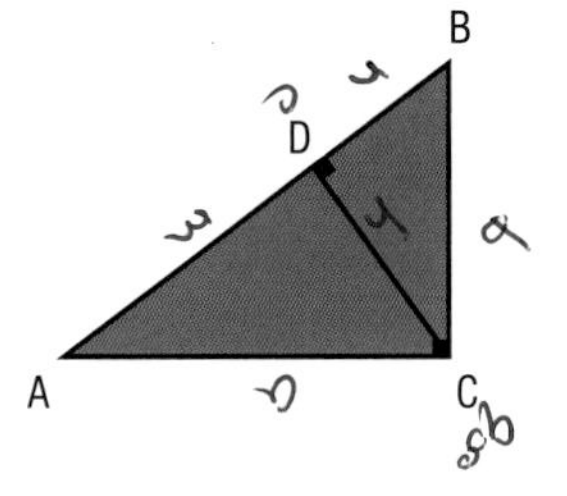

3 Complete the following table using the adjacent right triangle.

Length of segments

	a	b	c	m	n	h
a)	9	12				
b)				4		8
c)	10					6

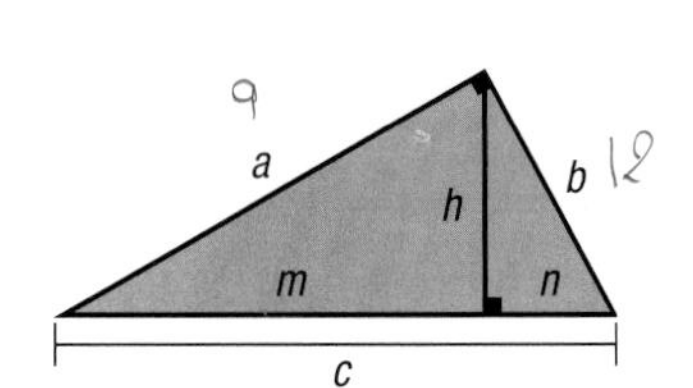

4 In the triangle below, what is the length of segment AD?

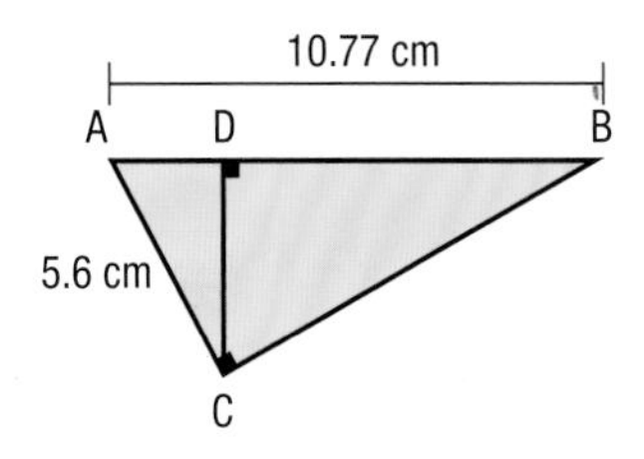

5 As illustrated in the adjacent figure, two people observe a kite. The angles of elevation are indicated.

a) Identify the geometric statement which allows you to conclude that:

1) $m\angle ABC = 90°$

2) $(m\,\overline{BD})^2 = m\,\overline{AD} \times m\,\overline{CD}$

b) What is the height of the kite?

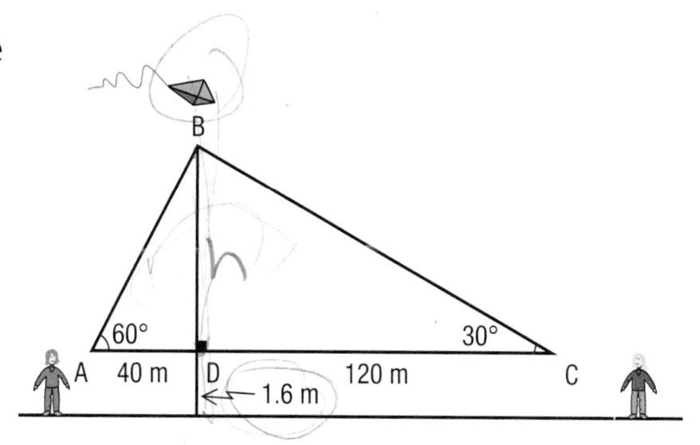

6 Two ladders leading to a tree house are shown in the adjacent figure. At what height from the ground is the tree house?

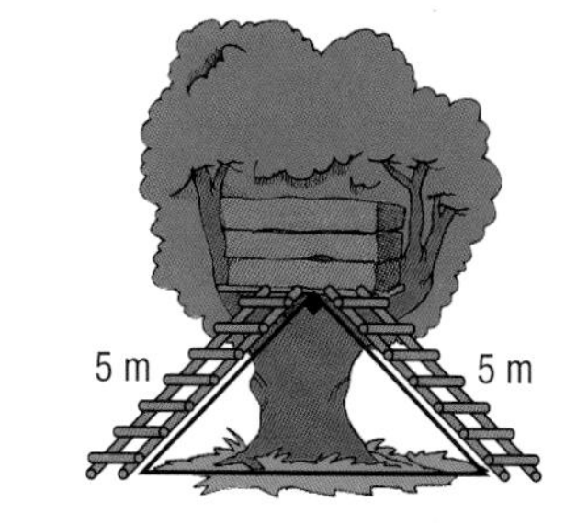

7 As shown in the adjacent figure, a carpenter needs to cut triangle GBE out of a rectangular piece of plywood. At what distance from point D should the cut be started?

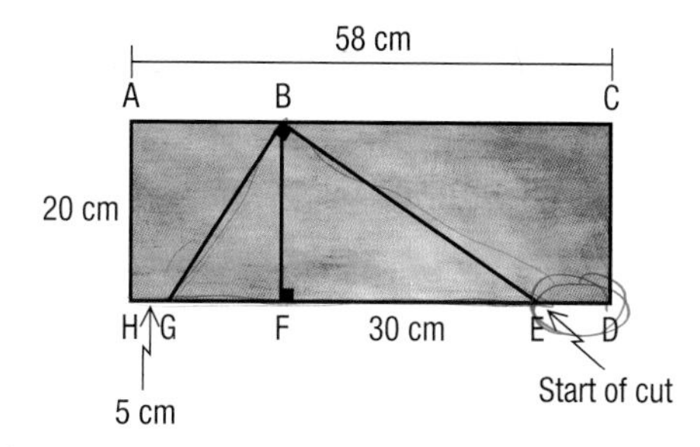

8 Triangle ABC, as shown, is right angled at vertex C. Determine the following lengths and identify the geometric statement on which you are basing your calculations.

a) $m\,\overline{BD}$

b) $m\,\overline{CB}$

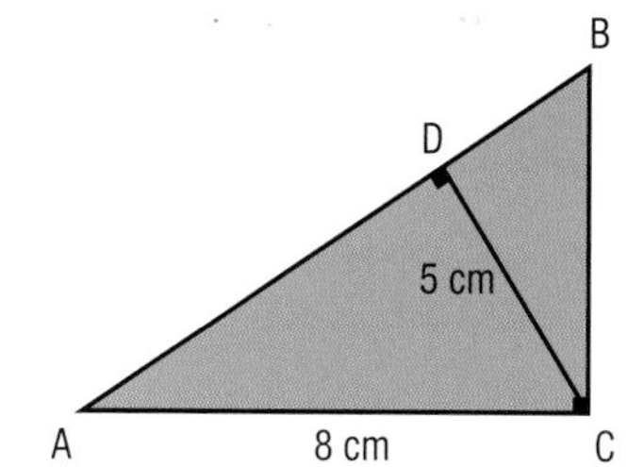

9 **EIFFEL TOWER** While visiting Paris, two tourists observed the Eiffel Tower from two different places. Wilford, in a boat on the Seine, is 635 m from the tower and he sees the top of the tower at an angle of elevation of 27°. Deymi, who is walking close to the tower, is 165 m from it. She sees the top of the tower at an angle of elevation of 63°. Use this information to determine the height of the tower.

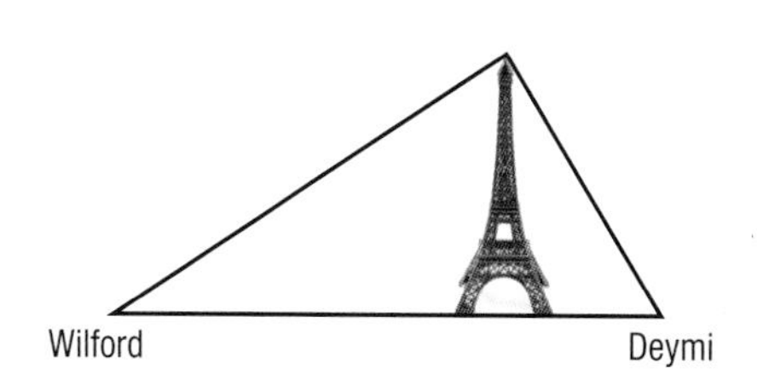

At the beginning of 2007, the Eiffel Tower had been visited more than 229 623 812 times.

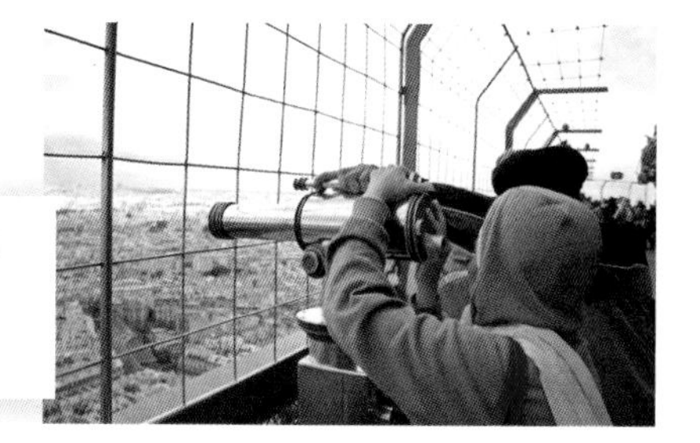

10 An entrepreneur attaches guy wires to a post. The following lists three situations.

Situation 1
Guy wires **A** and **B** are equal in length.

a) What distance from the base of the post should the two guy wires be anchored if the point of attachment on the post is 5 m above the ground?

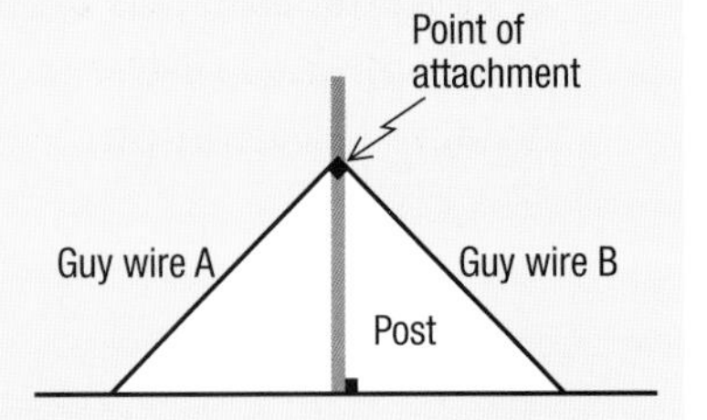

Situation 2
Guy wires **A** and **B** are not equal in length.

b) If the point of attachment on the post is 5 m above the ground and the distance between the base of the post and the anchor point of Guy wire **A** is 3 m, what is the distance between the base of the post and the anchor point of Guy wire **B**?

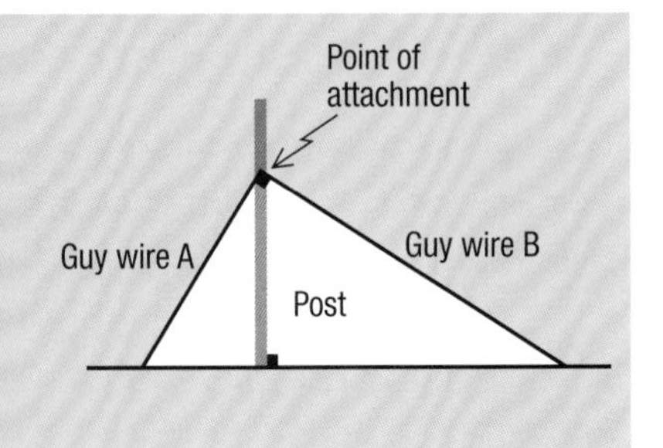

Situation 3
Guy wires **A** and **B** are not equal in length.
Guy wire **A** is 13 m long and the distance from its anchor point to the post is 12 m.

c) How high from the ground is the point where the guy wires are attached to the post?

d) What is the distance from the base of the post to the anchor point of Guy wire **B**?

e) What is the length of Guy wire **B**?

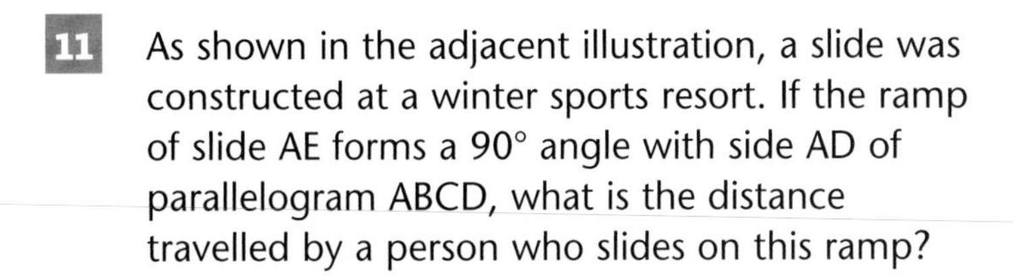

11 As shown in the adjacent illustration, a slide was constructed at a winter sports resort. If the ramp of slide AE forms a 90° angle with side AD of parallelogram ABCD, what is the distance travelled by a person who slides on this ramp?

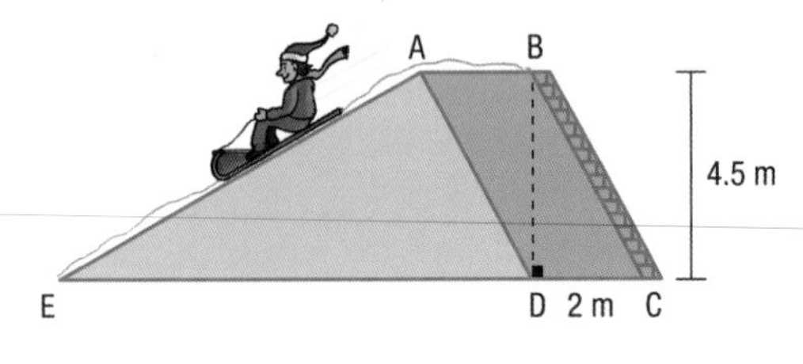

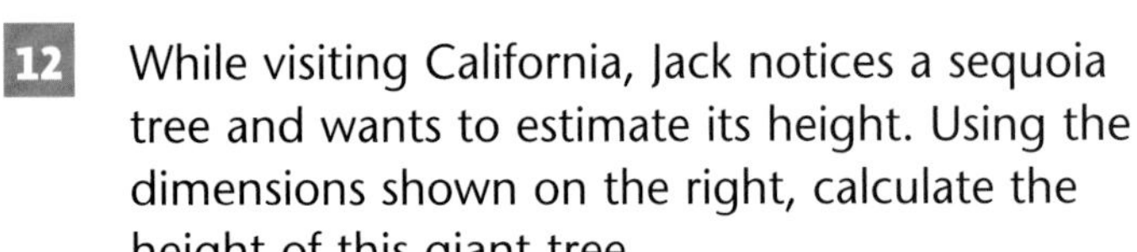

12 While visiting California, Jack notices a sequoia tree and wants to estimate its height. Using the dimensions shown on the right, calculate the height of this giant tree.

The largest known sequoia is "General Sherman" in Sequoia National Park in California, USA. It is 83 m high and its circumference is 30 m.

13 The owner of a stadium is preparing a celebration and wants to place a chain of electric lights along the blue lines of the seating as shown in the adjacent illustration. What will be the length of this chain?

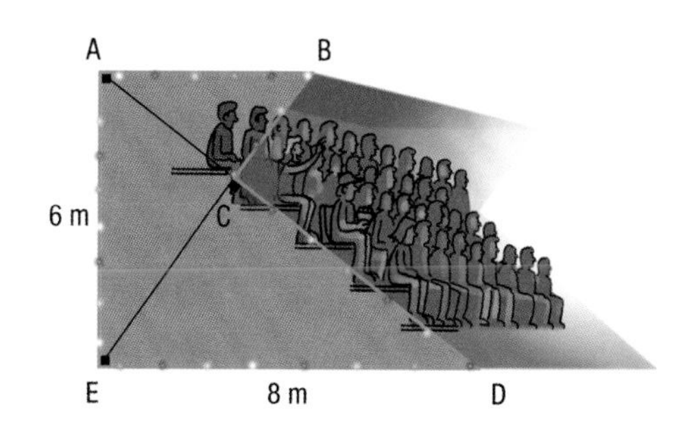

14 A field is in the shape of a parallelogram. Two drainpipes, represented by dotted lines, are added, as illustrated in the adjacent figure. If the drain costs $50/m, what is the cost of this drainage project?

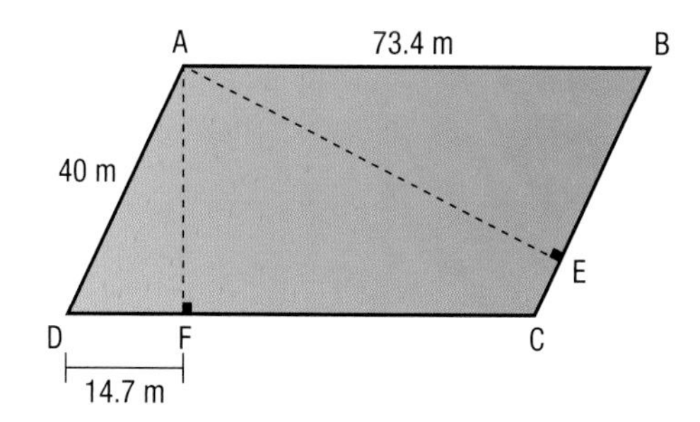

15 The adjacent illustration shows the side view of a warehouse. If the total height of the warehouse is the same as the length of the base of the wall, what is the height of the wall?

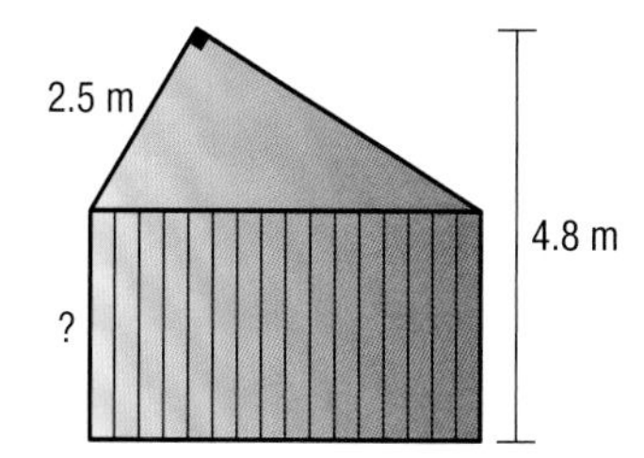

16 A circle with centre O is shown in the adjacent illustration. What is the area of the shaded section?

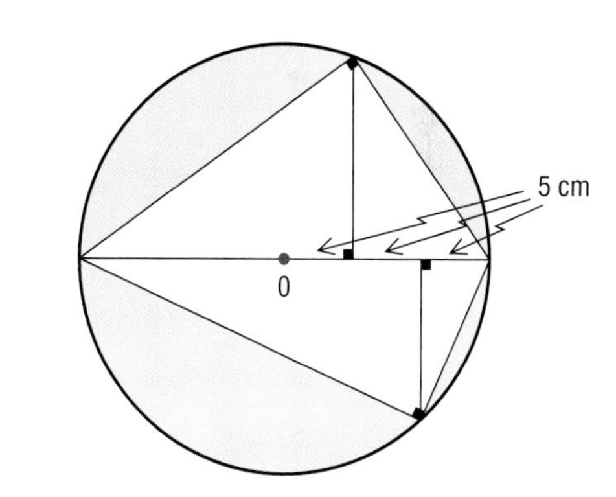

17 SANTOUR A santour is an Armenian instrument shaped like an isosceles trapezoid. Calculate the area of the top of the adjacent santour.

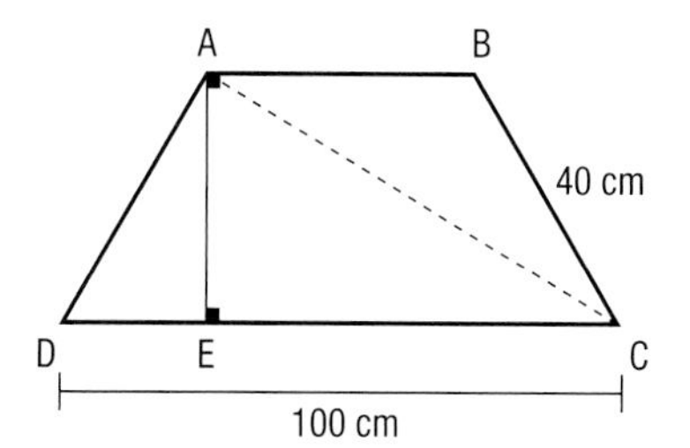

A santour is a stringed instrument which is struck with two finely curved batons.

Chronicle of the past

Ionian School

Since the beginning of time, mankind has sought answers to questions regarding the origins, the causes and the fundamental basis of all things. To answer these questions, people have turned to religion and philosophy. It is often said that philosophy is the mother of all sciences. In fact, a number of philosophers were also mathematicians. In the 6th century BCE, three philosophers from the ancient Greek city, Miletus, in Ionia, founded the Ionian school.

Ionia's contribution to the origins of philosophical and scientific thought has had an important influence on the Greek culture.

It seems that Thales was a merchant during the first part of his life. The fortune that he acquired allowed him to study and travel. He stayed for a while in Egypt where he became familiar with Egyptian mathematics and astronomy.

Thales of Miletus (ca. 625 BCE - ca. 546 BCE)

Thales is known for, among other things, the theorem that bears his name. The different configurations of this theorem enable the establishment of ratios of length and the calculation of various missing measurements using proportions. The following are two possible configurations of his theorem:

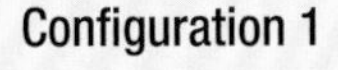

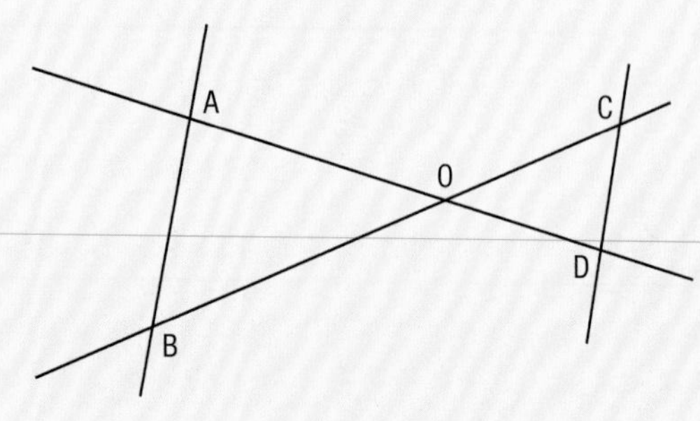

Configuration 2

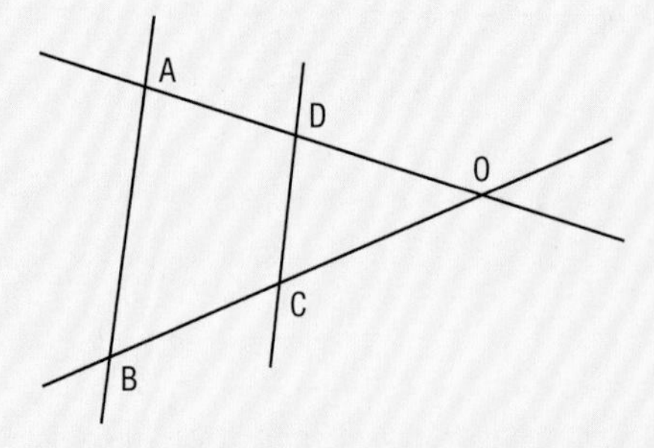

If lines AB and CD are parallel, then $\frac{m\overline{AO}}{m\overline{DO}} = \frac{m\overline{BO}}{m\overline{CO}}$.

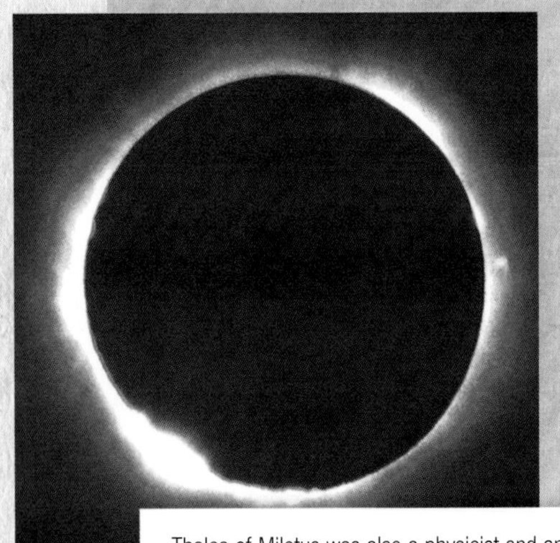

Thales of Miletus was also a physicist and an astronomer. He was the first to confirm that the moon was illuminated by the sun. He became a celebrity by predicting the solar eclipse in 585 BCE.

Anaximander of Miletus (ca. 610 BCE - ca. 546 BCE)

Anaximander, a student of Thales, later taught Pythagoras. Anaximander's interest in cosmology made him one of the fathers of astronomy, and this same interest led him to introduce the gnomon in Greece. Anaximander had gnomons built in Lacedemone (Sparta) to indicate solstices and equinoxes.

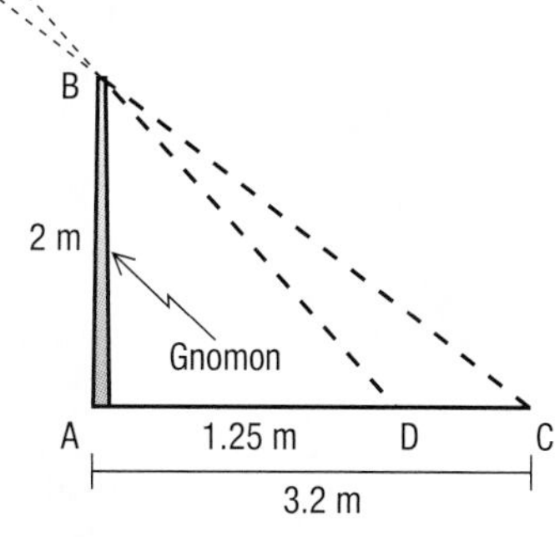

The shadow corresponding to segment AD is generated at noon, solar time, during the summer solstice, usually on June 21. The shadow corresponding to segment AC is generated by the gnomon at noon, solar time, during the winter solstice, usually on December 21.

A lunar crater was named in honor of Anaximander.

Anaximene (ca. 585 BCE - ca. 525 BCE)

Anaximene was one of the last disciples of the Ionian school. Like Anaximander, there are almost no written references of his work, and his contribution to mathematics is uncertain. Anaximene limited himself to popularizing the theories of his predecessors. He perfected the gnomon, explained the rainbow as light hitting condensed air (water) and was an inspiration to the Pythagoreans.

1. According to Thales' Configuration **1** theorem, if $m\,\overline{AO} = 3.9$ cm, $m\,\overline{DO} = 2.4$ cm and $m\,\overline{CO} = 2.9$ cm, determine the length of $\overline{BO}$.

2. By using Thales' Configuration **2** theorem:

   a) show that $\Delta\ ABO \sim \Delta\ DCO$

   b) can you state that m $\frac{m\,\overline{AO}}{m\,\overline{DO}} = \frac{m\,\overline{AB}}{m\,\overline{DC}}$? Explain your answer.

3. By using Anaximander's gnomon diagram, show that triangles ABC and ADB are similar.

In the workplace Carpenters

The trade

A carpenter's job is a skilled trade in the construction industry. This includes building walls and roofs; installing floor beams; fitting door frames and windows, mouldings, countertops, cabinets, shelves, and other elements of construction such as building staircases. Most of the work involves working with wood; however, working with metals could also be part of the trade.

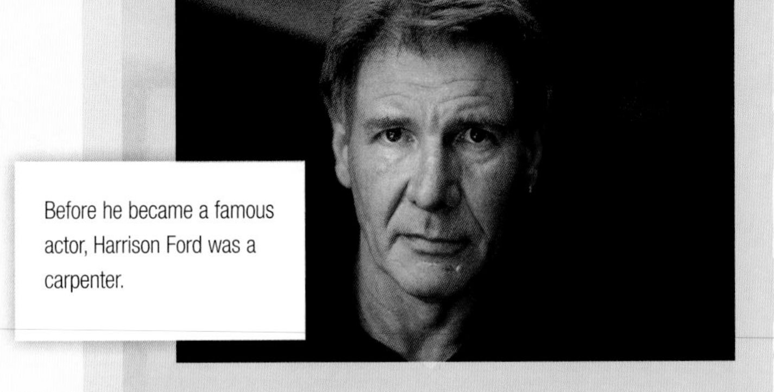

Before he became a famous actor, Harrison Ford was a carpenter.

"Measure seven times, but cut only once…" (Russian proverb).

Measure, saw, assemble

A carpenter must read plans and correctly interpret them in order to cut, shape, assemble, and join materials together. The worker must have knowledge of how to take precise measurements to avoid wasting materials, making poor cuts, or causing a weakness in the final structure.

Roof trusses

In Québec, slanted roofs are generally built with a triangular roof truss. Each roof truss is assembled with wooden frames and metal gussets. The placement of each wooden frame is extremely important for the sturdiness of the structure.

Figure 1: Roof truss

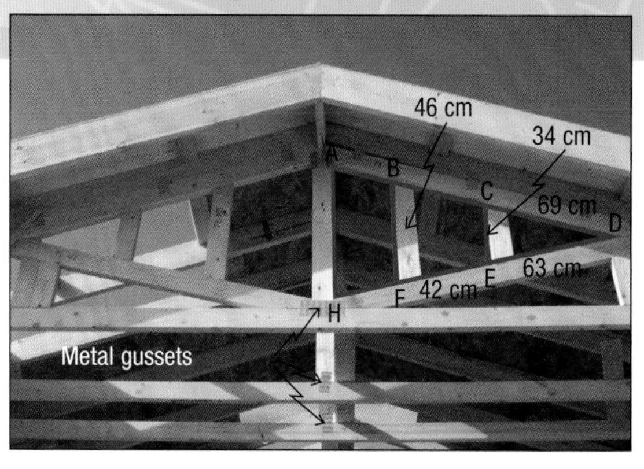

Wind bracing

Wind bracing is an element in construction used to protect a structure from a lateral deformation or from horizontal stresses induced by winds. For example, a carpenter uses a wind brace to strengthen the solidity of the walls of a house under construction. An attachment of wooden frames is installed on each side of the wall to avoid distortion or rafting caused by winds.

Figure 2: Wall in construction

1. In the roof truss shown in Figure **1**, $\overline{AH} \parallel \overline{BF} \parallel \overline{CE}$.

a) Show that Δ BDF $\sim$ Δ CDE.

b) Determine the length of the wooden beam shown by segment BF.

2. In Figure **2**, the rectangular wall BCEF measures 2.5 m by 3.9 m. Considering that m $\overline{AF}$ = m $\overline{DE}$, do the following:

a) Show that Δ ABF $\cong$ Δ CDE.

b) Determine the length of the wooden beam shown by segment CD.

overview

1 For each pair of triangles below, indicate the geometric statement which allows you to conclude that the triangles are congruent.

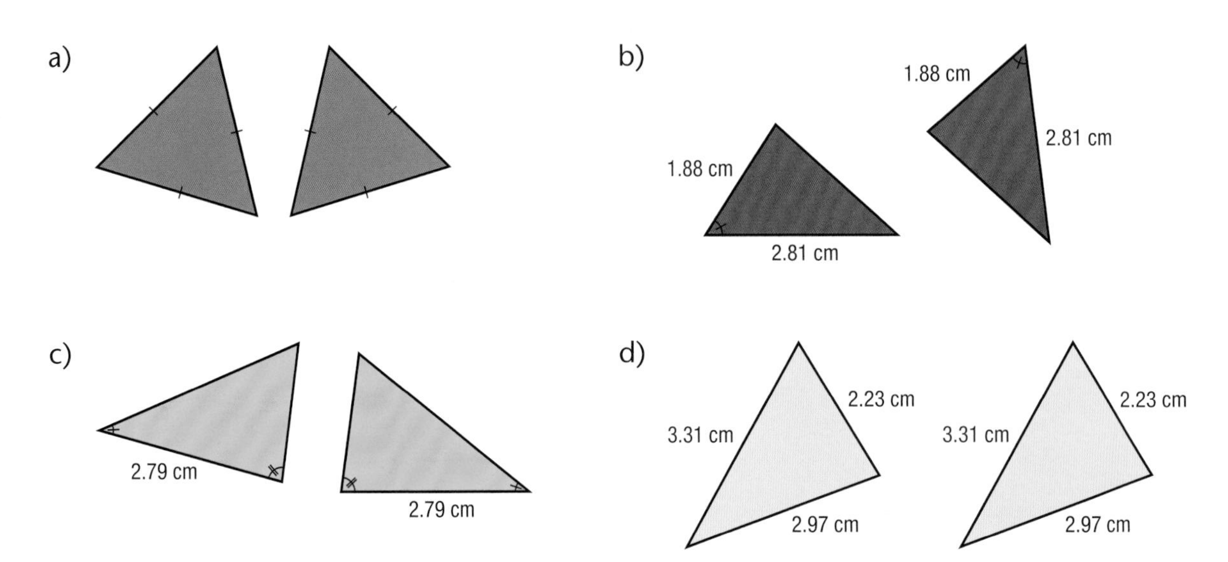

2 In each case, indicate whether the two triangles are congruent, similar or neither. Provide the geometric statement to support your claim.

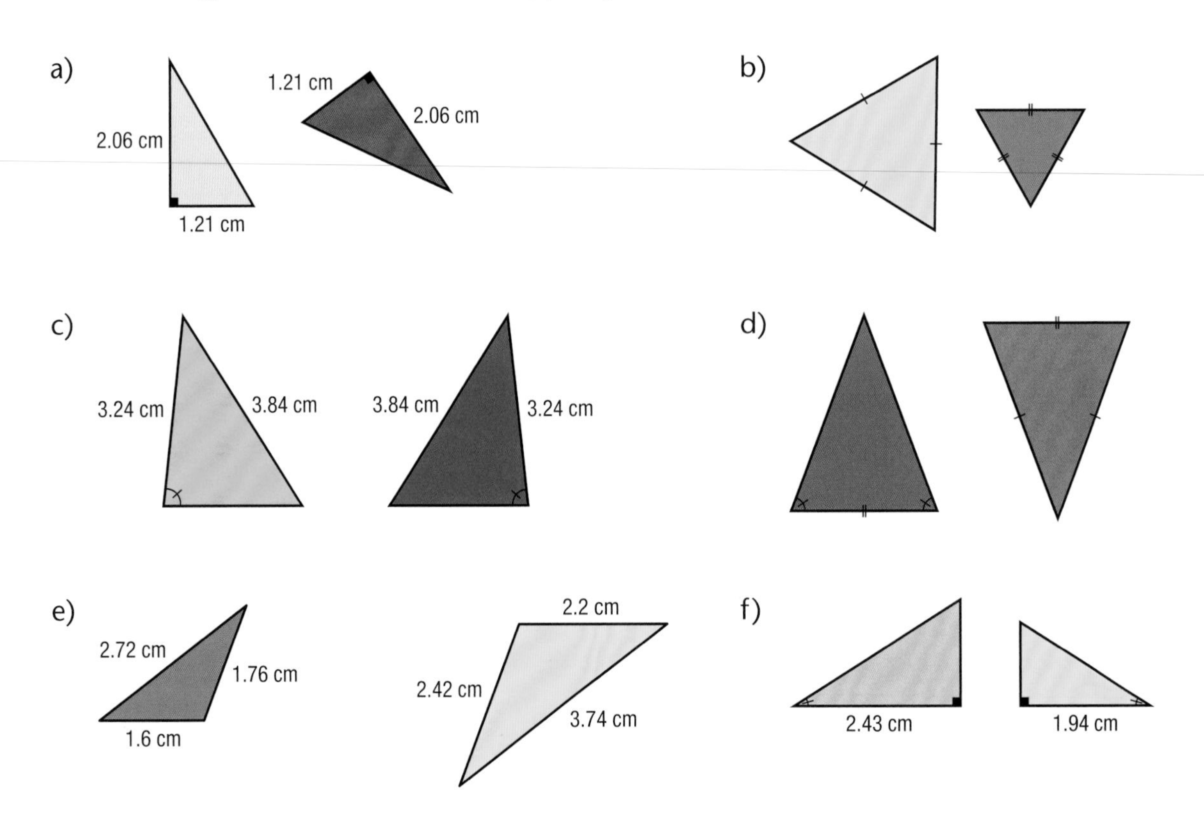

3 Decide whether the following statements are true or false. If false, provide a counter-example.

a) Two rhombuses with the same perimeter are congruent.

b) Two congruent rhombuses have the same area.

c) If the corresponding diagonals of two rhombuses are congruent, then the rhombuses are congruent.

d) Two rhombuses are congruent if their corresponding angles are equal.

4 Are the polygons below necessarily congruent? In each case, explain your answer.

a) two squares

b) two squares with a common side

c) two rectangles of which the corresponding diagonals are equal

d) two rectangles whose perimeters are equal

e) two parallelograms whose corresponding angles are equal

5 For each pair of triangles shown below, calculate the value of x.

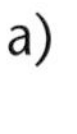

a)

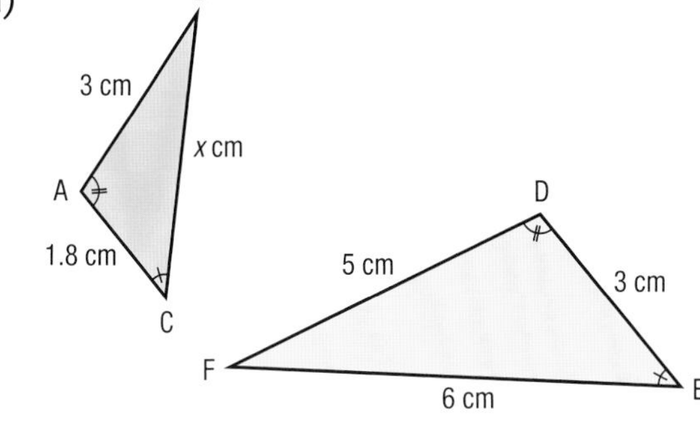

b)

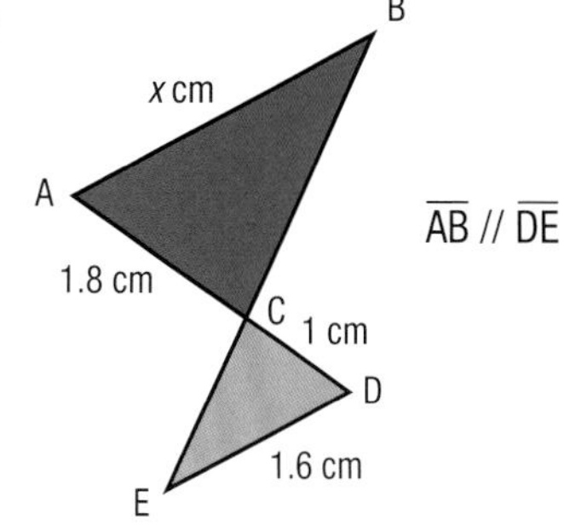

c)

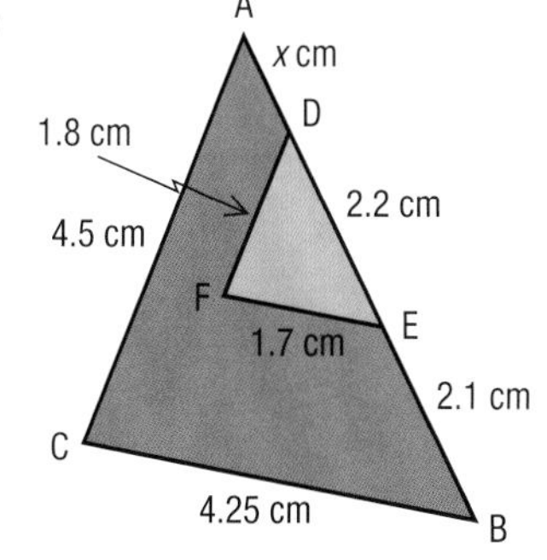

d)

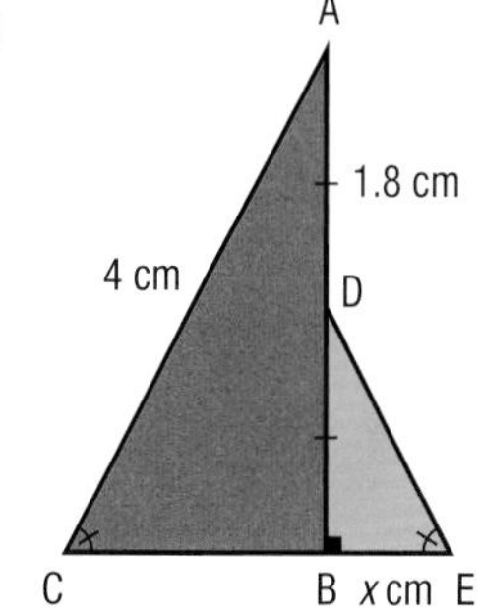

6 Using geometric statements concerning congruent triangles, show that the following statements are true.

a) If ABCD is a rectangle then diagonals AC and BD are congruent.

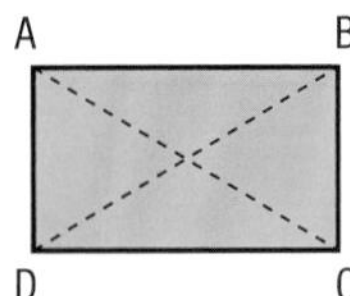

b) A quadrilateral in which two opposite sides are congruent and parallel is a parallogram.

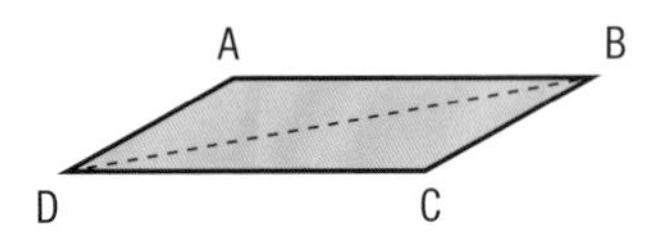

c) The diagonals of a parallelogram intersect at their midpoint.

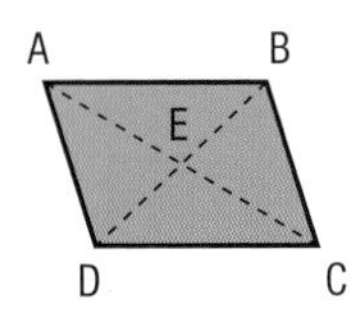

7 A carpenter needs to know if the floor of his shed, as illustrated, is a rectangle. He does not have a set square or a protractor. Nevertheless, he knows that the opposite sides are congruent. How can he be sure the floor is a rectangle?

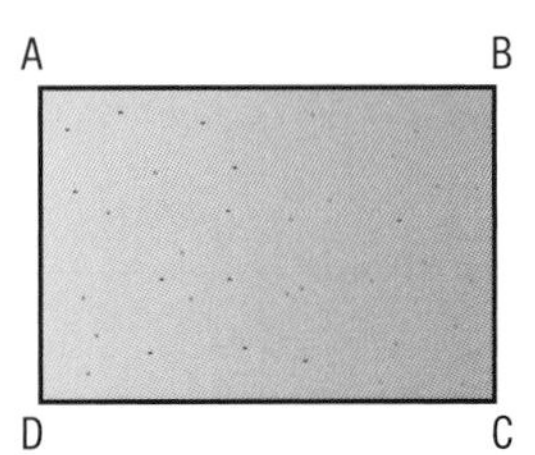

8 In the adjacent trapezoid ABCD, do the following:

a) Name the eight triangles that can be formed.

b) Name the three pairs of congruent triangles.

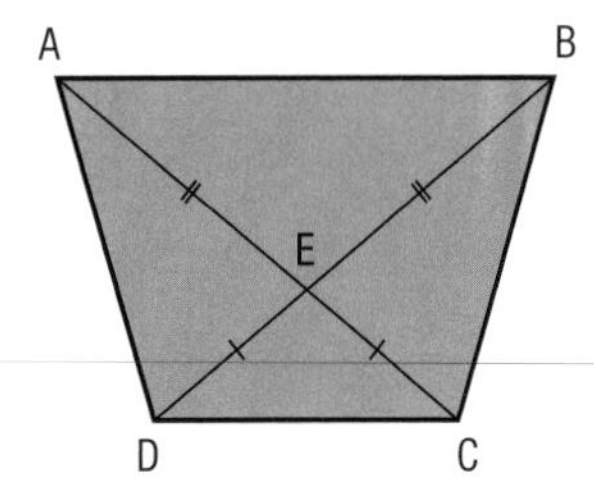

9 In the adjacent diagram, AB // CD // EF and GH // IJ // KL.

a) Find lengths of:

1) m $\overline{CE}$

2) m $\overline{HJ}$

3) m $\overline{CD}$

4) m $\overline{GH}$

b) Is it possible to find m $\overline{FH}$? Explain your answer.

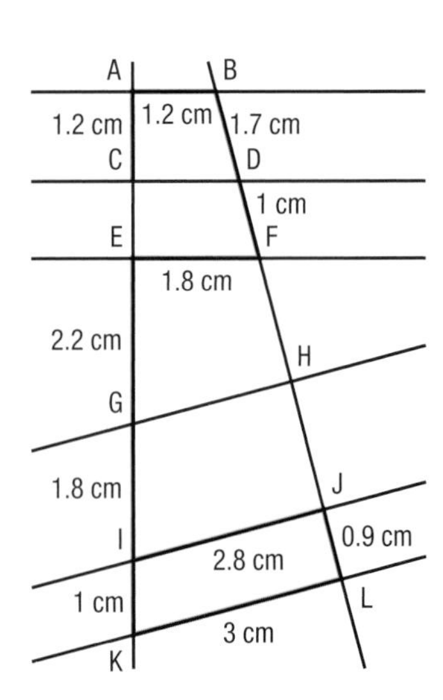

10 From a hot air balloon Elena is able to see her house and her school. From her house to the library, which is directly below the balloon basket, the distance is 1.5 km. If the distance from the library to the school is 1 km, at what altitude is the balloon basket?

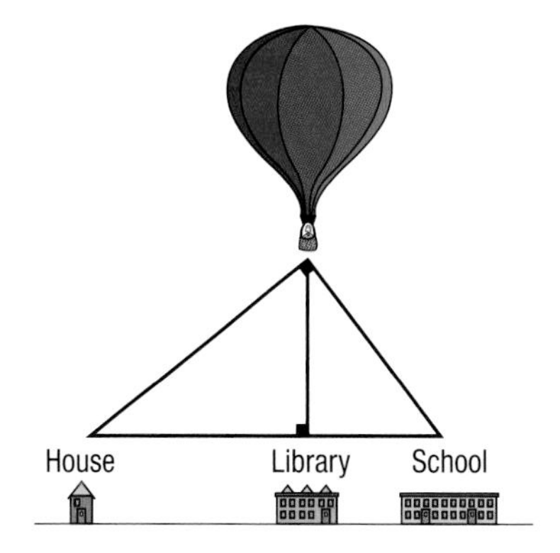

11 The adjacent illustration represents a house in the shape of a right prism with a triangular base. Calculate the volume of each of the five rooms.

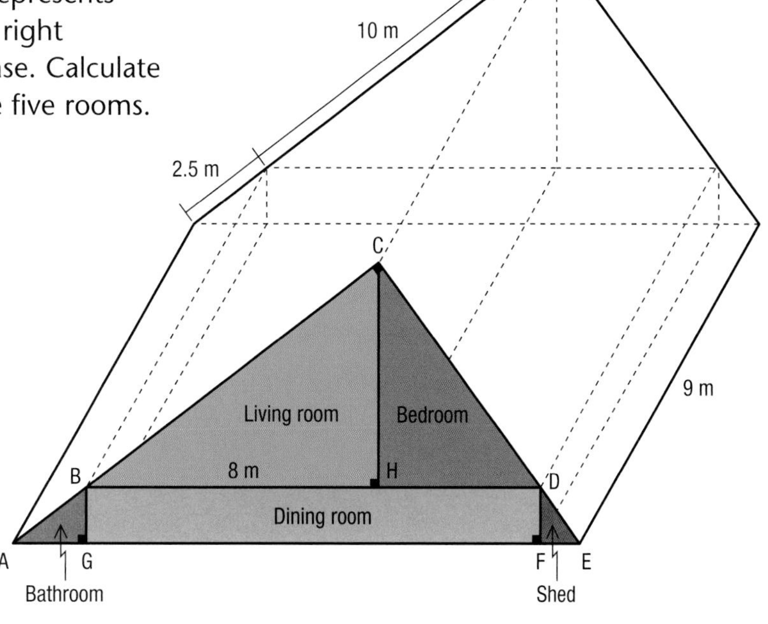

12 In the adjacent figure, $\overline{BE} \parallel \overline{CD}$. Find the length of segments AB, BE and CD. The dimensions are in centimetres.

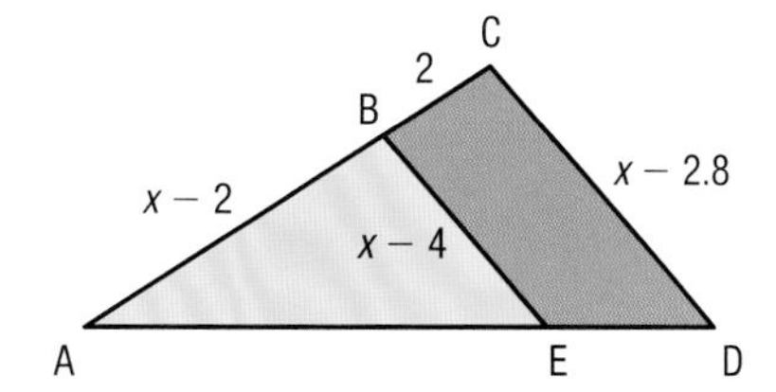

13 A horse trainer is standing at the edge of a fence of two adjoining rectangular paddocks as shown in the diagram below. Find the total area of the paddocks.

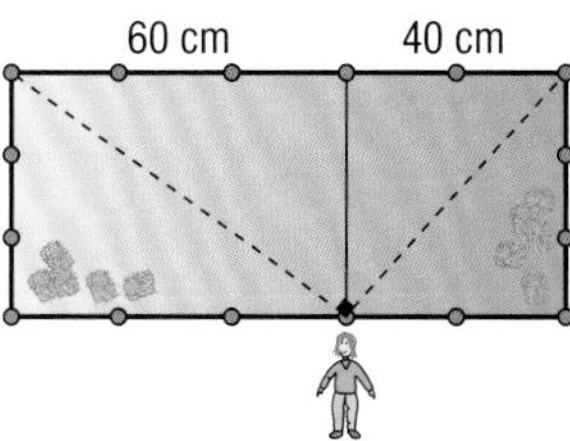

14 An electric lamppost casts a shadow of 480 cm. At the same moment, a person 1.7 m tall standing next to the lamppost casts a shadow of 90 cm. What is the height of the post?

15 In the adjacent figure the segment BE is a median of triangle ABC.

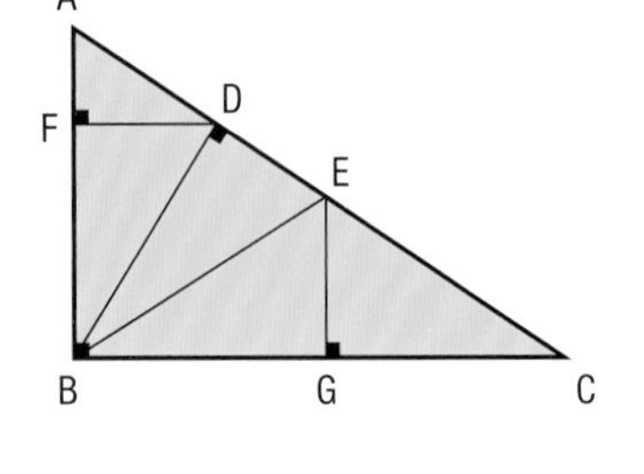

a) If possible, name a triangle similar to triangle:
   1) ABC
   2) EBG
   3) BED

b) If possible, name a triangle congruent to triangle:
   1) ECG
   2) FDB

16 Three right triangles were assembled to form the adjacent rectangle. Which geometric statement shows that:

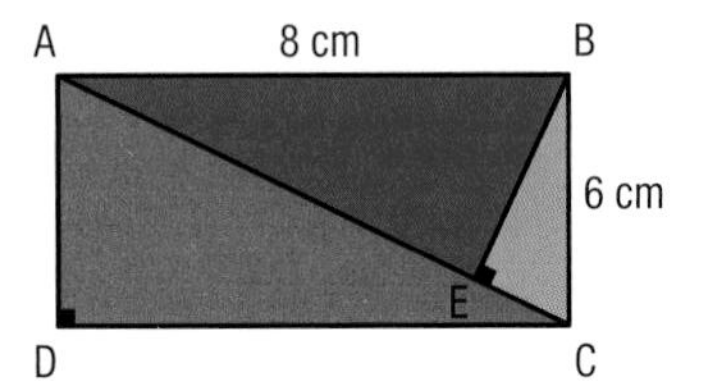

a) $\Delta ACD \cong \Delta ABC$?

b) $\Delta ACD \sim \Delta ABE$?

c) m $\overline{AC}$ = 10 cm?

d) m $\overline{AE}$ = 6.4 cm?

e) m $\overline{BE}$ = 4.8 cm?

17 The legs of the camping chairs below are attached to the four corners of a seat which is in the shape of a square. Among the triangles below indicate the ones that are congruent.

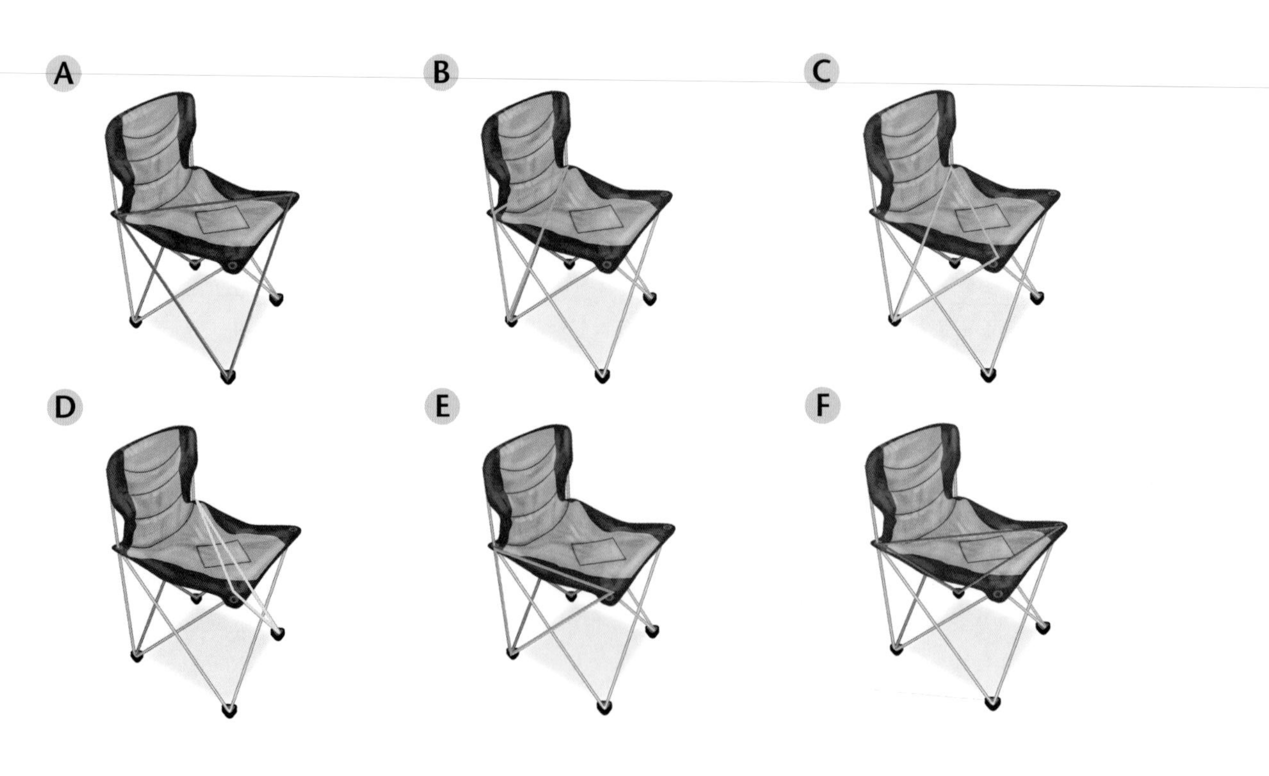

bank of problems

18 In the adjacent illustration, note the following:

- Triangle ABC is inscribed in the circle with centre O.
- Radius of the circle measures 6 cm.
- The segment BH corresponds to $\frac{3}{4}$ of the radius of the circle.

What is the length of segment OK?

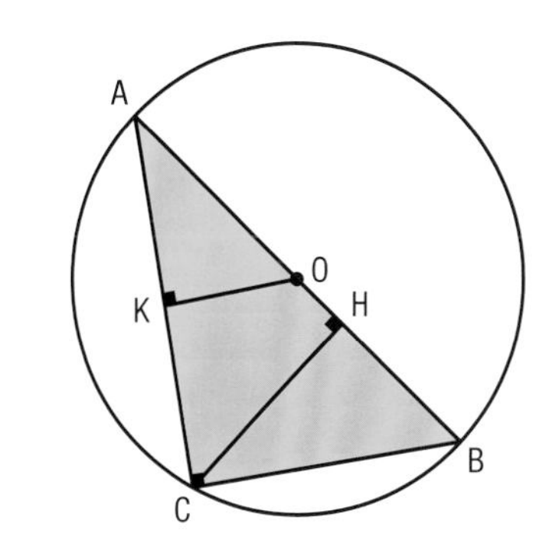

19 The adjacent illustration shows a bridge built of wooden sticks. Determine the lengths of *a*, *b*, and *c* and provide the geometric statements used for the basis of your calculations.

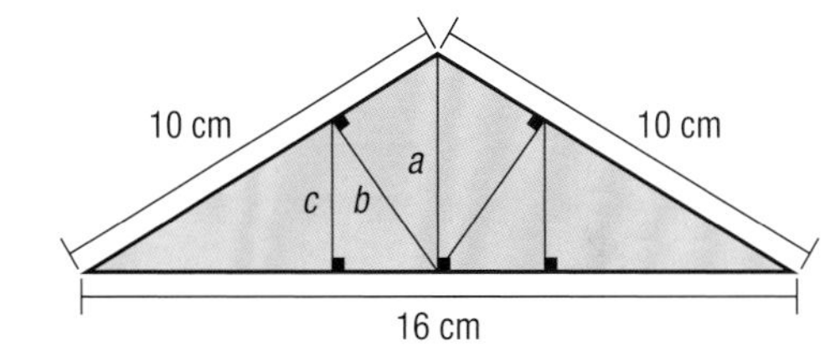

20 The metronome, illustrated below, has the shape of an equilateral triangle in which each side measures 20 cm. Determine the area of the grey region.

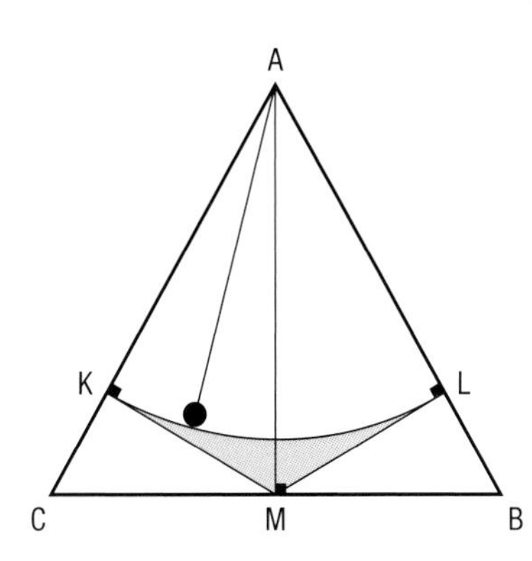

A metronome is a device that produces an audible or visual pulse to establish a tempo, the speed for the performance of musical compositions.

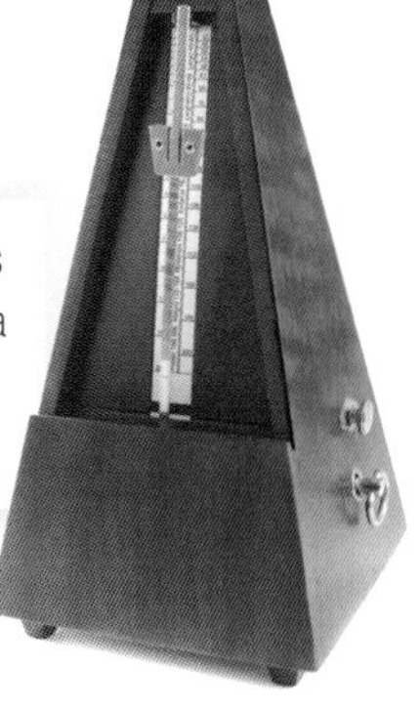

21 You can extend the diagonal of the rectangle ABCD so that m $\overline{AC}$ = m $\overline{CE}$. You can also extend side AD so that m $\overline{AD}$ = m $\overline{DF}$. Considering that $\overline{AF} \perp \overline{EF}$, determine the perimeter of the quadrilateral DGHF.

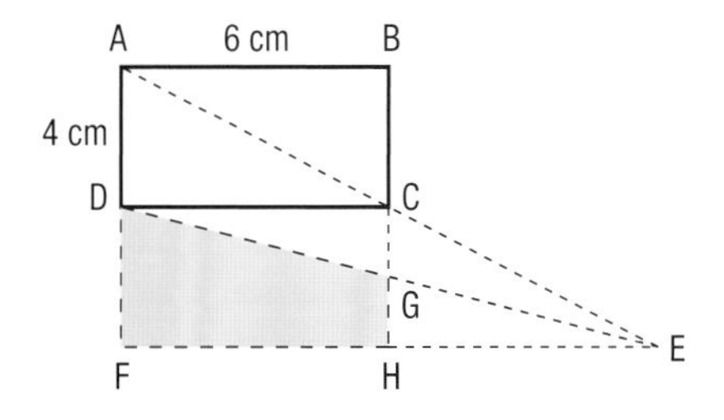

22 In the adjacent triangle ABC, determine the position of point D on side AC side so that segment DE is parallel to segment BC.

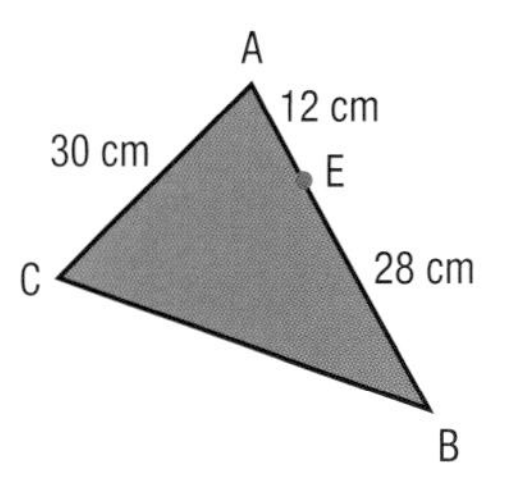

23 The adjacent illustration shows regular pentagon ABCDE. There are five right triangles drawn inside of it. Explain why these five triangles are congruent.

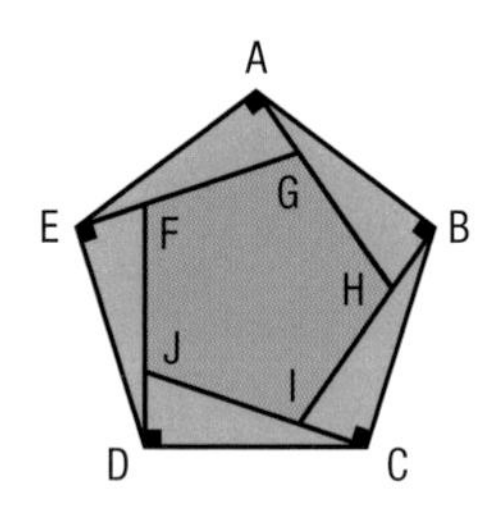

24 As shown in the adjacent illustration, when signalled, Clara and Anna walk in opposite directions. Clara walks at a speed of 3.75 km/h and Anna, at a speed of 8 km/h. Considering that they are separated by 22 km at the end of a 48 minute walk, how far apart were they at the start of their walk?

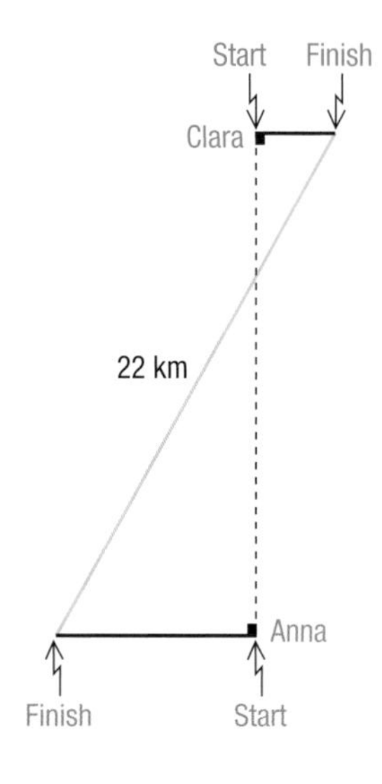

25 For the adjacent pentagon ABCDE, show that the diagonals AC and EB are congruent.

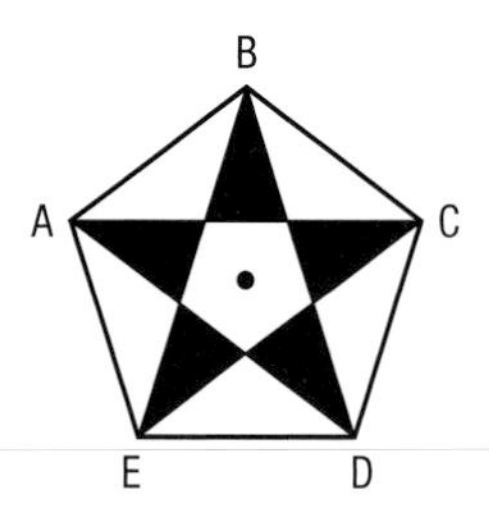

26 A company's logo is shown in the adjacent illustration. The company wants to incorporate the logo into a flowerbed design. Each section of the flowerbed will be decorated with different types of flowers. If in the diagram 1 cm $\triangleq$ 1 m, what is the real area of the blue region?

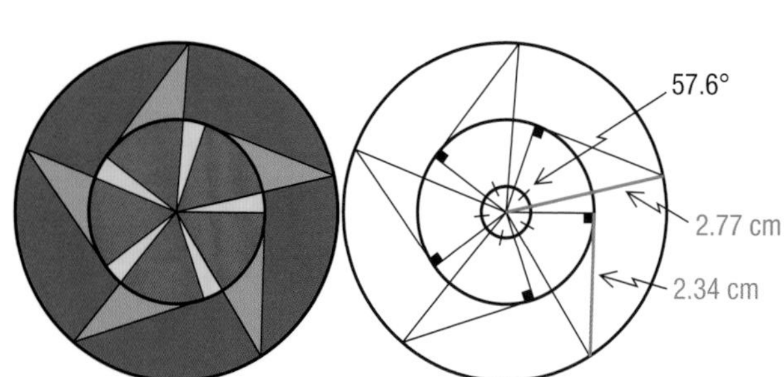

A logotype, more commonly known as a logo, is a unique symbol identifying companies, products, services, institutions, agencies, associations, events, or all other types of organizations. The purpose of a logo is to differentiate one organization from another.

LEARNING AND EVALUATION SITUATIONS

TABLE OF CONTENTS

VISION 1

VISION 2

VISION 3

Urbanization

Learning context

Throughout the world, cities are expanding to accommodate more and more people. Urbanization, the settling of a population in and around a city, is perhaps the most important social phenomenon in Québec since the beginning of the 20th century. In 1901, approximately 36% of Québec's population lived in urban areas. Thirty years later, this number had grown to approximately 63%. Urbanization can be triggered by economic, technological, geographical, demographical, political or social reasons.

Managing the geographical area of a city and its surroundings is not left to chance. In Québec, the management of urban land is accomplished using urban plans based on various aesthetic, economic, functional, environmental and social criteria.

View of a neighbourhood in Pachuca, Mexico.

This LES is related to sections 1.1 to 1.3.

LES 1

A new residential sector

An urban planner is a professional who plans the construction, growth and development of an urban community. He or she manages and orgnizes the use of city territory in accordance with an urban plan, and in doing so, takes into account the needs and wants of the citizens and the goals of sustainable development.

The trapezoid below represents a new residential sector.
The graph is scaled in metres.

As an urban planner you must prepare and present an urban plan for this new sector to the municipal council. According to this municipality's urban plan, you must ensure the following:

- This sector must be sub-divided into six lots.
- The area of four of these lots must be approximately 500 m^2.
- Fire hydrants must be installed at points A, B, C and D.

Your urban plan must include:

- a diagram of the sector, including the boundaries of each lot
- the exact surface area of each lot
- the equation of the line associated with each of the streets identified in the graph
- the coordinates of the points representing the fire hydrants

 LES 2

This LES is related to sections 1.1 to 1.4.

C2

Zoning

Working in the best interests of its population, a municipality may decide to divide its land into as many zones as it deems necessary. A municipality's zoning bylaws regulate the way or ways in which an area of land can be used based on environmental, functional, aesthetic or socioeconomic criteria. There are four types of zones: residential zones, agricultural zones, industrial zones and commercial zones.

Below is a municipality's zoning plan. The graph is scaled in metres.

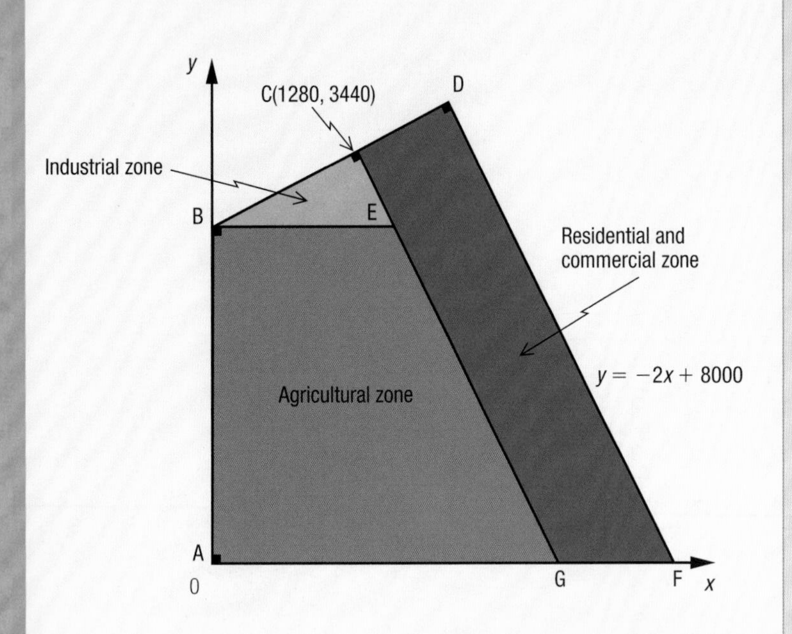

Your task is to verify whether the proposed plan respects the municipal zoning regulations. Among other things, these regulations stipulate the following:

- The land designated as industrial must occupy from 3% to 5% of the municipality's terriotory.
- The land designated as agricultural must occupy less than 62% of the municipality's territory.
- The land designated as residential and commerical must occupy more than 35% of the municipality's territory.

You must also determine the inequalities that graphically define each of this municipality's three zones.

Pollution

Learning context

Whether it is through recuperation, recycling, composting or the introduction of energy-saving measures, more and more Canadians are working together to help protect the environment. Since 2004, the Government of Canada has published national indexes of air quality, greenhouse gas emissions and freshwater quality. Statistics Canada and Health Canada work in collaboration to provide Canadians with updated, clear and tangible information on environmental conditions.

The national air quality index in Canada is based primarily on levels of exposure to ozone in the troposphere and to fine particles of matter, two primary components of smog. Human exposure to these two pollutants is of great concern. From 1990 to 2004, the ozone indicator registered a mean annual increase of 0.9%.

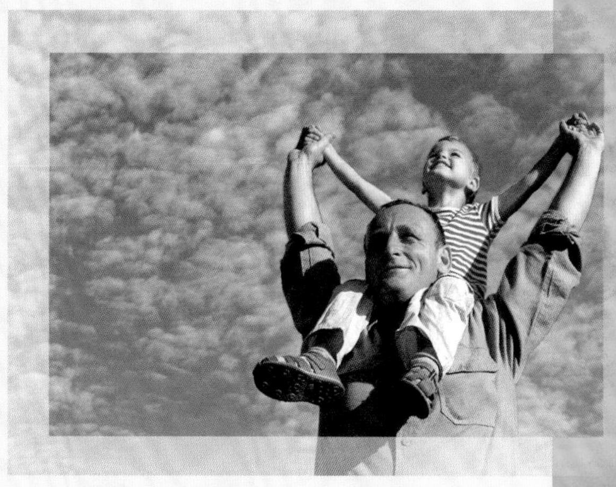

The greenhouse gas index refers to the total gas emissions at the national level. From 1990 to 2004, these emissions increased by approximately 27% across the country, exceeding the objective set by the Canadian government during the ratification of the Kyoto Protocol at the United Nations Framework Convention on Climate Change in 2002.

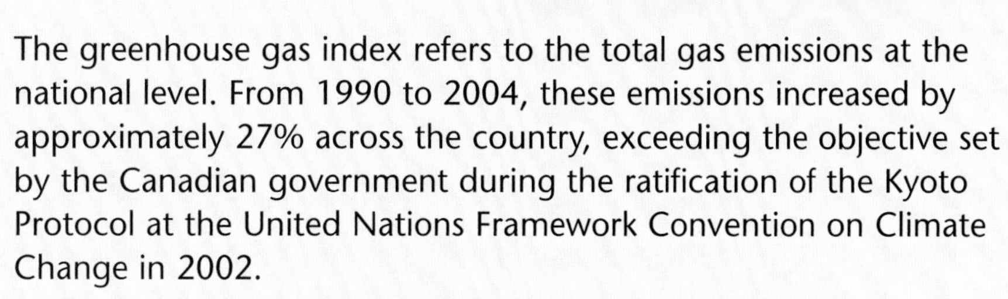

The water quality index emphasizes the capacity of Canada's bodies of water to support the needs of aquatic life. In a sample of 340 Canadian sites, freshwater quality was analyzed and considered good to excellent in 44% of cases, average in 34% of cases, and poor or bad in 22% of cases.

The Kyoto Protocol is an agreement signed by approximately 180 industrialized countries to commit to reducing greenhouse gas emissions. These gases are produced mainly by burning fossil fuels like coal, gasoline and diesel oil.

This LES is related to section 2.1.

Carbon exchange

In operation since January 2005, the Carbon Exchange is a market of negociation and exchange of credits for reducing greenhouse gas emissions. Quotas are first set by the participating countries which then issue notices to companies and industries enrolled in the program and specify the quantity of carbon dioxide (CO_2) they can emit over a given period of time. Those companies that reduce their greenhouse gas emissions can sell unused credits; companies producing too much pollution and exceeding the set emission limits must buy credits. One credit is equivalent to a ton of CO_2.

Listed below are data reguarding CO_2 emissions for companies registered with the Carbon Exchange for a single country:

Country

	Company															
	A	B	C	D	E	F	G	H	I	J	K	L	M	N	O	P
Quota of CO_2 (tons)	8	6	9	5	4	10	8	9	10	12	16	13	7	6	10	8
CO_2 emissions (tons)	11	5	10	6	4	11	5	5	11	14	18	7	12	6	7	9

Using the additional information and statistical calculations that will be supplied to you, analyze the data and draw conclusions about the amount of greenhouse gas emissions produced by the companies of four countries listed on the Carbon Exchange. Amongst other things, you must do the following:

- Make a diagram representing the quantities of CO_2 emitted by the companies in the same country.
- Identify the country that best respects the imposed quotas.
- Determine in which country Company **D** has reduced its greenhouse gas emissions the most.

O LES 4

This LES is related to sections 2.2 to 2.4.

C3

Water quality

The quality of fresh water in Canada is threatened by the way it is used in agriculture, in industry and by the general population. For example, fertilizer that drains or is dumped into the sewage system accelerates the growth of algae. The increase in algae reduces the quantity of oxygen normally available to plants and other organisms. When algae die, they decompose; this is a process that consumes even more oxygen. In turn, other aquatic organisms lack oxygen and die. The end result is that water becomes unfit for consumption.

BOD_5, or "Biochemical Oxygen Demand over five days," represents the quantity of oxygen (in mg/L) used by micro-organisms to decompose the organic matter found in a water sample maintained at 20°C, in the dark, for five days. Water from lakes and rivers contains a BOD_5 of several mg/L while household waste water may reach values as high as 300 mg/L.

The contingency table below lists the results of a study on the quality of various rivers in Québec.

Water quality

BOD_5 (mg/L) \ Number of bacteria (coliform colonies per litre of water)	18	19	20	21	22	23	24	25	26	27	28
12		1	2								
13			1		3						
14				1							
15					2						
16						1	3		2	1	
17					1	2					2
18								1		1	

As a biologist, you must study the effects of human pollution on various bodies of water in Québec. Your report must include an analysis and an interpretation of the relationship between the two variables presented in the contingency table above. You must also write a short text describing the effects of human pollution on aquatic ecosystems.

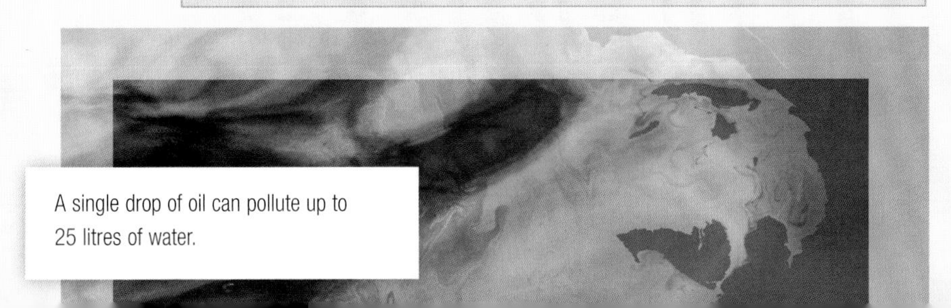

A single drop of oil can pollute up to 25 litres of water.

VISION 3

Overconsumption

Learning context

Deforestation, acid rain, degradation and pollution of various waterways, climate change, alteration of various wildlife habitats, and the thinning of the ozone layer are direct or indirect consequences of overconsumption in the environment. It is clear that overconsumption has become a global problem with potentially irreversible environmental consequences, in the foresecable future. The need for sustainable development calls for us to act as responsible consumers.

Companies also need to pay attention to overconsumption in order to minimize future production costs and negative environmental impacts. Air, water, soil and biodiversity are the victims of overconsumption. The misuse of chemicals has a negative impact on the environment with some of these chemical products or goods containing toxic and potentially dangerous materials. Following their initial usefulness, they often end up in dump sites, sewers, the water table or in the air that we breathe and become harmful to the environment and humans alike. For example, cyano bacteria or blue-green algae that grows mainly in the summer in waterways rich in phosphorus and nitrogen, come from the ingredients found in detergents and fertilizers.

LES 5

This LES is related to sections 3.1 and 3.2.

C2

Soil fertilization

Agriculturists are now aware that excessive use of fertilizers can be harmful to the environment. They must use these products in moderation.

The illustration below represents a grain farmer's fields. A given amount of fertilizer is spread over each of the fields in order to increase productivity. However, production costs and the environmental impact must be minimized.

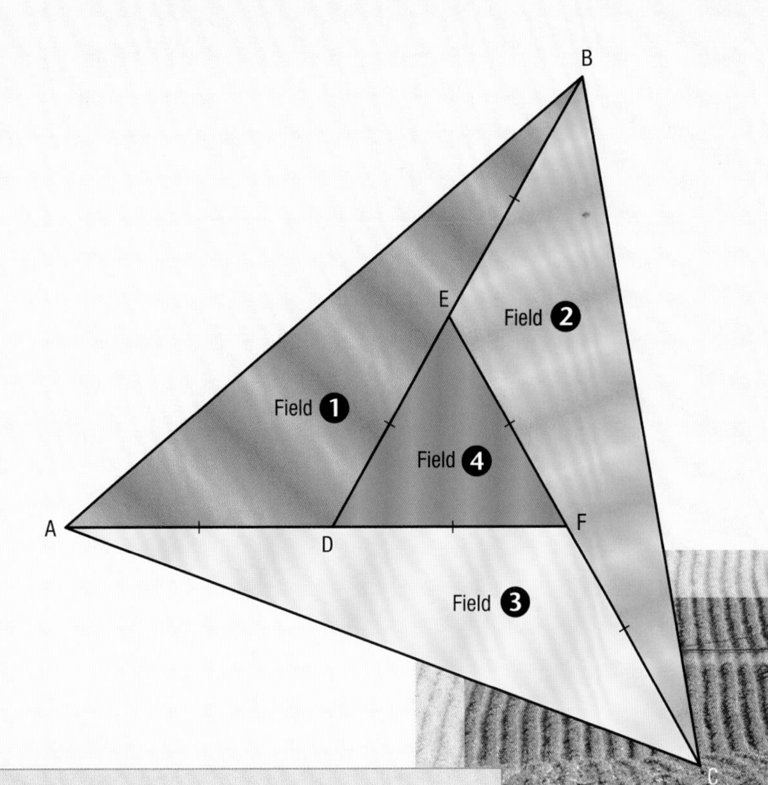

After doing various calculations, this grain farmer states that he will be able to fertilize all of his fields for less than $18,000. Using the additional data that you will be provided, support or refute this claim.

LES 6

This LES is related to sections 3.2 and 3.3.

C1

Roof trusses

Different situations urge people to make clear choices with regard to their consumption. The construction and renovation industry is such an example. In order to minimize their production costs, contractors must establish, as accurately as possible, the quantity of materials needed to construct a building or part of a building.

Below is a plan for the crossbeam structure of a roof truss to be used in the construction of a shed:

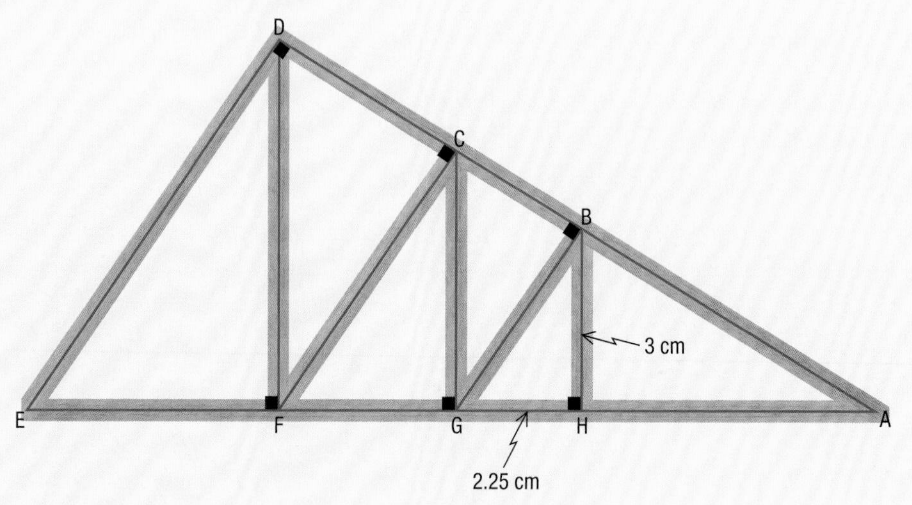

In this plan, note the following:

- The lengths indicated are 25 times smaller than the actual size.
- Segments BH, CG and DF as well as BG, CF and DE are parallel to each other.

As a construction entrepreneur, use the additional information you will be given and determine the quantity and lowest possible cost of the wood required to construct the roof trusses for the shed.

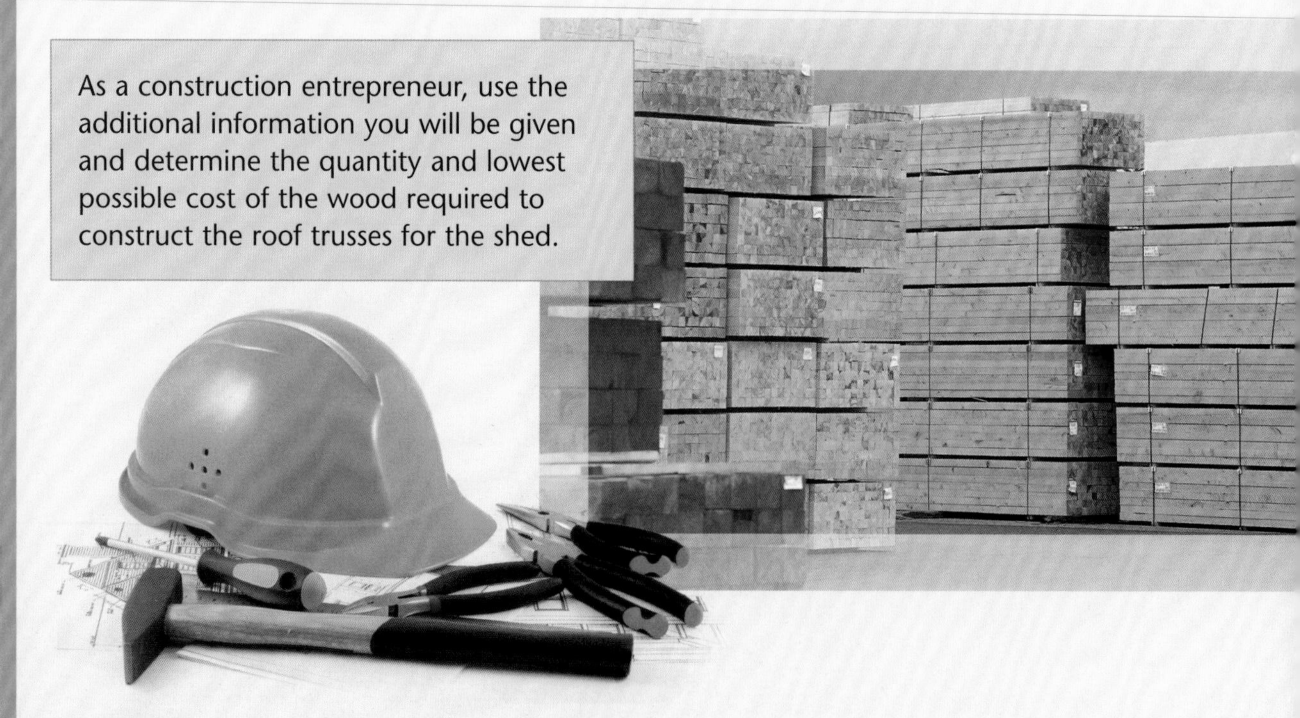

REFERENCE

TABLE OF CONTENTS

Graphing calculator

Sample calculations

It is possible to perform scientific calculations and to evaluate both algebraic and logical expressions.

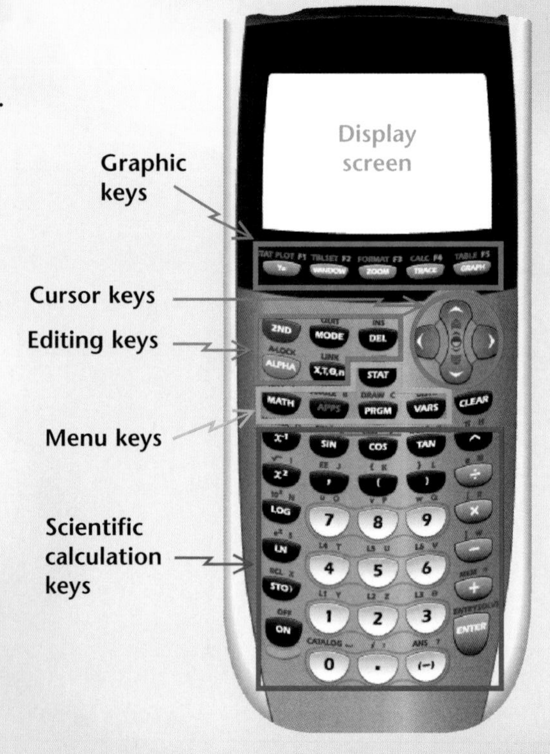

Scientific calculations

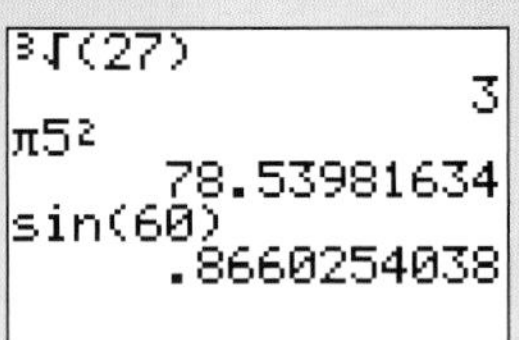

Logical expressions

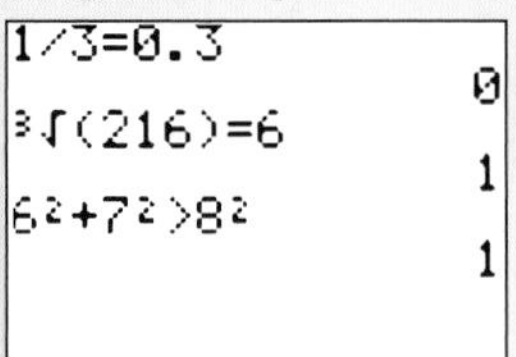
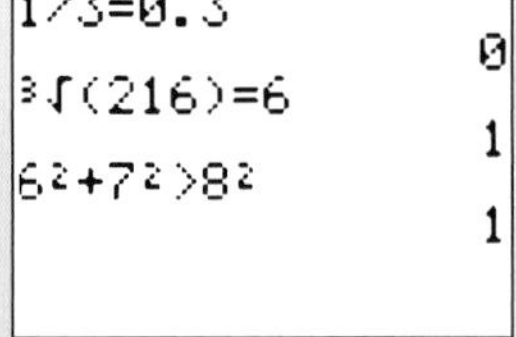

Algebraic expressions

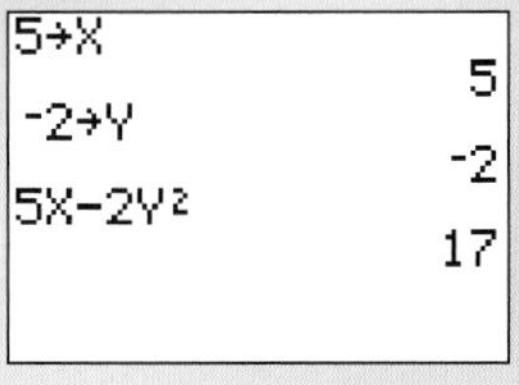

Probability

1. Display the probability menu.

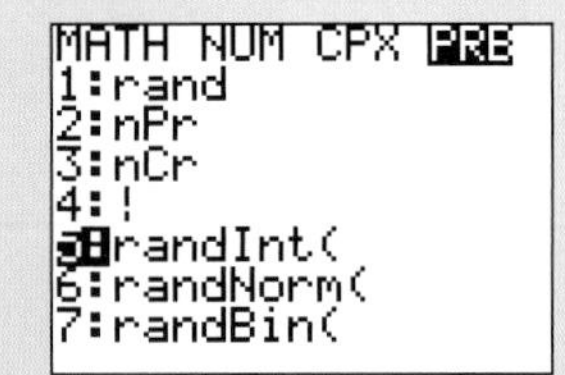

- Among other things this menu allows the simulation of random experiments. The fifth option generates a series of random whole numbers. Syntax: randInt (minimum value, maximum value, number of repetitions).

2. Display calculations and results.

```
randInt(0,1,5)
         {1 1 0 1 0}
randInt(1,6,7)
 {5 1 3 6 2 5 6}
```

- The first example simulates flipping a coin 5 times where 0 represents tails and 1 represents heads.The second example simulates seven rolls of a die with 6 faces.

Display a table of values

1. Define the rules.

```
Plot1 Plot2 Plot3
\Y1=2^X
\Y2=0.5X²
\Y3=
\Y4=
\Y5=
\Y6=
\Y7=
```

- This screen allows you to enter and edit the rules for one or more functions where Y is the dependent variable and X is the independent variable.

2. Define the viewing window.

- This screen allows you to define the viewing window for a table of values indicating the starting value of X and the step size for the variation of X.

3. Display the table.

X	Y1	Y2
0	1	0
1	2	.5
2	4	2
3	8	4.5
4	16	8
5	32	12.5
6	64	18

X=0

- This screen allows you to display the table of values of the rules defined.

Display a graphical representation

1. Define the rules.

```
Plot1 Plot2 Plot3
\Y1=0.5X+1
\Y2=4X-3
\Y3=
\Y4=
\Y5=
\Y6=
\Y7=
```

- If desired, the thickness of the curve (E.g. normal, thick or dotted) can be adjusted for each rule.

2. Define the viewing window.

```
WINDOW
 Xmin=-5
 Xmax=5
 Xscl=1
 Ymin=-8
 Ymax=8
 Yscl=1
 Xres=1
```

- This screen allows you to define the viewing window by limiting the Cartesian plane: Xscl and Yscl correspond to the step value on the respective axes.

3. Display the graph.

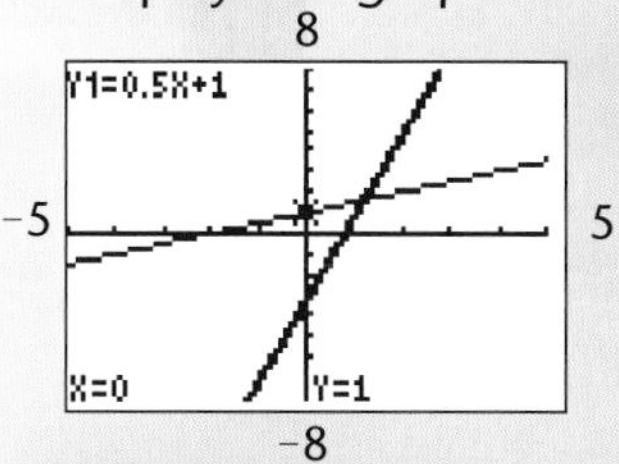

- This screen allows you to display the graphical representation of the rules previously defined. If desired, the cursor can be moved along the curves and the coordinates displayed.

Display a scatter plot and statistical calculations

1. Enter the data.

L1	L2	L3
1	9	
2	5	
3	6	
4	4	
5	4	
6	2	

L3(1)=

- This screen allows you to enter the data from a distribution. For a two-variable distribution, data entry is done in two columns.

2. Select the mode of representation.

- This screen allows you to choose the type of statistical diagram.

: scatter plot

: broken-line graph

: histogram

: box and whisker plot

3. Display the diagram.

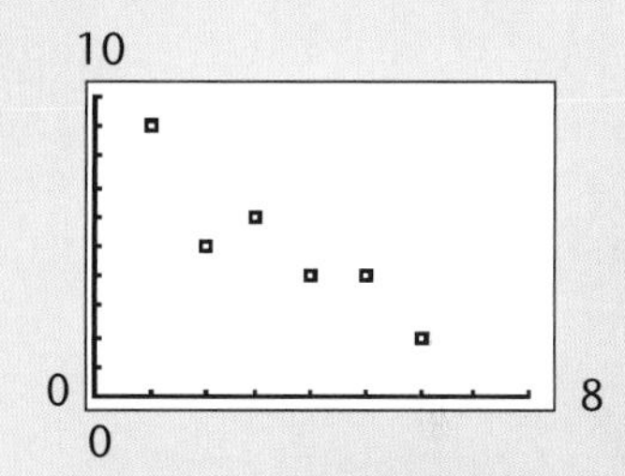

- This screen allows you to display the scatter plot.

4. Perform statistical calculations.

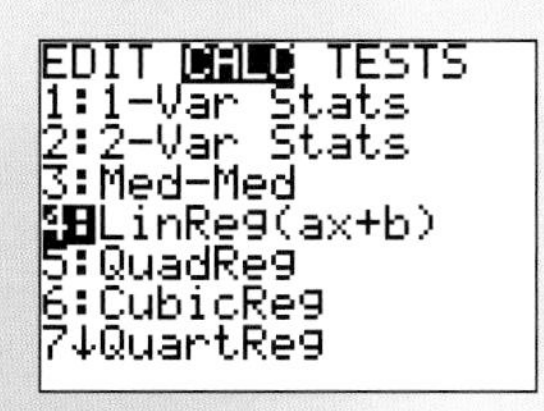

- This menu allows you to access different statistical calculations, in particular that of the linear regression.

5. Determine the regression and correlation.

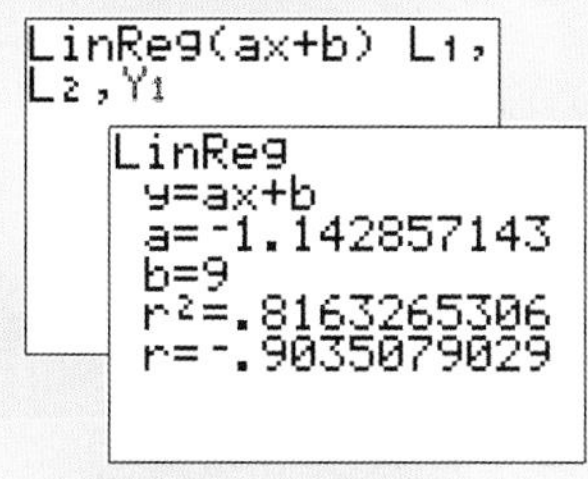

- These screens allow you to obtain the equation of the regression line and the value of the correlation coefficient.

6. Display the line.

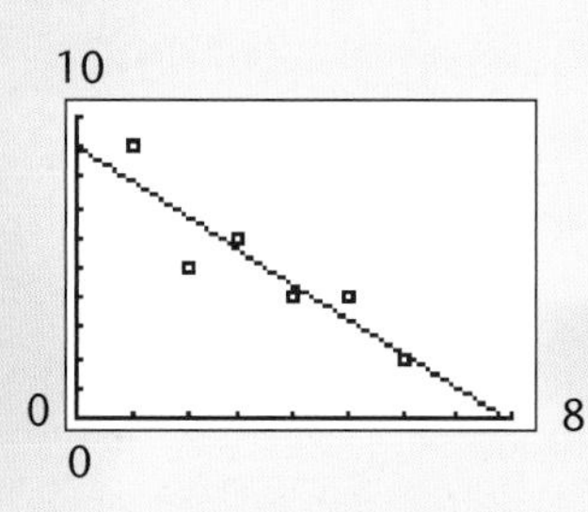

- The regression line can be displayed on the scatter plot.

Spreadsheet

A spreadsheet is software that allows you to perform calculations on numbers entered into cells. It is used mainly to perform calculations on large amounts of data, to construct tables and to draw graphs.

Spreadsheet Interface

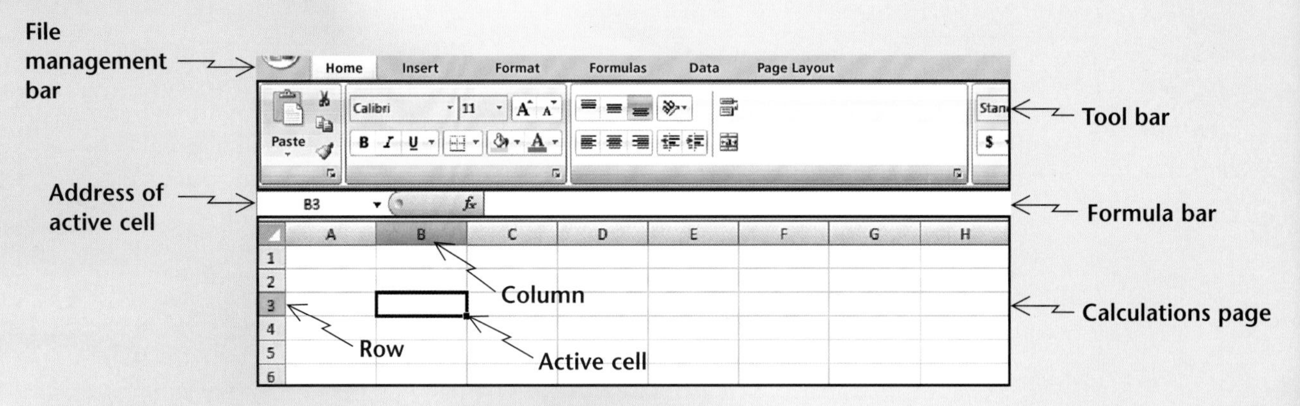

What is a cell?

A cell is the intersection of a column and a row. A column is identified by a letter and a row is identified by a number. Thus, the first cell in the upper right hand corner is identified as A1.

Entry of numbers, text and formulas in the cells

You can enter a number, text or a formula in a cell after clicking on it. Formulas allow you to perform calculations on numbers already entered in the cells. To enter a formula in a cell, just select it and begin by entering the "=" symbol.

Example:

Column **A** contains the data to be used in the calculations.

In the spreadsheet, certain functions are predefined to calculate the sum, the minimum, the maximum, the mode, the median, the mean and the mean deviation of a set of data.

◇	A	B	C	
1	Results			
2	27.4	Number of data	17	→ =COUNT(A2:A18)
3	30.15			
4	15	Sum	527	→ =SUM(A2:A18)
5	33.8			
6	12.3	Minimum	12.3	→ =MIN(A2:A18)
7	52.6			
8	28.75	Maximum	52.6	→ =MAX(A2:A18)
9	38.25			
10	21.8	Mode	33.8	→ =MODE(A2:A18)
11	35			
12	29.5	Median	30.15	→ =MEDIAN(A2:A18)
13	27.55			
14	33.8	Average	31	→ =AVERAGE(A2:A18)
15	15			
16	33.8	Mean deviation	8.417647059	→ =MEAN DEVIATION (A2:A18)
17	50			
18	42.3			
19				

How to construct a graph

Below is a procedure for drawing a graph using a spreadsheet.

1) Select the range of data.

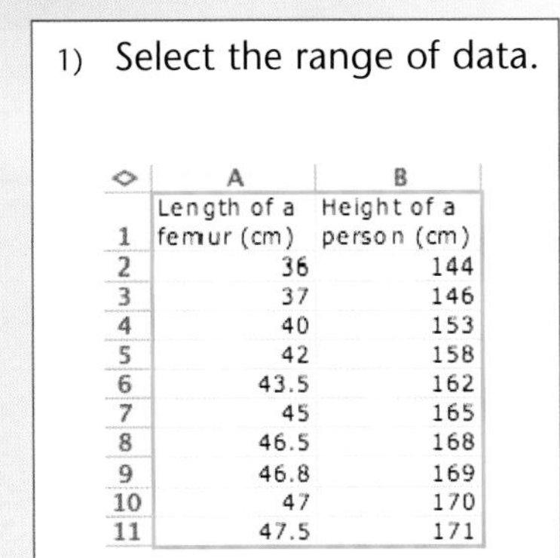

	A	B
1	Length of a femur (cm)	Height of a person (cm)
2	36	144
3	37	146
4	40	153
5	42	158
6	43.5	162
7	45	165
8	46.5	168
9	46.8	169
10	47	170
11	47.5	171

2) Select from the graph assistant.

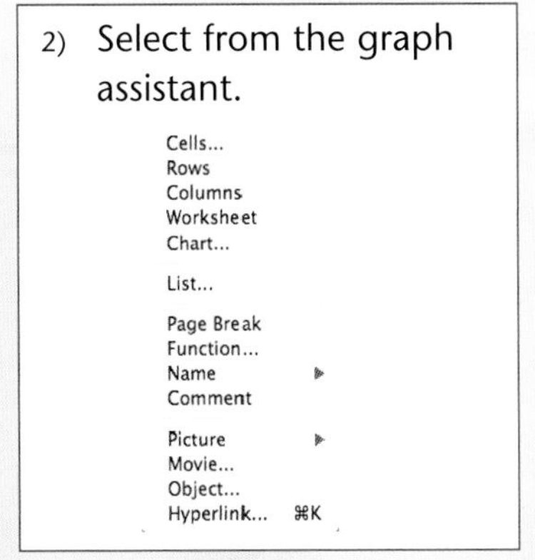

3) Choose the graph type.

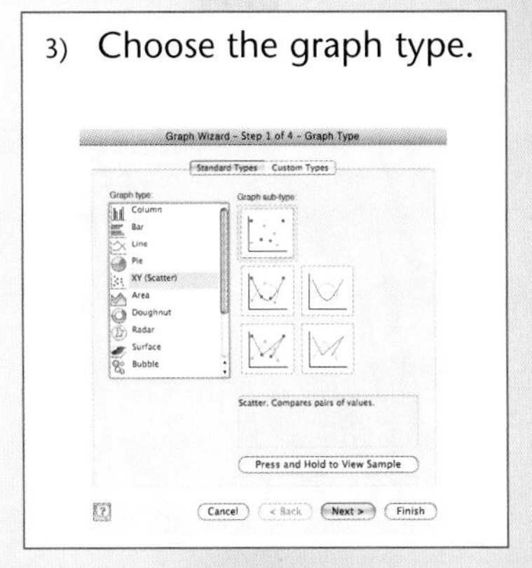

4) Confirm the data for the graph.

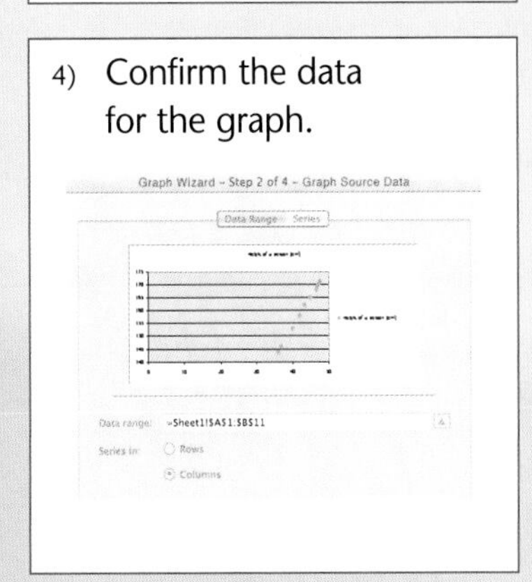

5) Choose graph options.

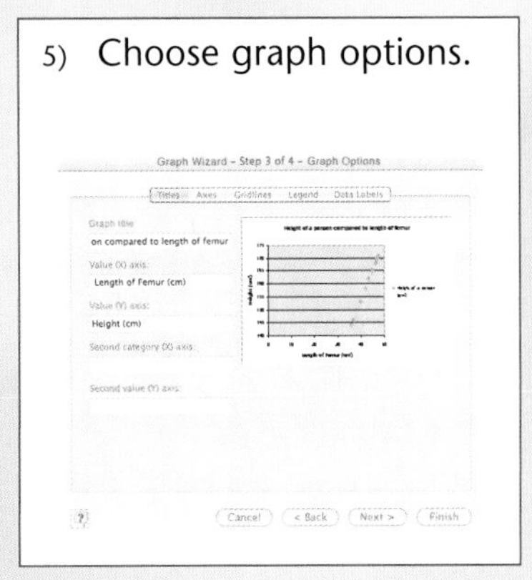

6) Choose the location of the graph.

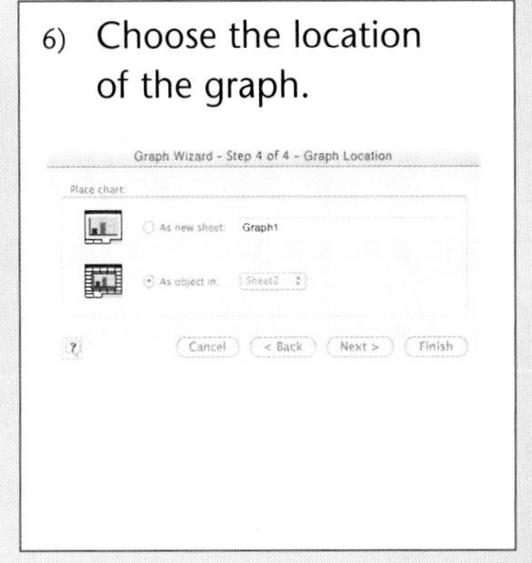

7) Draw the graph.

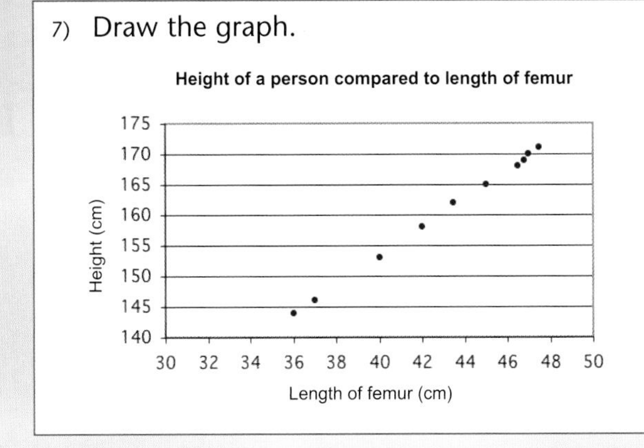

After drawing the graph you can modify different elements by double-clicking on the element to be changed: title, scale, legend, grid, type of graph, etc.

Below are different types of graphs you can create using a spreadsheet:

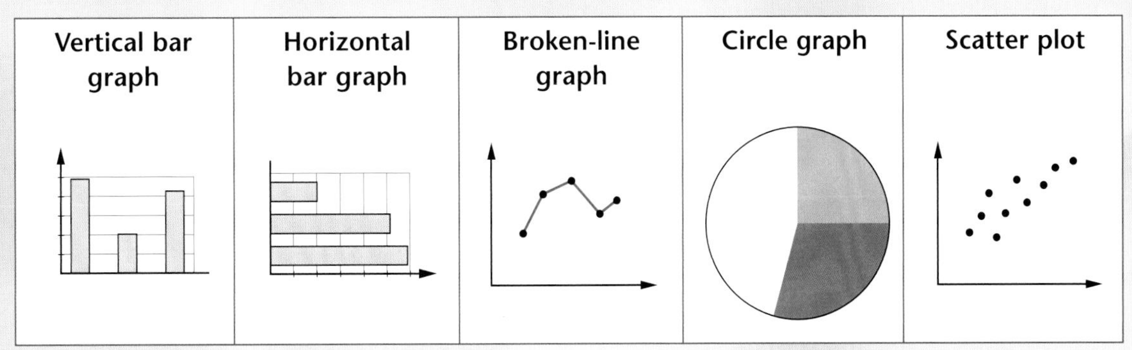

Dynamic geometry software

Dynamic geometry software allows you to draw and move objects in a workspace. The dynamic aspect of this type of software allows you to explore and verify geometric properties and to validate constructions.

The workspace and the tools

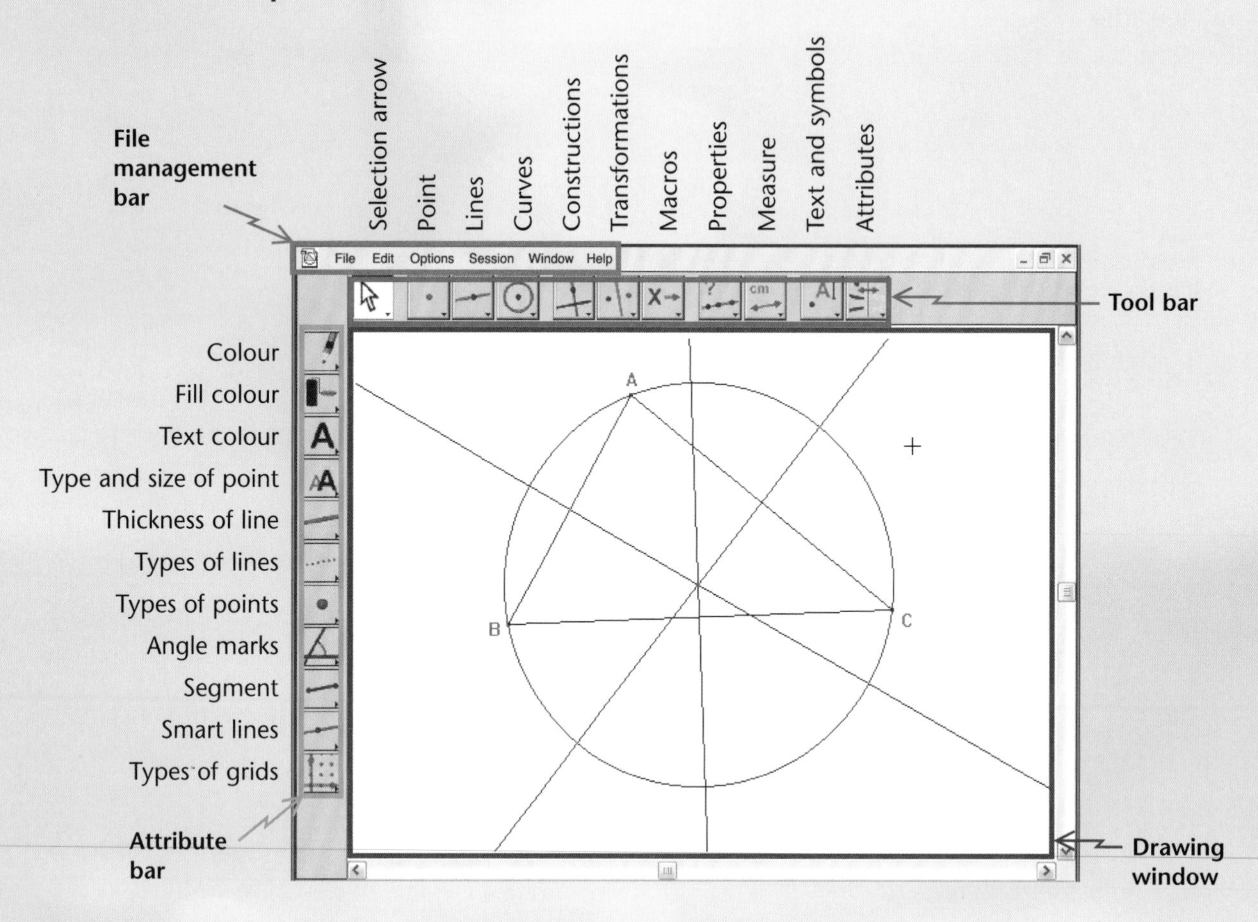

Cursors and their interpretations

+	Cursor used when moving in the drawing window.
	Cursor used when drawing an object.
What object?	Cursor used when there are several objects.
	Cursor used when tracing an object.
	Cursor used to indicate movement of an object is possible.
	Cursor used when working in the file management bar and in the tool bar.
	Cursor used when filling an object with a colour.
	Cursor used to change the attribute of the selected object.

Geometric explorations

1) A median separates a triangle into two other triangles. In order to explore the properties of these two triangles, perform the following construction. To verify that triangles ABD and ADC have the same area, calculate the area of each triangle. By moving the points A, B and C, notice that the areas of the two triangles are always the same.

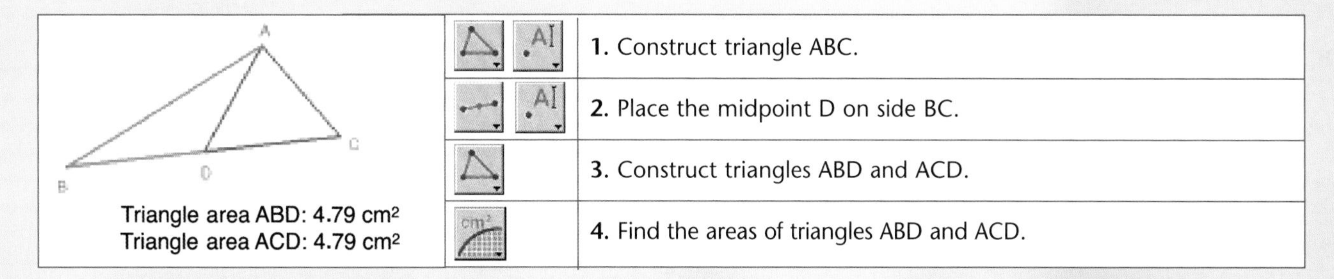

1. Construct triangle ABC.
2. Place the midpoint D on side BC.
3. Construct triangles ABD and ACD.
4. Find the areas of triangles ABD and ACD.

2) In order to determine the relation between the position of the midpoint of the hypotenuse in a right triangle and the three vertices of the triangle, perform the construction below. By moving points A, B, and C, note that the midpoint of the hypotenuse of a right triangle is equidistant from its three vertices.

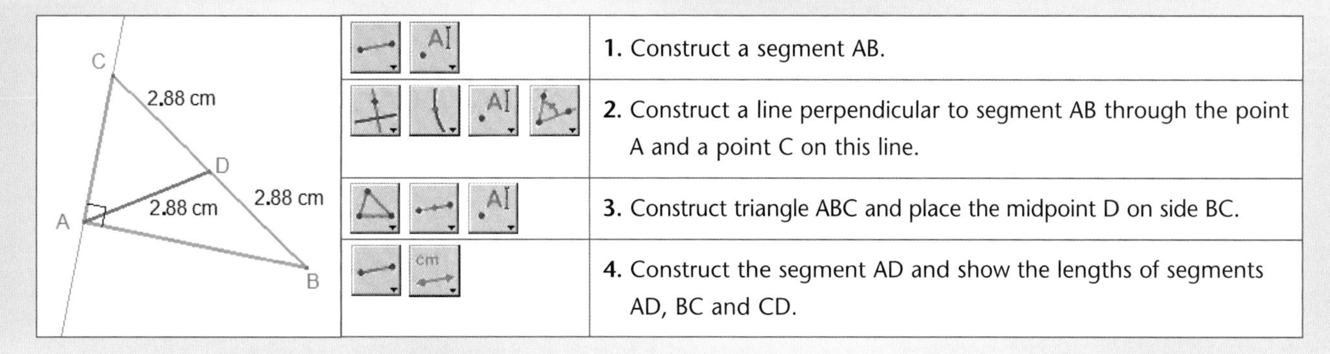

1. Construct a segment AB.
2. Construct a line perpendicular to segment AB through the point A and a point C on this line.
3. Construct triangle ABC and place the midpoint D on side BC.
4. Construct the segment AD and show the lengths of segments AD, BC and CD.

Graphical exploration

In order to discover the relation between the slopes of two perpendicular lines in the Cartesian plane, perform the construction below. By showing the product of the slopes and modifying the inclination of one of the lines, note a particular property of these slopes: the product of the slopes of these two perpendicular lines is -1.

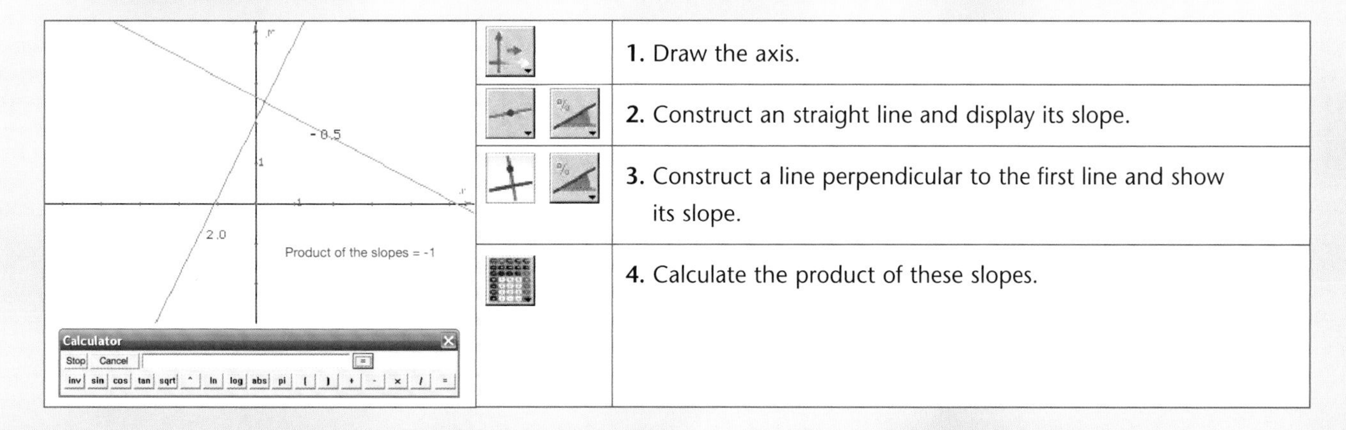

1. Draw the axis.
2. Construct an straight line and display its slope.
3. Construct a line perpendicular to the first line and show its slope.
4. Calculate the product of these slopes.

Notations and symbols

KNOWLEDGE

Notation & symbols	Meaning
{ }	Brace brackets, used to list the elements in a set
$\mathbb{N}$	The set of Natural numbers
$\mathbb{Z}$	The set of Integers
$\mathbb{Q}$	The set of Rational numbers
$\mathbb{Q}'$	The set of Irrational numbers
$\mathbb{R}$	The set of Real numbers
$\cup$	The union of sets
$\cap$	The intersection of sets
Ω	Read "omega," it represents the sample space in a a random experiment
$\varnothing$ or { }	The empty set (or the null set)
$=$	...is equal to...
$\neq$	...is not equal to... or ... is different from ...
$\approx$	...is approximately equal to...
$<$	...is less than...
$>$	...is greater than...
$\leq$	...is less than or equal to...
$\geq$	...is greater than or equal to...
$[a, b]$	Interval, including a and b
$[a, b[$	Interval, including a but excluding b
$]a, b]$	Interval, excluding a but including b
$]a, b[$	Interval excluding both a and b
∞	Infinity
(a, b)	The ordered pair a and b
$f(x)$	Is read as f of x or the value (image) of the function f at x
Δy	Variation or growth in y

Notation & Symbols	Meaning
( )	Parentheses show which operation to perform first
$-a$	The opposite of a
$\frac{1}{a}$ or a^{-1}	The reciprocal of a
a^2	The second power of a or a squared
a^3	The third power of a or a cubed
$\sqrt{a}$	The square root of a
$\sqrt[3]{a}$	The cube root of a
$\lvert a \rvert$	The absolute value of a
%	Percent
$a : b$	The ratio of a to b
$\approx$	...is approximately equal to...
π	Read "pi," it is approximately equal to 3.1416
°	Degree, unit of angle measure
$\overline{AB}$	Segment AB
m $\overline{AB}$	Measure of segment AB
$\angle$	Angle
m $\angle$	The measure of an angle
$\overset{\frown}{AB}$	The arc of the circle AB
m $\overset{\frown}{AB}$	The measure of the arc of the circle AB
//	... is parallel to...
$\perp$	... is perpendicular to ...
∟	Indicates a right angle in a geometric plane figure
$\triangle$	Triangle
$\cong$	...is congruent to...
$\sim$	...is similar to...
$\triangleq$	...corresponds to...
$P(E)$	The probability of the event E
Med	The median of a distribution

Geometric statements

KNOWLEDGE

	Statement	Example
1.	If two lines are parallel to a third line, then they are all parallel to each other.	If $l_1 // l_2$ and $l_2 // l_3$, then $l_1 // l_3$. 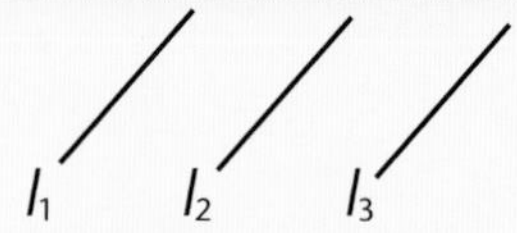
2.	If two lines are perpendicular to a third line, then the two lines are parallel to each other.	If $l_1 \perp l_3$ and $l_2 \perp l_3$, then $l_1 // l_2$. 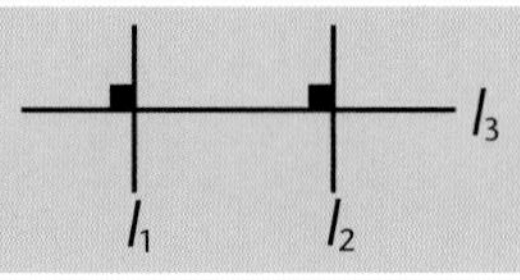
3.	If two lines are parallel, then every line perpendicular to one of these lines is perpendicular to the other.	If $l_1 // l_2$ and $l_3 \perp l_2$, then $l_3 \perp l_1$. 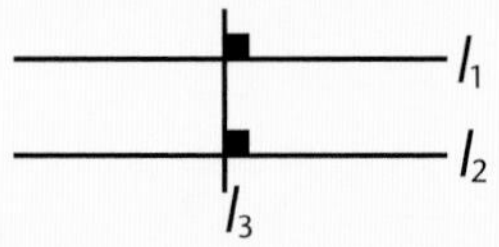
4.	If the exterior arms of two adjacent angles are collinear, then the angles are supplementary.	The points A, B and D are collinear. $\angle$ ABC & $\angle$ CBD are adjacent and supplementary. 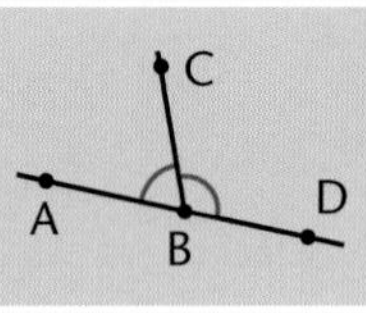
5.	If the exterior arms of two adjacent angles are perpendicular, then the angles are complementary.	$\overline{AB} \perp \overline{BD}$ $\angle$ ABC and $\angle$ CBD are adjacent and complementary. 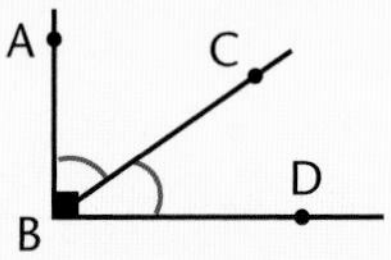
6.	Vertically opposite angles are congruent.	$\angle 1 \cong \angle 3$ $\angle 2 \cong \angle 4$ 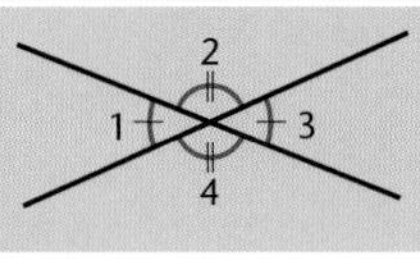
7.	If a transversal intersects two parallel lines, then the alternate interior, alternate exterior and corresponding angles are respectively congruent.	If $l_1 // l_2$, then angles 1, 3, 5 and 7 are congruent as are angles 2, 4, 6 and 8. 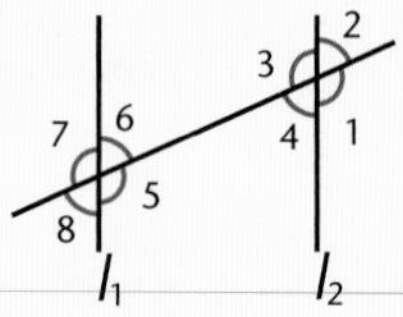
8.	If a transversal intersects two lines resulting in congruent corresponding angles (or alternate interior angles or alternate exterior angles), then those two lines are parallel.	In the figure for statement 7, if the angles 1, 3, 5 and 7 are congruent and the angles 2, 4, 6 and 8 are congruent, then $l_1 // l_2$.
9.	If a transversal intersects two parallel lines, then the interior angles on the same side of the transversal are supplementary.	If $l_1 // l_2$, then $m\angle 1 + m\angle 2 = 180°$ and $m\angle 3 + m\angle 4 = 180°$. 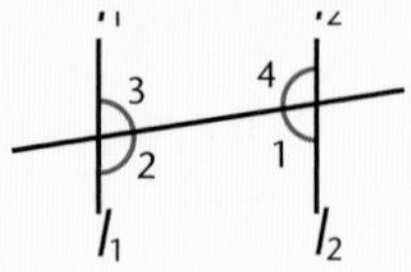

	Statement	Example
10.	The sum of the measures of the interior angles of a triangle is 180°.	$m\angle 1 + m\angle 2 + m\angle 3 = 180°$
11.	Corresponding elements of congruent plane or solid figures have the same measurements.	$\overline{AD} \cong \overline{A'D'}$, $\overline{CD} \cong \overline{C'D'}$, $\overline{BC} \cong \overline{B'C'}$, $\overline{AB} \cong \overline{A'B'}$ $\angle A \cong \angle A'$, $\angle B \cong \angle B'$, $\angle C \cong \angle C'$, $\angle D \cong \angle D'$
12.	In an isosceles triangle, the angles opposite the congruent sides are congruent.	In the isosceles triangle ABC: $\overline{AB} \cong \overline{AC}$ $\angle C \cong \angle B$
13.	The axis of symmetry of an isosceles triangle represents a median, a perpendicular bisector, an angle bisector and an altitude of the triangle.	Axis of symmetryof triangle ABC, Median from point A Perpendicular bisector of the side BC Bisector of angle A Altitude of the triangle
14.	The opposite sides of a parallelogram are congruent.	In the parallelogram ABCD: $\overline{AB} \cong \overline{CD}$ and $\overline{AD} \cong \overline{BC}$
15.	The diagonals of a parallelogram bisect each other.	In the parallelogram ABCD: $\overline{AE} \cong \overline{EC}$ and $\overline{DE} \cong \overline{EB}$
16.	The opposite angles of a parallelogram are congruent.	In the parallelogram ABCD: $\angle A \cong \angle C$ and $\angle B \cong \angle D$
17.	In a parallelogram, the sum of the measures of two consecutive angles is 180°.	In the parallelogram ABCD: $m\angle 1 + m\angle 2 = 180°$ $m\angle 2 + m\angle 3 = 180°$ $m\angle 3 + m\angle 4 = 180°$ $m\angle 4 + m\angle 1 = 180°$
18.	The diagonals of a rectangle are congruent.	In the rectangle ABCD: $\overline{AC} \cong \overline{BD}$
19.	The diagonals of a rhombus are perpendicular.	In the rhombus ABCD: $\overline{AC} \perp \overline{BD}$
20.	The measure of an exterior angle of a triangle is equal to the sum of the measures of the interior angles at the other two vertices.	$m\angle 3 = m\angle 1 + m\angle 2$

	Statement	Example
21.	In a triangle the longest side is opposite the largest angle.	In triangle ABC, the largest angle is A, therefore the longest side is BC. 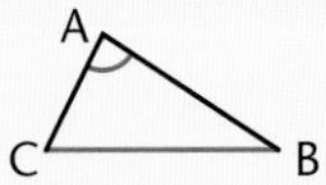
22.	In a triangle, the smallest angle is opposite the smallest side.	In triangle ABC, the smallest angle is B, therefore the smallest side is AC. 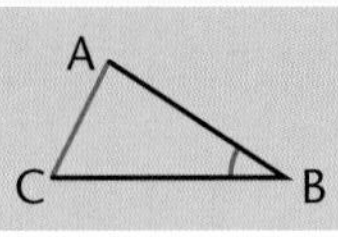
23.	The sum of the measures of two sides in a triangle is larger than the measure of the third side.	$2 + 5 > 4$ $2 + 4 > 5$ $4 + 5 > 2$ 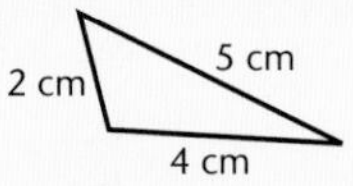 2 cm, 5 cm, 4 cm
24.	The sum of the measures of the interior angles of a quadrilateral is 360°.	$m\angle 1 + m\angle 2 + m\angle 3 + m\angle 4 = 360°$ 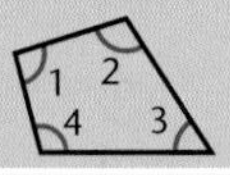
25.	The sum of the measures of the interior angles of a polygon with n sides is $n \times 180° - 360°$ or $(n - 2) \times 180°$.	$n \times 180° - 360°$ or $(n - 2) \times 180°$ 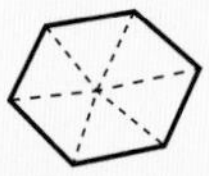 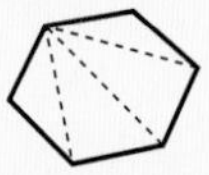
26.	The sum of the measures of the exterior angles (one at each vertex) of a convex polygon is 360°.	$m\angle 1 + m\angle 2 + m\angle 3 +$ $m\angle 4 + m\angle 5 + m\angle 6 = 360°$ 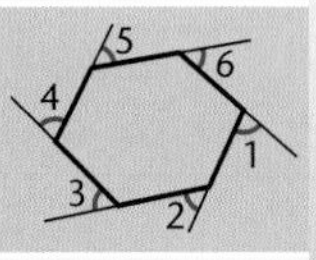
27.	The corresponding angles of similar plane figures or of similar solids are congruent and the measures of the corresponding sides are proportional.	The triangle ABC is similar to triangle A'B'C': $\angle A \cong \angle A'$ $\angle B \cong \angle B'$ $\angle C \cong \angle C'$ $\frac{m\overline{A'B'}}{m\overline{AB}} = \frac{m\overline{B'C'}}{m\overline{BC}} = \frac{m\overline{A'C'}}{m\overline{AC}}$ 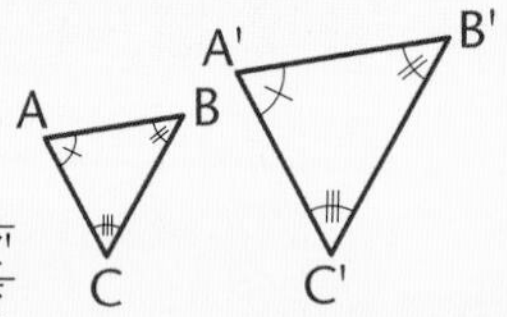
28.	In similar plane figures, the ratio of the areas is equal to the square of the ratio of similarity.	In the above figures, $\frac{m\overline{A'B'}}{m\overline{AB}} = \frac{m\overline{B'C'}}{m\overline{BC}} = \frac{m\overline{A'C'}}{m\overline{AC}} = k$ ← Ratio of similarity $\frac{\text{area of triangle A'B'C'}}{\text{area of triangle ABC}} = k^2$
29.	Three non-collinear points define one and only one circle.	There is only one circle which contains the points A, B and C. 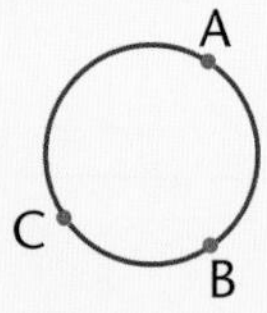
30.	The perpendicular bisectors of any chords in a circle intersect at the centre of the circle.	l_1 and l_2 are the perpendicular bisectors of the chords AB and CD. The point of intersection M of these perpendicular bisectors is the centre of the circle. 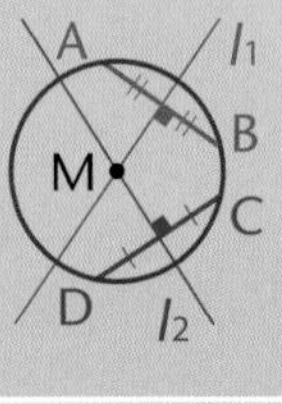

	Statement	Example
31.	All the diameters of a circle are congruent.	$\overline{AD}$, $\overline{BE}$ and $\overline{CF}$ are diameters of the circle with centre O. $\overline{AD} \cong \overline{BE} \cong \overline{CF}$
32.	In a circle, the measure of the radius is one-half the measure of the diameter.	$\overline{AB}$ is a diameter of the circle with centre O. $m\ \overline{OA} = \frac{1}{2}\ m\ \overline{AB}$
33.	In a circle, the ratio of the circumference to the diameter is a constant represented by π.	$\frac{C}{d} = \pi$
34.	In a circle, a central angle has the same degree measure as the arc contained between its sides.	In the circle with centre O, $m\ \angle\ AOB = m\ \overset{\frown}{AB}$ is stated in degrees.
35.	In a circle, the ratio of the measures of two central angles is equal to the ratio of the arcs intercepted by their sides.	$\frac{m\angle\ AOB}{m\ \angle COD} = \frac{m\ \overset{\frown}{AB}}{m\ \overset{\frown}{CD}}$
36.	In a circle, the ratio of the areas of two sectors is equal to the ratio of the measures of the angles at the centre of these sectors.	$\frac{\text{Area of the sector AOB}}{\text{Area of the sector COD}} = \frac{m\angle\ AOB}{m\ \angle COD}$

Glossary

Algebraic term - see Term.

Altitude of a triangle
Segment from one vertex of a triangle, perpendicular to the line containing the opposite side. The length of such a segment is also called a height of the triangle.

Angles

Classification of angles according to their measure

Name	Measure	Representation
Zero	0°	
Acute	Between 0° & 90°	
Right	90°	
Obtuse	Between 90° & 180°	
Straight	180°	
Reflex	Between 180° & 360°	
Perigon	360°	

Alternate interior angles, p. 151

Central angle
Angle formed by two radii in a circle. The vertex of the angle is the centre of the circle.

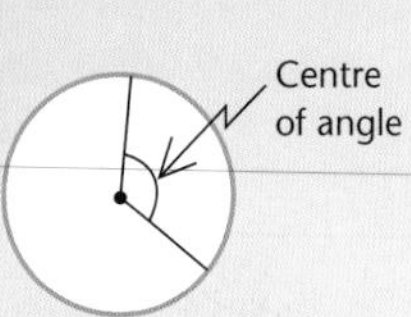

Corresponding angles, p. 151

Apothem of a regular polygon
Segment (or length of segment) from the centre of the regular polygon perpendicular to any of its sides. It is determined by the centre of the regular polygon and the midpoint of any side.

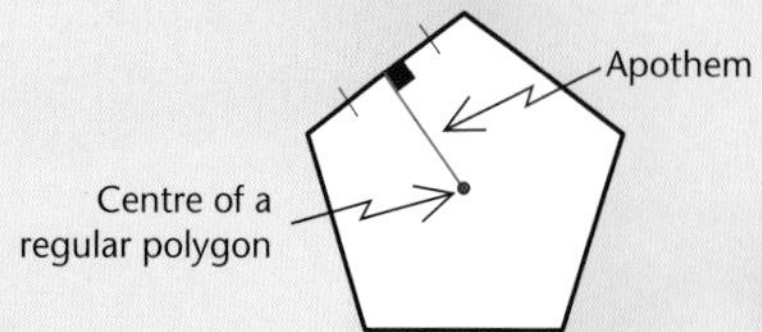

Arc of a circle
Part of a circle defined by two points on the circle.

Area
The surface of a figure. Area is expressed in square units.

Area of a circle, p.150

Area of a parallelogram, p.150

Area of a rectangle, p.150

Area of a regular polygon, p.150

Area of a rhombus, p.150

Area of a right circular cone

$A_{\text{right circular cone}} = \pi r^2 + \pi ra$

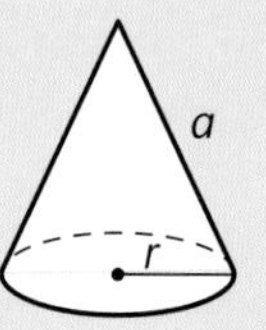

Area of a sector

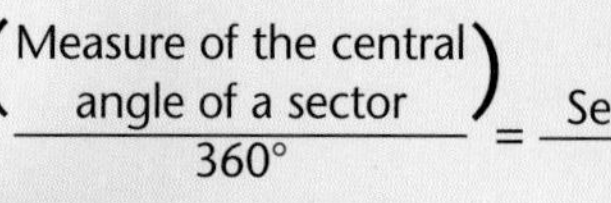

$$\frac{\left(\text{Measure of the central angle of a sector}\right)}{360°} = \frac{\text{Sector area}}{\pi r^2}$$

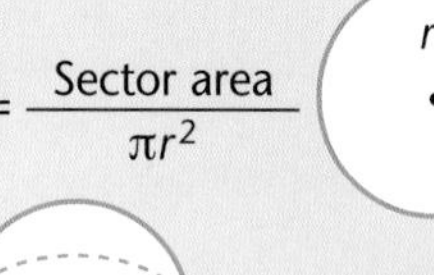

Area of a sphere

$A_{\text{sphere}} = 4\pi r^2$

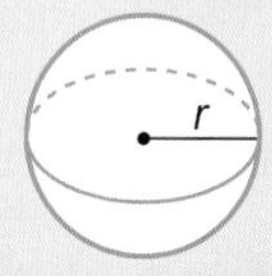

Area of a square, p. 150

Area of a trapezoid, p. 150

Area of a triangle, p. 150

Box-and-whisker plot, p. 72

Capacity
Volume of a fluid which a solid can contain.

Cartesian plane
A plane formed by two scaled perpendicular lines. Each point is located by its distance from each of these lines respectively.

Census
A search for information on an entire population.

Central angle - see Angles.

Circle
The set of all points in a plane at an equal distance from a given point called the centre.

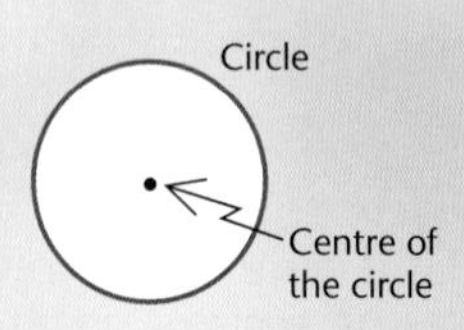

Circumference
The perimeter of a circle. In a circle whose circumference is C, diameter is d and radius is r:
$C = \pi d$ and $C = 2\pi r$.

Class (interval)
In statistics, an interval of the form [lower limit, upper limit[.

Correlation coefficient, p. 110, 123

Conjecture, p. 161

Constant function - see Function.

Contingency table, p. 93

Continuous quantitative variable
A variable for which the data collected are numbers which can take any value on a real interval.

Coordinates of a point
Each of the two numbers used to describe the position of a point in a Cartesian plane.

Correlation, p. 93

Counter-example, p. 161

Cube root
The inverse of the operation which consists of cubing a number is called finding the cube root. The symbol for this operation is $\sqrt[3]{\ }$.

E.g. 1) $\sqrt[3]{125} = 5$
2) $\sqrt[3]{-8} = -2$

Degree of a monomial
The sum of the exponents of the monomial.
E.g. 1) The degree of the monomial 9 is 0.
2) The degree of the monomial $-7xy$ is 2.
3) The degree of the monomial $15a^2$ is 2.

Degree of a polynomial in one variable
The largest exponent of that variable in the polynomial.
E.g. The degree of the polynomial $7x^3 - x^2 + 4$ is 3.

Dependent variable - see Variable.

Diameter
Segment (or length of segment) which is determined by two points on a circle passing through the centre of the circle.

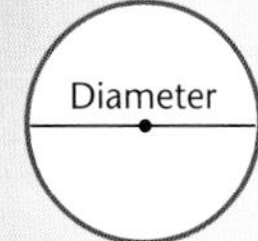

Dilatation
A geometric transformation defined by a centre, initial point, corresponding image and scale factor. A dilatation results in an enlargement or reduction of the initial figure.

Direct variation function - see Function.

Discrete quantitative variable
A variable for which the data collected are numbers which can not take all values on a real interval.

Distance between two points, p. 15

Distribution table
A table which summarizes data gathered in a statistical survey.

Edge
Segment formed by the intersection of any two faces of a solid.

Equation
Mathematical statement of equality involving one or more variables.
E.g. $4x - 8 = 4$

Equivalent equations
Equations having the same solution.
E.g. $2x = 10$ and $3x = 15$ are equivalent equations, because the solution of each is 5.

Exponentiation
Operation which consists of raising a base to an exponent.
E.g. In 5^8, the base is 5 and the exponent is 8.

Face
Plane or curved surface bound by edges.

Factoring
Writing an expression as a product of factors.
E.g. The factorization of $6a^2 + 15a$ can be expressed as $3a(2a + 5)$.

Frequency
In statistics, the number of times a data value occurs, or the number of data values in a given category.

Favourite Meal	Frequency
Pizza	10
Lamb	12
Salad	7

Function
A relation between two variables in which each value of the independent variable is associated to at most one value of the dependent variable.

Direct variation function
A function in which a constant change in the independent variable results in a constant, non-zero change to the dependent variable. Its graph is an oblique line through the origin of the Cartesian plane.

First-degree polynomial function
A function whose rule can be written as a first-degree polynomial.
E.g. $f(x) = 7.1x + 195$

Partial variation function
A function in which a constant change in the independent variable results in a constant, non-zero change to the dependent variable. Its graph is an oblique line which does not pass through the origin of the Cartesian plane.

Polynomial function
A function whose rule can be written as a polynomial.
E.g. $f(x) = 3x^2 + 7$

Zero-degree polynomial function (constant function)
A function in which a constant change in the independent variable results in no change in the dependent variable. Its graph is a horizontal line parallel to the x-axis.
E.g. $f(x) = -5$

H

Half-plane, p. 49

Height of a triangle
See Altitude of a triangle.

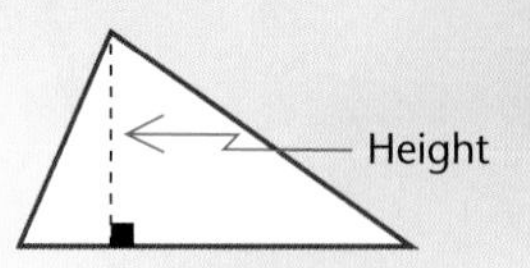

Hypotenuse, p.151

Image
In geometry, figure obtained by a geometric transformation performed on an initial figure.

Independent variable - see variable

Inequality
A mathematical statement which compares two numerical expressions with an inequality symbol (which may include variables).
E.g. 1) $4 < 4.2$
2) $-10 \leq -5$
3) $4a > 100$

First-degree inequality
in one variable, p. 7
in two variables, p. 49

Initial figure
Figure on which a geometric transformation is performed.

Integer
Any number belonging to the set $\mathbb{Z} = \{...,-2, -1, 0, 1, 2, 3, ...\}$.

Interquartile range
Difference between the 1st and 3rd quartiles of a one-variable distribution.

Interval
A set of all the real numbers between two given numbers called the endpoints. Each endpoint can be either included or excluded in the interval.
E.g. The interval of real numbers from -2 included to 9 excluded is $[-2, 9[$.

Irrational number
A number which cannot be expressed as a ratio of two integers, and whose decimal representation is non-periodic and non-terminating.

L

Laws of Exponents

Law	
Product of powers:	$a^m \times a^n = a^{m+n}$
Quotient of powers: For $a \neq 0$	$\frac{a^m}{a^n} = a^{m-n}$
Power of a product:	$(ab)^m = a^m b^m$
Power of a power:	$(a^m)^n = a^{mn}$
Power of a quotient: $b \neq 0$	$\left(\frac{a}{b}\right)^m = \frac{a^m}{b^m}$

Legs (or arms) of a right triangle, p. 151

Like terms - see Terms.

Linear correlation, p. 94, 95, 123, 124

Line
- Mayer's, p. 112
- Median-median, p. 111
- Regression, p. 111, 123

Lines
- parallel, p. 27, 40
- perpendicular, p. 27

M

Maximum of a distribution
Largest number in a data set.

Maximum of a function
The largest value of the dependent variable in the function.

Mean, p. 71

Mean deviation, p. 81

Measure
- **Central tendency,** p. 71
- **Dispersion,** p. 71, 81
- **Position,** p. 81

Median of a distribution, p. 71

Median of a triangle
Segment determined by a vertex and the midpoint of the opposite side.
E.g. The segments AE, BF and CD are the medians of triangle ABC.

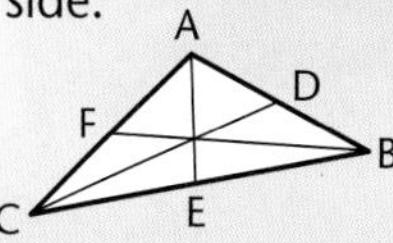

Metric relations (in a right triangle), p. 181

Minimum of a distribution
The smallest number in a data set.

Minimum of a function
The smallest value of the dependent variable in the function.

Mode, p. 71

Monomial
Algebraic expression formed by one number or a product of numbers and variables.
E.g. 9, $-5x^2$ and 4xy are monomials.

N

Natural number
Any number belonging to the set
$\mathbb{N} = \{0, 1, 2, 3, ...\}$.

Numerical coefficient of a term
Numerical value multiplied by the variable or variables of a term.
E.g. In the algebraic expression $x + 6xy - 4.7y$, the numerical coefficients of the first, second and third terms are 1, 6 and -4.7 respectively.

O

Origin of a Cartesian plane
The point of intersection of the two axes in a Cartesian plane. The coordinates of the origin are (0, 0).

P

Parallelogram, p. 149

Percentile, p. 81

Partial variation function - see Function.

Perimeter
The length of the boundary of a closed figure. It is expressed in units of length.

Perpendicular bisector
A perpendicular line passing through the midoint of a segment. It is also an axis of symmetry for the segment.
E.g.

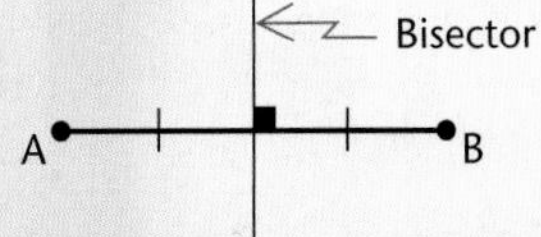

Polygon
A closed plane figure with three of more sides.

Polygons

Number of Sides	Name of Polygon
3	Triangle
4	Quadrilateral
5	Pentagon
6	Hexagon
7	Heptagon
8	Octagon
9	Nonagon
10	Decagon
11	Undecagon
12	Dodecagon

Polyhedron
A solid determined by plane polygonal faces.
E.g.

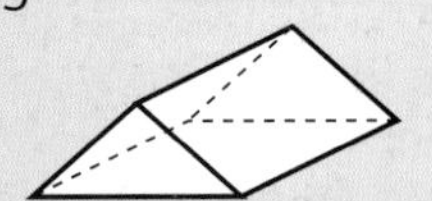

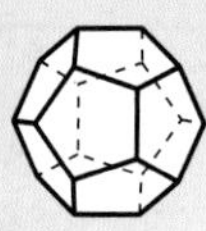

Polynomial
An algebraic expression containing one or more terms. E.g. $x^3 + 4x^2 - 18$

Polynomial function - see Function.

Point of division, p. 16

Population
A set of living beings, objects or facts which are the object of a statistical survey.

Prism
A polyhedron with two congruent parallel faces called "bases." The parallelograms determined by the corresponding sides of these bases are called the "lateral faces."

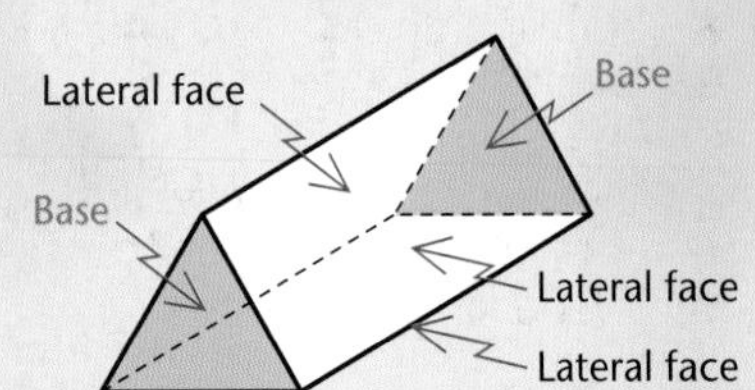

Regular prism
A prism whose bases are regular polygons.
E.g. A regular heptagonal prism

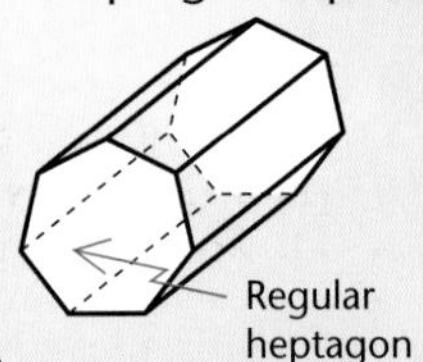

Right prism
A prism whose lateral faces are rectangles.
E.g. A right trapezoidal prism

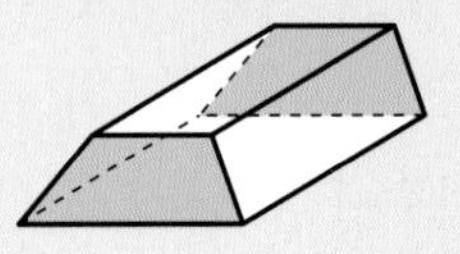

Proof, p. 161

Proportion
A statement of equality between two ratios or two rates
E.g. 1) 3:11 = 12:44
2) $\frac{7}{5} = \frac{14}{10}$

Pyramid
A polyhedron with one polygonal base, whose lateral faces are triangles with a common vertex called the apex.
E.g. Octagonal pyramid

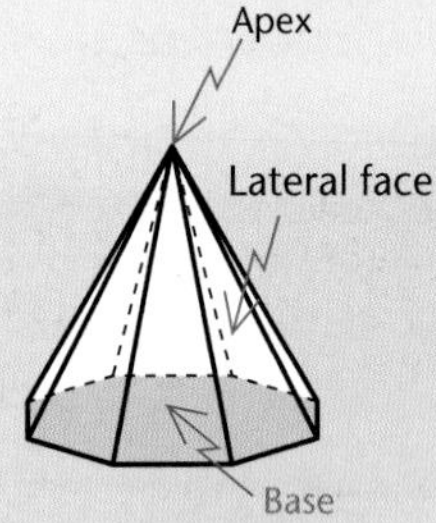

Regular pyramid
A pyramid whose base is a regular polygon.
E.g. A regular hexagonal pyramid

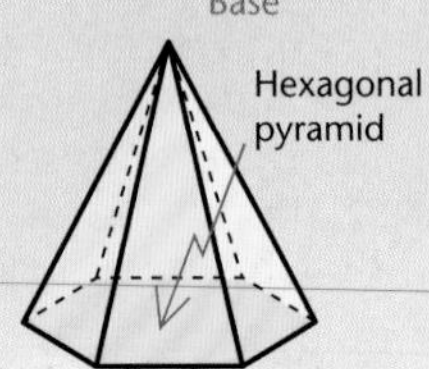

Right pyramid
A pyramid such that the segment from the apex, perpendicular to the base, intersects it at the centre of the polygonal base.
E.g. A right rectangular pyramid

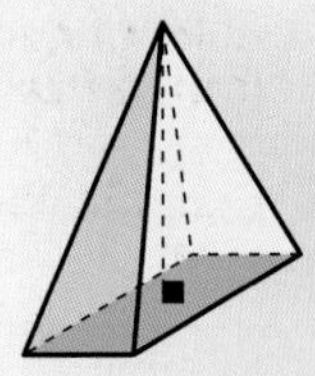

Pythagorean theorem, p. 151

Q

Quadrant
Each of the four regions defined by the axis of a Cartesian plane. The quadrants are numbered 1 to 4 as shown below.

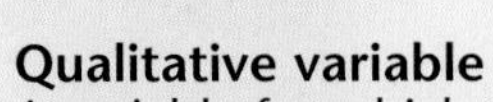

Qualitative variable
A variable for which the data collected are words or codes.

Quartiles, p. 72

Radius
A radius is a segment (or length of a segment) which is determined by the centre of a circle and any point on the circle.

Range, p. 71

Rate
A way of comparing two quantities or two sizes expressed in different units and which requires division.

Rate of change
In a relation between two variables, a comparison between two corresponding variations.

$$\text{Rate of change} = \frac{\left(\text{variation of the dependent variable}\right)}{\left(\text{variation of the independent variable}\right)}$$

Ratio
A way of comparing two quantities or two sizes expressed in the same units and which requires division.

Rational number
A number which can be written as the quotient of two integers where the denominator is not zero. Its decimal representation can be terminating or non-terminating and periodic.

Ratio of similarity
Ratio of corresponding segments resulting from a dilatation.

Real number
A number belonging to the union of the set of rational numbers and the set of irrational numbers.

Rectangle, p. 149

Reflection
A geometric transformation which maps an initial point to an image point such that a given line (called the reflection line) is the perpendicular bisector of the segment determined by the point and its image. The reflection of a figure is the reflection of all of its points.

Relation
A relationship between two variables.

Removing a common factor
Writing an expression as a product of two factors, one of which is common to the terms of the original expression.
E.g. $8a^2 - 44a = 4a(2a - 11)$

Rhombus, p. 150

Right circular cone
Solid made of two faces, a circle and a sector. The circle is the base and the sector forms the lateral face.

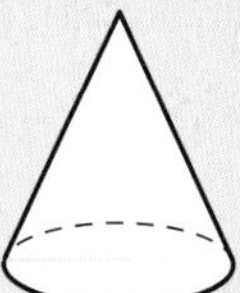

Right circular cylinder
Solid made of three faces, two congruent circles and a rectangle. The circles form the bases and the rectangle forms the lateral face.

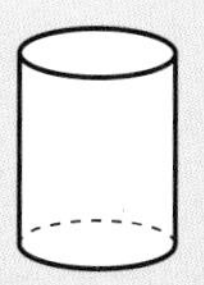

Root - see Square root, cube root.

Rotation
A geometric transformation which maps an object to an image using a centre, an angle and a direction of rotation.

Rule
An equation which translates a relationship between variables.

Rules for transforming equations
Rules that allow you to obtain equivalent equations.
You can preserve the value of the equation:
- by adding the same amount to both sides of the equation
- by subtracting the same amount from both sides of the equation
- by multiplying both sides of the equation by the same amount or
- by dividing both sides of the equation by the same non-zero amount

Rules for transforming inequalities
Rules that allow you to obtain equivalent inequalities.
The value of an inequalities is preserved:
- by adding the same amount to both sides of the inequality
- by subtracting the same amount from both sides of the inequality
- by multiplying or dividing both sides of the inequalities by the same strictly positive amount

The direction of an inequality is reversed:
- by multiplying or dividing both sides of the inequality by the same strictly negative amount

S

Sample
Subset of a population.

Sample survey
A search for information on a representative subset of a population in order to draw conclusions about the entire population.

Scatter plot, p. 94

Scientific notation
A notation which facilitates the reading and writing of numbers which are very large or very small.

E.g. 1) 56 000 000 = 5.6×10^7
2) 0.000 000 008 = 8×10^{-9}

Section of a solid
Face obtained by a plane which cuts a solid.

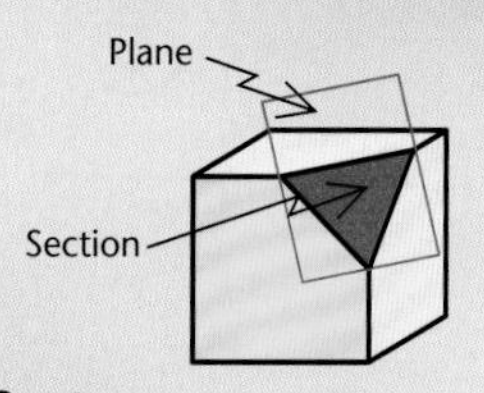

The section obtained by the intersection of this plane and this cube is a triangle.

Sector
Part of a circle defined by two radii.

Similar figures
Two figures are similar if and only if a dilation enlargement or reduction of one results in a figure congruent to the other.

Slant height of a right circular cone
Segment (or length of a segment) defined by the apex and any point on edge of the base.

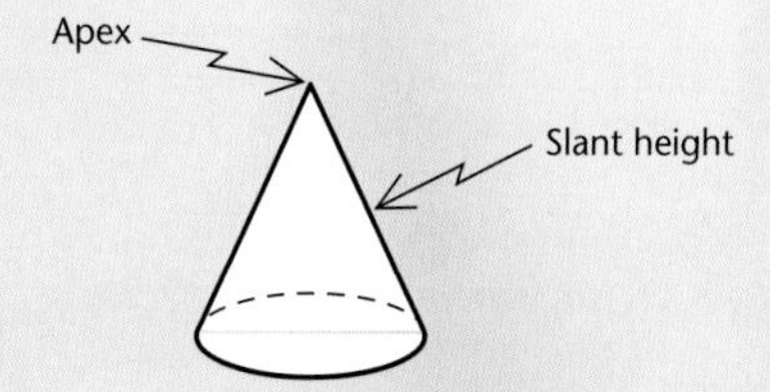

Slant height of a regular pyramid
Segment from the apex perpendicular to any side of the polygon forming the base of the pyramid. It corresponds to the altitude of a triangle which forms a lateral face.

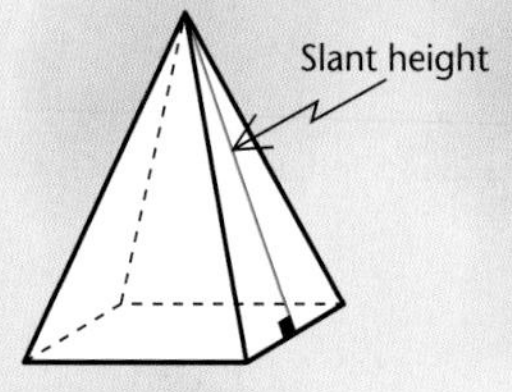

Slope, p. 15, 26, 27

Solid
Portion of space bounded by a closed surface.
E.g.

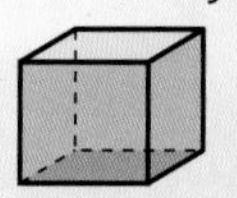

Sphere
The set of all points in space at a given distance (radius) from a given point (centre).

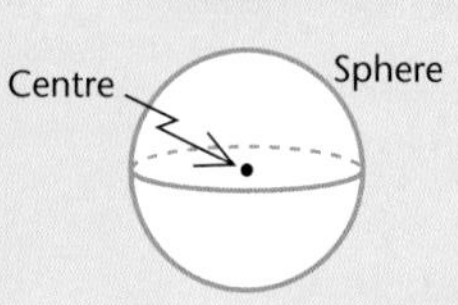

Stem-and-leaf plot, p. 80

Square, p. 150

Square root
The inverse of the operation which consists of squaring a positive number is called finding the square root. The symbol for this operation is $\sqrt{\ }$.
E.g. The square root of 25, written $\sqrt{25}$, is 5.

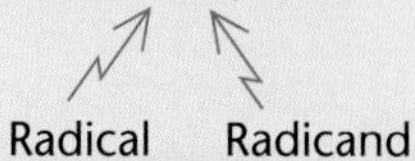
Radical Radicand

Statistical variable
In statistics, the variable being researched.

Surface area - see Area.

System of equations
Solving by comparaison, p. 6
Solving by elimination, p. 39
Solving by substitution, p. 39

T

Terms

Like terms
Constants or terms composed of the same variables raised to the same exponents.
E.g. 1) $8ax^2$ and ax^2 are like terms
2) 8 and 17 are like terms

Algebraic term
A term can be composed of one number or of a product of numbers and variables.
E.g. 9, x and $3xy^2$ are terms

Theorem, p. 161

Translation
A geometric transformation which maps an initial point to an image point given a specified direction and length.

Trapezoid, p. 149

Triangles, p. 149
congruent, p. 160
similar, p. 171